STATES, WAVES AND PHOTONS:
A Modern Introduction to Light

JOSEPH W. SIMMONS and MARK J. GUTTMANN
La Salle College

STATES, WAVES AND PHOTONS: A Modern Introduction to Light

ADDISON-WESLEY PUBLISHING COMPANY
Reading, Massachusetts · Menlo Park, California · London · Don Mills, Ontario

This book is in the

ADDISON-WESLEY SERIES IN PHYSICS

Consulting Editor: DAVID LAZARUS

Printed in the United States of America. Published simultaneously in Canada. Library of Congress Catalog Card No. 73-102998.

To Henry J. Bolger, C.S.C., *1901–1964*

Chairman, Department of Physics, University of Notre Dame, 1936–1963

If you were there, you understand why . . .

PREFACE

It is the nature of any realm of human endeavor to establish what must be called a body of accomplishment. For example, the paintings of Rembrandt and the mechanics of Newton, respectively, belong to the realms of art and physical science. Few, if any, would criticize the viewpoint that these achievements should rightfully be called art and science. To put this another way, any attempt to define art or science would have to be broad enough to include Rembrandt's paintings under the term "art" and the laws of Newton under the term "physics" and, except in some emotional sense, to exclude them from the other category. Inevitably, such a body of accomplishment expands, usually in quite unpredictable ways. These ways are such that what was once a good delineation of a field becomes inadequate. Two hundred years ago it would perhaps have been possible to give a definition of art which would completely exclude the whole of modern nonrepresentational painting. Yet, whether we like these paintings or not, it is generally considered that this modern art is indeed art. The point to be taken from the above is that there is a quite definite relation between the accomplishments made in a given field and the definition of that field. To be concrete, what physics is and what physicists do must always be a question which cannot be completely answered except by hindsight.

Most intermediate optics texts relegate to a final chapter or two those aspects of physics which have forced us in this century to alter drastically our "definition" of physics. As a result, after having been introduced to the photon concept and quantum mechanics in either atomic physics or a quantum mechanics course, the student does quite well until a nagging thought appears: "What happened to the electromagnetic wave?" Part of the pedagogic difficulty may be that the student has formed an incomplete definition of physics in general and optics in particular. If the student receives a thoroughly rigorous course in classical optics, he emerges feeling that what he sees *is* an electromagnetic wave. Certainly, after exposure to the triumphs of the classical theory and its experimental confirmation, it is awkward for him to start again with a photon description, leaving no bridge except the correspondence principle and some mention of the existence of quantum electrodynamics. The present text is an attempt to solve this pedagogic problem by an *a priori* inclusion of quantum effects as a fundamental part of "optics". Such an attempt leads naturally to the basic formalism of quantum mechanics within the confines of a two-state system. This formalism is exploited in Chapters 2 through 5 for two principal reasons:

1) it is a valid description of the photon as a quantum mechanical entity;
2) it introduces the student to a nomenclature and approach that characterize much of the physics of our time.

The latter reason is not insignificant for those engaged in undergraduate teaching. Such teachers are well aware of the difficulty of devising curricula that are in touch with the physics of their generation. For example, the space devoted to the representation and production of polarization states might seem disproportionately large, unless understood within the framework of an undergraduate curriculum.

The physics programs at most schools are rather different from those of a decade ago. This must be, but it is well known that the upper division course in optics has often been a casualty. However, after 6 years of utilizing the approach presented here, the authors have found the optics course transformed from a luxury to a dynamic and integral part of the curriculum. At present this text forms the basis for a one-semester course (three hours per week plus laboratory work) given in the second semester of the junior year at La Salle. There is more than enough material in this book for such a course, and a selection must be made. Although other choices are certainly possible, our method of presenting the material may be of interest. The material of Chapter 1 is covered in the third semester of the general physics course and the junior course begins with Chapter 2 and the understanding that the student is familiar with the elements of matrix algebra. The students also have had a strong course in electromagnetic theory. To accomplish the central purpose of the course (as outlined in the following paragraph) within the allotted time, we have generally omitted Sections 6.4 and 6.5, on the Cornu spiral, and Sections 7.9 through 7.16, dealing with coherence. With the exception of the detailed calculations in Appendix VI, the remaining material in the text is covered without difficulty.

The principal motivation which guided us in the writing of the text was to show how polarization states, the electromagnetic field, and the spin of the photon are interrelated. Thus, an abrupt change of pace is necessary in Chapter 7 when the classical electromagnetic field is introduced so that it can later be related to the angular momentum of a photon in a K-space representation in Chapters 8 and 9. Then, coming full circle, the eigenstates of the spin operators for the electromagnetic field in K space are the ket vectors of the Jones calculus of the early chapters.

By the very nature of its approach, the text serves the purpose of an introduction to the concepts and techniques of quantum mechanics. For this reason it is subtitled as "A Modern Introduction to Light" rather than "An Introduction to Modern Optics". The distinction is nontrivial and explains the absence of many important topics, e.g., optical instruments, computer lens design, nonlinear optics, etc.

Finally, the authors would like to thank for their suggestions and criticisms the many students who have suffered through preliminary versions of this text.

Philadelphia, Pa.
February 1970

J. W. S.
M. J. G.

CONTENTS

CHAPTER 1

MATRIX OPTICS

Geometric optics is the traditional beginning for a study of optics. There are a number of reasons, more or less compelling, for following this tradition. If the student has been previously exposed to geometric optics, he will be meeting familiar material within the framework of an aesthetically satisfying conceptual scheme. Furthermore, this venerable area of optics is an ideal introduction to the matrix algebra techniques which are often utilized in this book. Last but least, as mentioned above, this is the way optics books usually start.

The prerequisites required for mastery of the material in this chapter are slight. It is assumed that the student is familiar with the rectilinear propagation of light, the law of reflection, and Snell's law. The matrix algebra required is not more advanced than that given in Sections 1 and 3 of Appendix I.

1.1 A STATEMENT OF THE PROBLEM

The problem that will be considered in this chapter is illustrated in Fig. 1.1. Between the wavy lines is some optical system, S, which consists of various reflecting and refracting surfaces. System S is such that it possesses a rotational axis of symmetry, which will be called the *system axis* or *optic axis*. The intersection of the optic axis with any of the reflecting or refracting boundaries is termed the *vertex* of the boundary.

Now consider a ray which passes above point 1 on the axis before entering the system and after emerging from S passes over point 2. The entrant ray and the emergent ray are each defined by their height above the axis and their inclination relative to the horizontal. We will assume that there exists a linear relation between the entrant and emergent rays of the form

$$\begin{aligned} y_2 &= s_{11}y_1 + s_{12}\theta_1, \\ \theta_2 &= s_{21}y_1 + s_{22}\theta_1, \end{aligned} \tag{1.1}$$

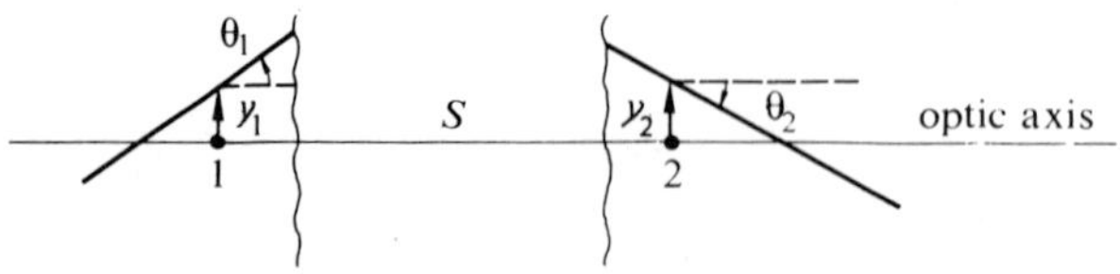

Fig. 1.1. Entrant and emergent rays for an optical system.

where the coefficients s_{ij} depend only on the system S and not on the particular ray. Later it will be seen that this is equivalent to assuming that all angles are such that the approximation

$$\sin\theta = \tan\theta = \theta \tag{1.2}$$

may be made. This assumption is known as the paraxial approximation, and geometric optics in this approximation is called Gaussian optics.

Equation (1.1) may conveniently be written as

$$|R_2) = \mathsf{S}|R_1), \tag{1.3}$$

where

$$|R_2) = \begin{pmatrix} y_2 \\ \theta_2 \end{pmatrix}, \qquad |R_1) = \begin{pmatrix} y_1 \\ \theta_1 \end{pmatrix}, \qquad \mathsf{S} = \begin{pmatrix} s_{11} & s_{12} \\ s_{21} & s_{22} \end{pmatrix}.$$

Following the notation of Appendix I, the symbol $|R)$ indicates a column vector. The information about the entrant and emergent rays has been included in the column vectors $|R_1)$ and $|R_2)$, and all pertinent information about the system has been isolated in the 2×2-matrix S.

If we have a composite system as shown in Fig. 1.2, then in the paraxial approximation we have

$$|R_2) = \mathsf{S}_A|R_1), \qquad |R_3) = \mathsf{S}_B|R_2),$$

and

$$|R_3) = \mathsf{S}_B\mathsf{S}_A|R_1).$$

Thus, if the individual system matrices are known, the matrix of the composite system is the product of the component system matrices. Consequently, if we are to represent any optical array by a matrix, we must find matrices that represent the simplest possible subsystems of such an array. This amounts to determining the matrices that represent refraction, reflection, and translation. When we have these elementary matrices, any Gaussian array may be

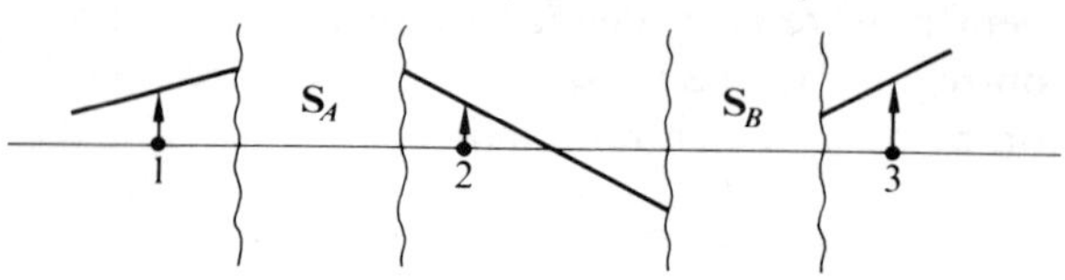

Fig. 1.2. A composite optical system.

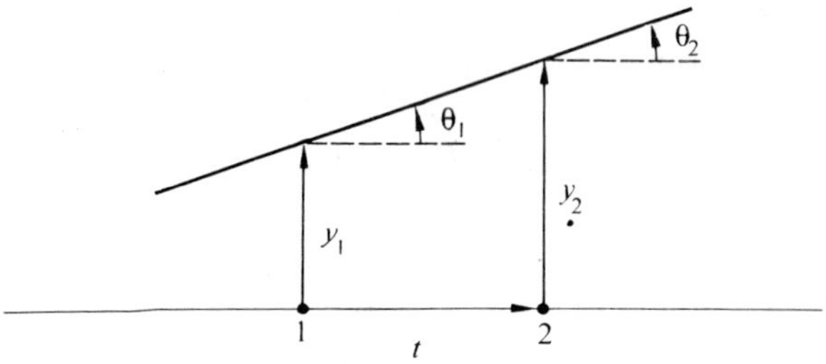

Fig. 1.3. Simple translation of a ray.

represented as a product of the elementary matrices. The matrix elements of this product can then provide information about focal lengths, image-object relations, etc. The determination of the matrix for an optical system and the extraction from the matrix elements of all pertinent information are the problems to be examined in this chapter.

1.2 THE MATRIX OPERATORS

The elementary matrix operators mentioned in the last section will now be derived in the paraxial approximation. It is assumed that reflecting and refracting surfaces possess rotational symmetry and are spherical, and that the refracting media are homogeneous.

a) The Translation Matrix

It is a simple matter to construct a matrix which describes a light ray in a homogeneous medium. It is evident from Fig. 1.3 that the relation between the coordinates of the ray above points 1 and 2 are

$$y_2 = y_1 + t \tan\theta, \qquad \theta_2 = \theta_1. \tag{1.4}$$

This relation is not linear, since $\tan\theta$ appears rather than θ. These equations can, however, be put in the form of the Gaussian approximation if we limit ourselves to those rays for which θ is small enough to use Eq. (1.2). If this is done, Eqs. (1.4) may be written in compact matrix form, where the translation matrix T_{21} is given by

$$\mathsf{T}_{21} = \begin{pmatrix} 1 & t \\ 0 & 1 \end{pmatrix}, \tag{1.5}$$

with t considered as a positive quantity.

Here, and in all that follows, heights above the axis are considered as positive and angles are taken as positive if the ray has a positive slope, i.e., if y increases from left to right.

In obtaining the matrix T_{21} no consideration was given to the direction in which light was travelling along the ray, and T_{21} correctly relates the ray at point 2 to that at point 1 regardless of the direction in which light travels along the ray. If $|R_2)$ is known and $|R_1)$ is desired, the equation $|R_2) = \mathsf{T}_{21}\,|R_1)$ may be inverted to yield $|R_1) = \mathsf{T}_{21}^{-1}\,|R_2)$. The inverse matrix is computed to be

$$\mathsf{T}_{12} = \mathsf{T}_{21}^{-1} = \begin{pmatrix} 1 & -t \\ 0 & 1 \end{pmatrix}. \tag{1.6}$$

Thus, the form of Eq. (1.5) may be used to relate any two points i and j separated by a distance d with

$$\mathsf{T}_{ij} = \begin{pmatrix} 1 & d \\ 0 & 1 \end{pmatrix}, \tag{1.7}$$

where d is positive if point j is to the left of point i and negative if it is to the right.

b) The Refraction Matrix

In Fig. 1.4 two distinct regions are separated by a spherical boundary whose radius of curvature is r. As usual, the index of refraction, n, is defined as the ratio of the velocity of light *in vacuo*, c, to the velocity of light in the medium, v: $n = c/v$. The two regions are characterized by their indices of refraction, n_1 and n_2. It is also convenient to define a relative index of refraction n_{12} by $n_{12} = n_1/n_2$. Since the radius r is the normal direction at the point of refraction, we write, using Snell's law,

$$n_1 \sin \phi_1 = n_2 \sin \phi_2, \tag{1.8}$$

which in the Gaussian approximation becomes

$$n_1\phi_1 = n_2\phi_2. \tag{1.9}$$

From Fig. 1.4 we see that

$$\phi_1 = \psi + \theta_1, \qquad \phi_2 = \psi + \theta_2. \tag{1.10}$$

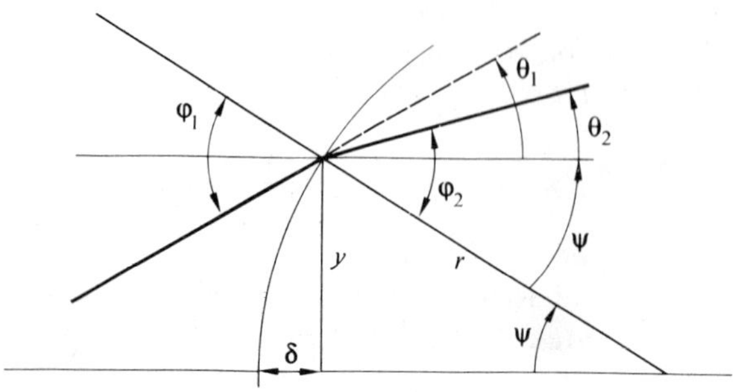

Fig. 1.4. Refraction of a ray at a spherical surface.

To the same approximation, we have $\psi = yK$, where the curvature of the surface is defined as $K = 1/r$. Since the height of the ray does not change in crossing the boundary, $y = y_1 = y_2$ and Eqs. (1.8), (1.9), and (1.10) may be solved to yield

$$\begin{aligned} y_2 &= y_1, \\ \theta_2 &= (n_{12} - 1)Ky_1 + n_{12}\theta_1. \end{aligned} \tag{1.11}$$

If ψ is small, then δ may be neglected.

If $|R_1)$ and $|R_2)$ are the column vectors that represent the incident and the refracted rays, we can write

$$|R_2) = \mathsf{R}_{21}\,|R_1), \tag{1.12}$$

where

$$\mathsf{R}_{21} = \begin{pmatrix} 1 & 0 \\ (n_{12} - 1)K & n_{12} \end{pmatrix}. \tag{1.13}$$

We have assumed that the curvature is positive when the surface is curved as shown, i.e., convex when viewed from the left. In practice it is best to enter all curvatures as positive symbols and to enter positive or negative values only in calculations.

As in the development of the translation matrix, we have made no mention of the direction in which the ray travels, since this was again unnecessary. If the ray on the right of the surface is known, the ray on the left may be found by inverting Eq. (1.12):

$$|R_1) = \mathsf{R}_{21}^{-1}\,|R_2).$$

Here $\mathsf{R}_{21}^{-1} = \mathsf{R}_{12}$ is found to be

$$\mathsf{R}_{12} = \begin{pmatrix} 1 & 0 \\ (n_{21} - 1)K & n_{21} \end{pmatrix}.$$

Thus, the same refraction matrix is used regardless of which way the ray is traced, so long as the proper relative index of refraction is used.

c) The Reflection Matrix

In Fig. 1.5 a ray is incident from the left on a spherical reflecting surface of curvature K. The same sign convention is used here as for refracting surfaces. The law of reflection governs the relation between $|R_1)$ and $|R_2)$.

Since the radius of curvature is normal to the surface,

$$|\phi_1| = |\phi_2|. \tag{1.14}$$

Since both θ_1 and ψ are shown as positive angles,

$$|\phi_1| = \psi + \theta_1. \tag{1.15}$$

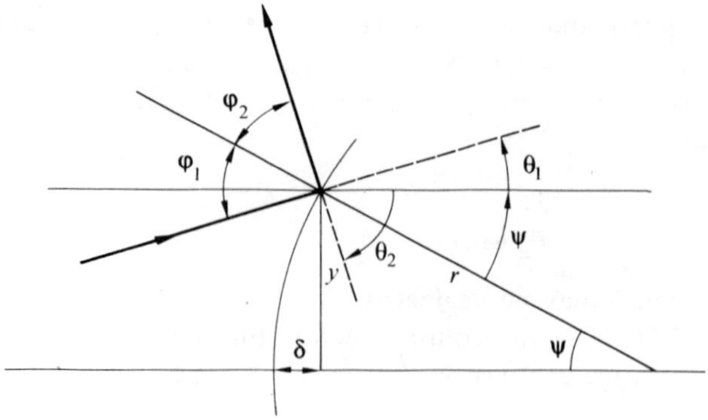

Fig. 1.5. Reflection of a ray at a spherical surface.

However, θ_2 is drawn as a negative angle, so we have

$$\begin{aligned}|\phi_2| &= |\theta_2| - \psi \\ &= -\theta_2 - \psi.\end{aligned} \tag{1.16}$$

Again taking $y = y_1 = y_2$, neglecting δ, and setting $\psi = yK$, Eqs. (1.14), (1.15), and (1.16) yield

$$y_2 = y_1, \qquad \theta_2 = -2Ky_1 - \theta_1. \tag{1.17}$$

The rays $|R_1)$ and $|R_2)$ are thus related by the reflection or mirror matrix M_{21} through $|R_2) = \mathsf{M}_{21}\,|R_1)$, with

$$\mathsf{M}_{21} = \begin{pmatrix} 1 & 0 \\ -2K & -1 \end{pmatrix}. \tag{1.18}$$

In the derivation of Eq. (1.18) both the incident and reflected rays were taken on the left-hand side of the surface. The solid rays on the left and their dashed extensions on the right represent the same rays, having the same heights above the axis at the surface and the same inclinations. Equation (1.18) would have been obtained if we had assumed the light to be incident from the right along the dashed rays. Therefore, the same matrix relates incident and reflected rays, for both convex and concave mirrors, regardless of on which side of the mirror the rays intersect.

The optical systems which we will study will consist of various refracting and reflecting surfaces separated by different thicknesses of homogeneous materials of different indices of refraction. Any such system has a matrix which can be written as a product of the elementary matrices we have developed. In order to show how these elementary matrices are used to obtain a matrix describing a system, we will now derive a matrix for a "thick lens" system. The significance of the matrix elements for this system will become clear in Section 1.4.

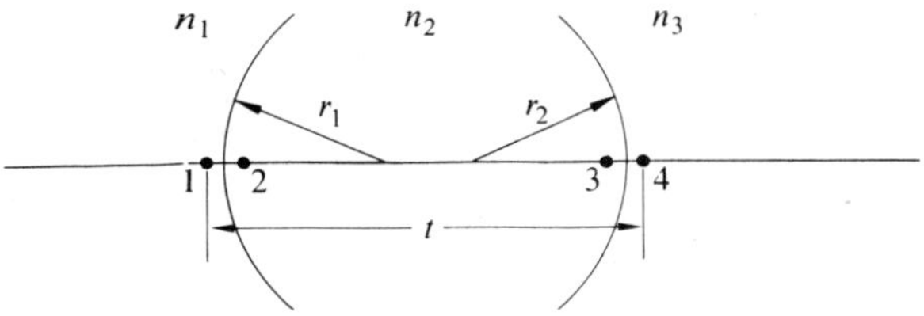

Fig. 1.6. A single thick lens.

Example 1. In Fig. 1.6 three media with indices n_1, n_2, and n_3 are separated by two refracting surfaces of radii r_1 and r_2. The surfaces are a distance t apart. The matrix V_{41} relates a ray passing over point 1 just before the first surface to the emergent ray which passes over point 4 just after it leaves the last surface. This matrix is given as the product of two refracting matrices and one translation matrix,

$$\mathsf{V}_{41} = \mathsf{R}_{43}\mathsf{T}_{32}\mathsf{R}_{21}.$$

Note that as we proceed through an optical system from left to right we must multiply the appropriate matrices from right to left. The elementary matrices are given by

$$\mathsf{R}_{21} = \begin{pmatrix} 1 & 0 \\ (n_{12} - 1)K_1 & n_{12} \end{pmatrix}, \qquad \mathsf{T}_{32} = \begin{pmatrix} 1 & t \\ 0 & 1 \end{pmatrix},$$

$$\mathsf{R}_{43} = \begin{pmatrix} 1 & 0 \\ (n_{23} - 1)K_2 & n_{23} \end{pmatrix}.$$

The reader may verify that the product yields

$$\mathsf{V}_{41} = \begin{pmatrix} v_{11} & v_{12} \\ v_{21} & v_{22} \end{pmatrix},$$

where

$$\begin{aligned} v_{11} &= 1 + (n_{12} - 1)K_1 t, \\ v_{12} &= n_{12} t, \\ v_{22} &= n_{13} + tK_2(n_{13} - n_{12}), \\ v_{21} &= (n_{13} - n_{23})K_1 + (n_{23} - 1)K_2 + \frac{(n_2 - n_3)(n_1 - n_2)}{n_2 n_3} K_1 K_2 t. \end{aligned}$$

It should be noted that if the system is as shown in Fig. 1.6, K_1 is a positive quantity and K_2 is negative according to the sign convention adopted for convex and concave surfaces.

An extremely useful relation may be obtained by noting that the determinants of the T-, R-, and M-matrices are, respectively, 1, n_{ij}, and -1. The determinant of a product of matrices is the product of the individual determinants. It is left to the reader to show that this implies that the determinant of the matrix for an optical system represented by the matrix S is

$$\text{Determinant (S)} = (n_{\text{initial}}/n_{\text{final}})(-1)^N, \tag{1.19}$$

where N is the number of reflecting surfaces in the system. Since

$$\text{Det (S)} = s_{11}s_{22} - s_{12}s_{21},$$

given any three elements of S, the fourth may be found if the initial and final media are known and if the number of reflecting surfaces is also known. We will often write Det (S) $= D$, or $|\mathsf{S}|$.

1.3 THE S- AND V-MATRICES

A matrix associated with an optical system is also associated with two points on the optic axis and with two planes perpendicular to the axis which pass through these points, e.g., points 1 and 2 in Fig. 1.1. Such a matrix may be termed a *system* matrix S. These reference points and planes are not unique. Much as the origin of a coordinate system may be placed wherever convenient, so the reference points of an optical system may be taken wherever we wish. Moving the reference points will, of course, modify the appearance of the system matrix, but cannot alter the behavior of the actual physical optical system.

If the state of a ray is known as it passes through the first reference plane, the matrix of the system allows us to find the state of the ray as it passes through the second plane. It is often useful to pick out as initial and final planes those which touch the initial and final vertices of the optical system. The matrix between these planes will be called the vertex matrix and is given the symbol V. Any matrix associated with the system may be termed a system matrix S, with the V-matrix being a special case of S.

If for a given system an S-matrix is known between two points 1 and 2, then it is possible to find the system matrix S′, with elements s_{ij} between any other two points 1′ and 2′, as indicated in Fig. 1.7.

The quantities z and z' are taken as positive if measured from left to right. The system matrix connecting 1′ and 2′ is then given by

$$\mathsf{S}' = \mathsf{T}_{z'}\mathsf{S}\mathsf{T}_z = \begin{pmatrix} 1 & z' \\ 0 & 1 \end{pmatrix}\begin{pmatrix} s_{11} & s_{12} \\ s_{21} & s_{22} \end{pmatrix}\begin{pmatrix} 1 & z \\ 0 & 1 \end{pmatrix},$$

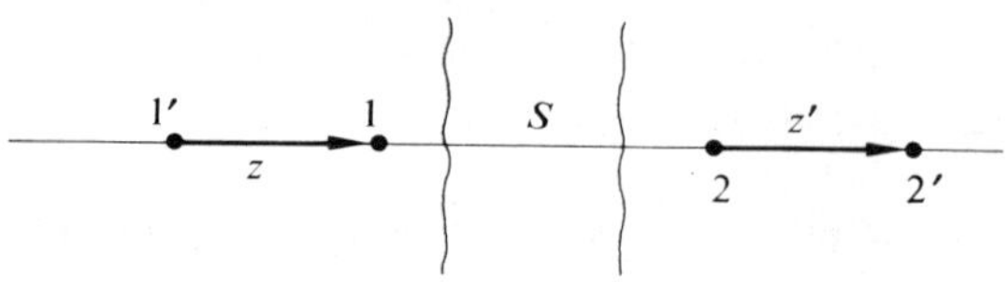

Fig. 1.7. Translation of reference points for an optical system.

where the points 1′ and 1 are connected through T_z and the points 2 and 2′ through $\mathsf{T}_{z'}$. The result of the matrix multiplication is

$$\mathsf{S}' = \begin{pmatrix} s_{11} + z's_{21} & zs_{11} + z's_{22} + zz's_{21} + s_{12} \\ s_{21} & s_{22} + zs_{21} \end{pmatrix}. \tag{1.20}$$

The matrix given in Eq. (1.20) is of basic importance to what follows. It can, for example, permit an evaluation of image-object relationships. The reader should be sure that he understands its derivation and the meaning of each symbol in this equation; in particular the meaning of positive and negative values for z and z'.

1.4 SIGNIFICANCE OF THE INDIVIDUAL ELEMENTS s_{ij}

The four elements, s_{ij}, of a system matrix contain the total information which describes that system in the paraxial approximation. This information is shared among the elements. From Eq. (1.19) the determinant of a system matrix cannot be zero. This implies that at most two elements of the matrix can be zero, and if two are zero, they must be the two diagonal elements or the two off-diagonal elements. The implications of vanishing matrix elements will now be investigated.

a) $s_{11} = 0$

If s_{11} is zero, Eqs. (1.1) imply

$$\begin{aligned} y_2 &= s_{12}\theta_1, \\ \theta_2 &= s_{21}y_1 + s_{22}\theta_1. \end{aligned} \tag{1.21}$$

Thus, if s_{11} is zero, the height of a ray at 2 is independent of its original height at 1. All rays which pass through the initial plane with the same inclination, θ_1, pass through a common point in the final plane. This is illustrated in Fig. 1.8.

That plane in which all parallel entering rays pass through a common point is called the *second focal plane* of the system, FP_2. The point of intersection of the second focal plane and the optic axis is called the *second focal point*, Fp_2.

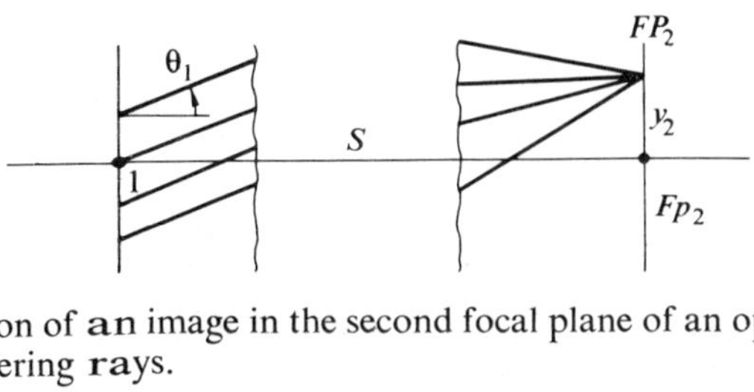

Fig. 1.8. Formation of an image in the second focal plane of an optical system for a set of parallel entering rays.

In general, s_{11} will not be zero for a system matrix and the point 2 will not be the second focal point, Fp_2. If s_{11} is not zero, let us assume that Fp_2 is some distance z' to the right of point 2. Then, point 2′ must be Fp_2 and in Eq. (1.20) s'_{11} must be zero. This yields

$$z' = -s_{11}/s_{21}. \tag{1.22}$$

Thus, given a system matrix S between any two points 1 and 2, the second focal point, Fp_2, is a distance $-s_{11}/s_{21}$ to the right of point 2. If $-s_{11}/s_{21}$ happens to be a negative quantity, the second focal point is to the left of point 2.

b) $s_{22} = 0$

When s_{22} is zero, we have from Eqs. (1.1)

$$\begin{aligned} y_2 &= s_{11}y_1 + s_{12}\theta_1, \\ \theta_2 &= s_{21}y_1. \end{aligned} \tag{1.23}$$

All rays which pass through the initial plane at a height y_1 above the axis emerge from the system at an inclination θ_2 which is independent of their inclination θ_1. This is illustrated in Fig. 1.9.

That plane to the left of the system such that the emergent inclination is independent of the entering inclination, depending only on the entrance height, is called the *first focal plane* of the system, FP_1. The point where this plane intersects the optic axis is called the *first focal point*, Fp_1. Alternatively, it is the plane for which parallel rays entering the system from the right pass through a common point when emerging on the left.

As before, the system matrix will usually not have $s_{22} = 0$. In this case we assume that Fp_1 is some distance z to the left of point 1. Then, demanding that $s'_{22} = 0$ in Eq. (1.20), we obtain

$$z = -s_{22}/s_{21}. \tag{1.24}$$

Therefore, if the elements of the system matrix S, between points 1 and 2, are known, Fp_1 is a distance $-s_{22}/s_{21}$ to the left of point 1. As before, if $-s_{22}/s_{21}$ is negative, Fp_1 is to the right of point 1.

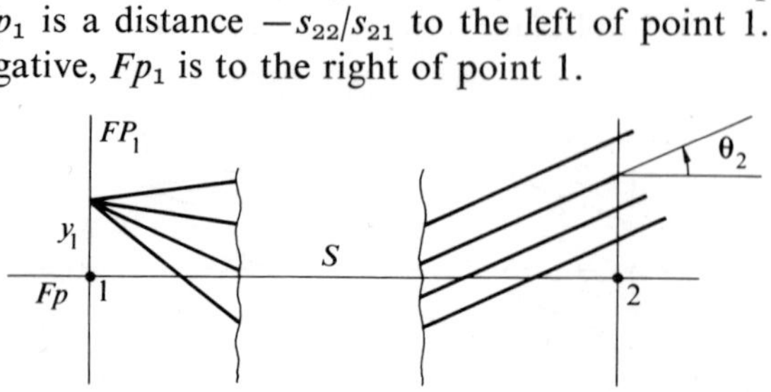

Fig. 1.9. Rays diverging from a point in the first focal plane of an optical system leave as a parallel system of rays.

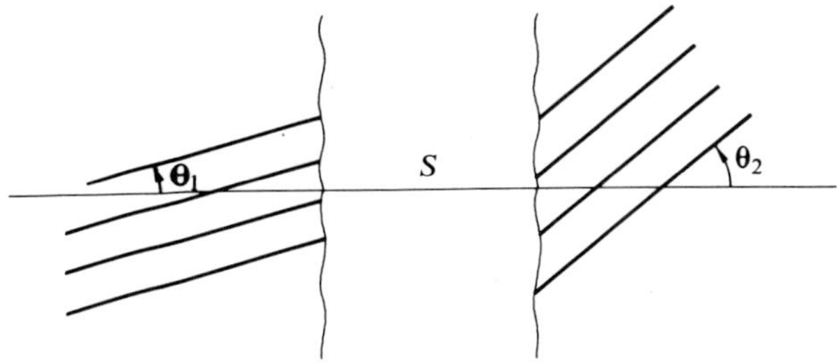

Fig. 1.10. Behavior of a system of parallel rays in a telescopic system.

c) $s_{21} = 0$

In the first two cases, if s_{11} or s_{22} was not zero, it was possible to translate to a system in which s'_{11} or s'_{22} was zero. This is not possible for s_{21}, since in Eq. (1.20) it is seen that $s'_{21} = s_{21}$ and translation of the reference points has no effect on this element. This is the only element which translation does not affect, and s_{21} must therefore be determined only by the system and not by the location of the reference points 1 or 2.

The special significance of s_{21} will be examined later, but at this point we consider the very special case $s_{21} = 0$. From Eq. (1.1) we then have

$$\begin{aligned} y_2 &= y_1 s_{11} + \theta_1 s_{12}, \\ \theta_2 &= \theta_1 s_{22}, \end{aligned} \tag{1.25}$$

and the emergent angle of any ray depends only on the entrance inclination. In Fig. 1.10 illustrating this case the points 1 and 2 are not shown; for if s_{21} is zero between one pair of reference points, it is zero between any pair. A system for which s_{21} is zero is called a *telescopic system*. The *angular magnification* of such a system is defined to be θ_2/θ_1 and is given by

$$\text{Angular magnification} = s_{22}. \tag{1.26}$$

It is important to note that although angular magnification may be defined for any ray in any system, it may be different for different rays in a given system. Only for a telescopic system is the phrase "angular magnification of the system" meaningful.

d) $s_{12} = 0$

If s_{12} is zero, Eqs. (1.1) yield

$$\begin{aligned} y_2 &= s_{11} y_1, \\ \theta_2 &= s_{21} y_1 + s_{22} \theta_1, \end{aligned} \tag{1.27}$$

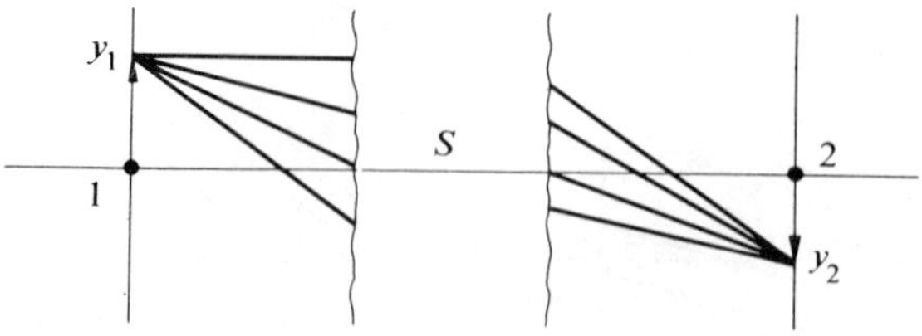

Fig. 1.11. Conjugate planes of an optical system.

which, as illustrated in Fig. 1.11, implies that the height of a ray passing through the second plane depends only on the height of the ray when it left the first plane and is independent of θ_1. All rays leaving a common point in the first plane converge at a common point in the second plane. Such points and planes are said to be conjugate points and planes, respectively. If one of the points is a source of light, its image is simply its conjugate point and the corresponding planes are frequently called object and image planes. From Eqs. (1.20), setting $s'_{12} = 0$, it is seen that, given any system matrix between points 1 and 2, pairs of conjugate points exist a distance z to the left of 1 and z' to the right of 2 related by

$$\frac{s_{11}}{z'} + \frac{s_{22}}{z} + \frac{s_{12}}{zz'} = -s_{21}. \tag{1.28}$$

When two planes associated with an optical system are conjugate, one may define the *lateral magnification* between these planes. Referring to Fig. 1.11, the definition is given as

$$\text{Lateral magnification} = y_2/y_1 = s_{11}. \tag{1.29}$$

A good illustration of how the foregoing results apply is given below for a thin lens.

Example 2. By a thin lens we understand the optical system of Example 1 with the distance t small enough to be ignored. We further assume that $n_1 = n_3 = 1$ and set $n_2 = n$. The vertex matrix V_{41} in Example 1 then becomes

$$\mathsf{V}_{41} = \begin{pmatrix} 1 & 0 \\ -(n-1)(K_1 - K_2) & 1 \end{pmatrix},$$

and from Eqs. (1.22) and (1.24) the first and second focal points are located distances f to the right and left of the lens given by

$$\frac{1}{f} = (n-1)(K_1 - K_2). \tag{1.30}$$

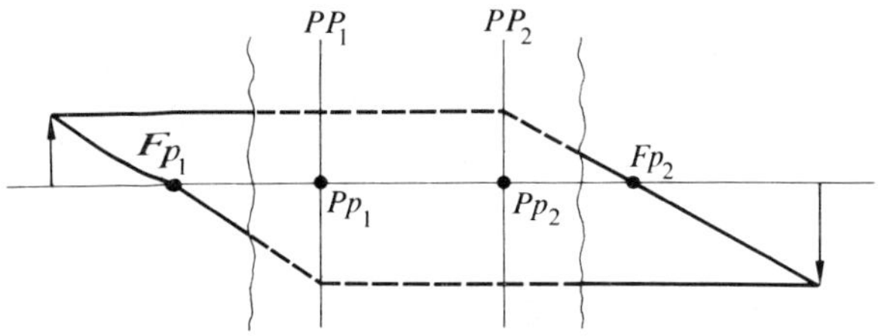

Fig. 1.12. The principal planes and points of an optical system.

The relation between object and image distances measured from the lens is given by Eq. (1.28) as

$$\frac{1}{z} + \frac{1}{z'} = \frac{1}{f}, \tag{1.31}$$

which is the well-known thin lens equation.

1.5 THE PRINCIPAL PLANES AND LENS EQUATIONS

Any ray which enters an optical system from the left, parallel to the optic axis, passes through the second focal point. Any ray which passes through the first focal point leaves the system traveling parallel to the axis. Typical rays of each kind are shown in Fig. 1.12. These rays may be refracted and reflected many times in passing through the system and we do not know the exact path which they follow. In Fig. 1.12 the rays have been extended into the system as dotted lines until the entering ray intersects the emergent ray. The entire effect of the system on each ray can in this way be represented by one deviation, located at the planes PP_1 and PP_2 which intersect the axis at the points Pp_1 and Pp_2. These planes and points are called, respectively, the first and second *principal planes* and *points* of the system.

From Fig. 1.12 it is seen that if the locations of the principal planes and focal points are known, the image location for any object may be obtained by a geometric construction. The order from left to right of object, image, focal points, and principal planes need not be as given in Fig. 1.12. For example, in Fig. 1.13 the object is placed inside the first focal point and the order of the principal planes is reversed. Ray A, which enters parallel to the axis, must always be extended until it meets the second principal plane and

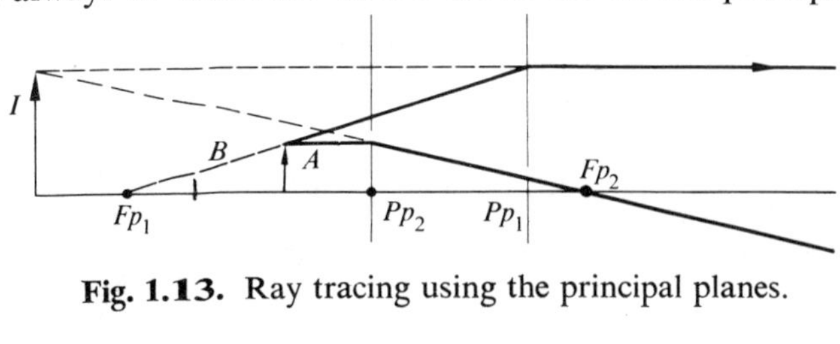

Fig. 1.13. Ray tracing using the principal planes.

then bent to pass through Fp_2. Ray B, although it does not actually pass through Fp_1, behaves as if it had and, hence, must be extended until it strikes PP_1 and then bent to leave the system parallel to the axis. Rays A and B never do pass through a real image point. Extended backward, however, they intersect at point I. Such an image is called a *virtual image*. Although the rays do not pass through I, upon leaving the system they behave as if they were diverging from I.

The examples given are for a system which must possess an even number of reflecting surfaces. The rays enter from the left and leave on the right. Ray tracing for a mirror system, or a system with an odd number of reflecting surfaces, proceeds in exactly the same manner. However, when the final rays are found, they must be extended backward to emerge from the left of the system.

If we assume that none of the four points Fp_1, Fp_2, Pp_1, and Pp_2 are coincident, and since their locations can in principle be independently varied, there are 24 different orders in which they may occur along the axis. If the different locations of the object are then considered, we find in general 120 different cases in which an image location may be determined by the ray tracing techniques illustrated in Figs. 1.12 and 1.13. The remaining 118 cases are left for the problems!

We now locate the principal points of an optical system in terms of the matrix elements which determine that system. Consider an object which is very close to the first principal plane. Its image is found by ray tracing in Fig. 1.14. It is seen that the image is very close to the second principal plane and that in the limit as the object approaches PP_1 the image approaches PP_2 and y_2 approaches y_1. Therefore, when an object is located in PP_1, the image is located in PP_2 and the lateral magnification is unity. The principal planes are often called the planes of unit lateral magnification.

Assume that the system matrix, S, is given between two points 1 and 2, as in Fig. 1.7. Furthermore assume that PP_1 is located at 1′ and PP_2 at 2′. Since the principal planes are conjugate, $s'_{12} = 0$ in Eq. (1.20), and since they are planes of unit magnification, $s'_{11} = 1$. These yield

$$s_{11} + z's_{21} = 1, \tag{1.32}$$

$$zs_{11} + z's_{22} + zz's_{21} + s_{12} = 0. \tag{1.33}$$

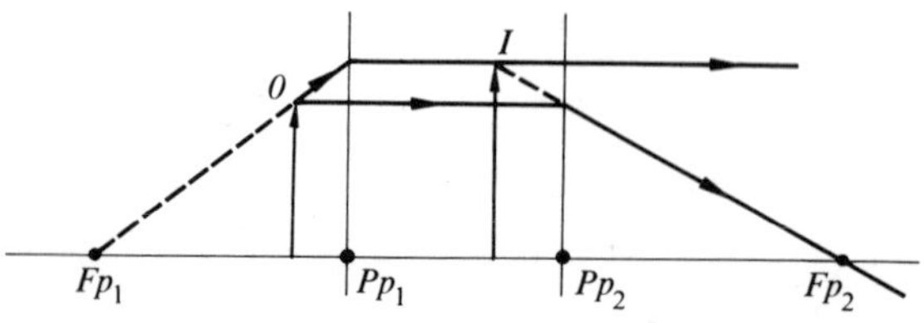

Fig. 1.14. Geometric construction leading to the definition of the principal planes as planes of unit lateral magnification.

Eq. (1.32) implies that, given any system matrix between points 1 and 2, the second principal point is located a distance to the right of 2 given by

$$z' = \frac{1 - s_{11}}{s_{21}} = Pp_2. \tag{1.34}$$

If Eq. (1.34) is placed in Eq. (1.33), we find the first principal point located a distance to the left of 1 given by

$$z = \frac{\text{Det (S)} - s_{22}}{s_{21}} = Pp_1, \tag{1.35}$$

where Det (S) $= s_{11}\, s_{22} - s_{12}\, s_{21}$.

Given an arbitrary system matrix, the locations of the focal points of the system are given by Eqs. (1.22) and (1.24). Combining these equations with Eqs. (1.34) and (1.35), we learn that Pp_1 is a distance f_1 to the right of Fp_1 given by

$$f_1 = -\text{Det (S)}/s_{21}, \tag{1.36}$$

and the distance right from Pp_2 to Fp_2, called f_2, is

$$f_2 = -1/s_{21}. \tag{1.37}$$

The quantities f_1 and f_2 are known, respectively, as the *first* and *second focal lengths* of the system. It is easily seen that for any system

$$f_1/f_2 = \text{Det (S)}. \tag{1.38}$$

From Eq. (1.19) it is seen that the magnitude of the focal lengths will be equal only if the initial and final indices of refraction are equal.

It is of great practical importance to note that the second focal lengths of *any* optical system, whose system matrix S relates *any* two points 1 and 2, is always given by the negative reciprocal of the element s_{21}. This is the special significance of this element which was mentioned earlier. We have seen how to trace two very special rays through a system, one which enters parallel to the axis and one which leaves parallel. We now determine how to trace an arbitrary ray.

The matrix which connects the principal points of a system will be called the P-matrix of the system. Since the principal planes are conjugate planes of unit magnification, the element p_{12} must be zero and we may immediately write, using Eq. (1.37),

$$\mathsf{P} = \begin{pmatrix} 1 & 0 \\ -1/f_2 & \text{D} \end{pmatrix}. \tag{1.39}$$

If we operate with P on an arbitrary ray striking PP_1, we see that it must leave PP_2 at the same height above the axis as it had at PP_1. Only the angle at which it leaves PP_2 must be determined. This angle is found by imagining

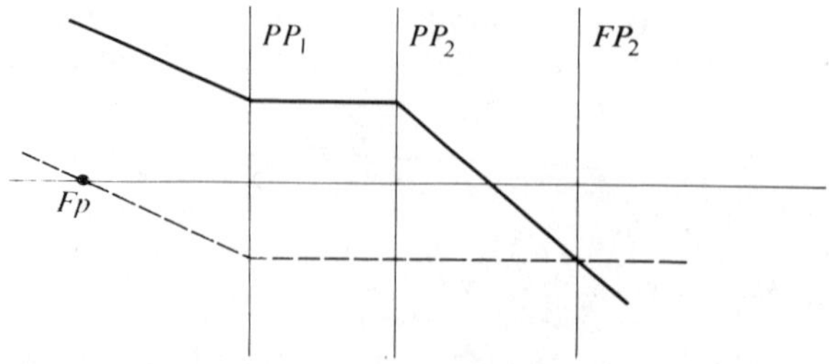

Fig. 1.15. Ray tracing for an arbitrary ray.

that the arbitrary ray is one of a set of parallel rays striking PP_1 which must pass through a common point in the second focal plane. One of these rays must pass through the first focal point and, hence, leave the second principal plane parallel to the axis. Where this ray intersects the plane FP_2, so does the arbitrary ray. This is illustrated in Fig. 1.15.

The matrix connecting the principal points is given by Eq. (1.39). Using this matrix, we may find the matrix connecting a point O, a distance z to the left of Pp_1, and a point I, a distance z' to the right of Pp_2, and thereby derive the Gaussian form of the equation relating image and object distances.

The matrix connecting O and I is

$$\mathsf{M} = \mathsf{T}_{z'}\mathsf{P}\mathsf{T}_{z} = \begin{pmatrix} 1 - z'/f_2 & z + z'D - \dfrac{zz'}{f_2} \\ -\dfrac{1}{f_2} & D - z/f_2 \end{pmatrix}. \tag{1.40}$$

If we demand that O and I be conjugate, we stipulate

$$z + z'D - \frac{zz'}{f_2} = 0. \tag{1.41}$$

Dividing by zz' and remembering that $D = f_1/f_2$, we obtain

$$\frac{f_1}{z} + \frac{f_2}{z'} = 1. \tag{1.42}$$

If the final and initial media have the same index of refraction and the system contains an even number of reflecting surfaces, $D = 1$ and we obtain the Gaussian form of the lens equation,

$$\frac{1}{z} + \frac{1}{z'} = \frac{1}{f}, \tag{1.43}$$

where $f_1 = f_2 = f$.

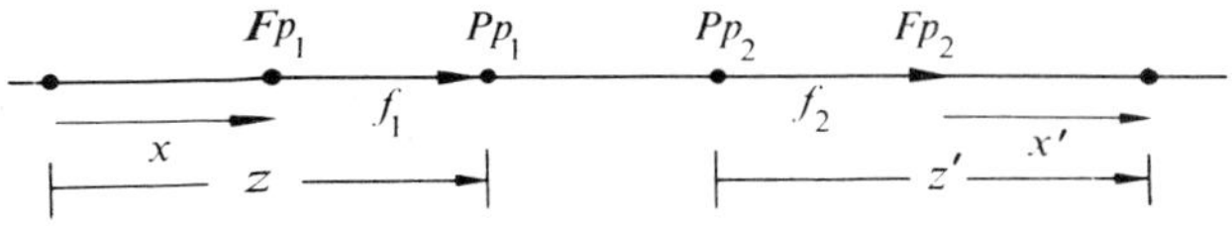

Fig. 1.16. Geometry for the Newtonian form of the lens equation.

If the system contains an odd number of reflecting surfaces, the Gaussian form of the mirror equation is obtained,

$$\frac{1}{z} - \frac{1}{z'} = \frac{1}{f}, \tag{1.44}$$

where $f_1 = -f_2 = f$. Known as the "thin mirror" equation, this is the equation obtained in elementary treatments of a single reflecting surface. By contrast, a thick mirror system would be any system containing more than a single reflecting surface.

Thus, any Gaussian system obeys either the thin lens equation or the thin mirror equation if object and image distances are measured from the principal points.

An alternative form of Eq. (1.43), first obtained by Newton, is found if distances are measured not from the principal planes but from the focal points. In Fig. (1.16) let $x = z - f_1$, $x' = z' - f_2$. Substituting into Eq. (1.43) leads to

$$xx' = f_1 f_2, \tag{1.45}$$

or, if $D = \pm 1$,

$$xx' = \pm f^2. \tag{1.46}$$

1.6 THE NODAL POINTS

The matrix of an optical system connecting two reference points contains only four independent elements, and only three are independent if the initial and final indices are specified. If the locations of the principal points and focal points are given, the matrix elements may be calculated. Experimentally, the focal points may be determined by locating the image of a distant object, which, from Eq. (1.42), must lie in the focal plane. The experimental location of the principal planes is not always so easy. A third set of points, the *nodal points*, which may be readily determined by experiment, is therefore defined. It will be seen that the nodal points, Np_1 and Np_2, and the principal points coincide when the determinant of the system is unity.

We have seen that any ray striking PP_1, at height y above the axis, must emerge from PP_2 at the same height. In particular, the ray which strikes Pp_1 must travel along the optic axis and emerge at Pp_2. Using the P-matrix, we see that the emergent angle of this ray is equal to the incident angle

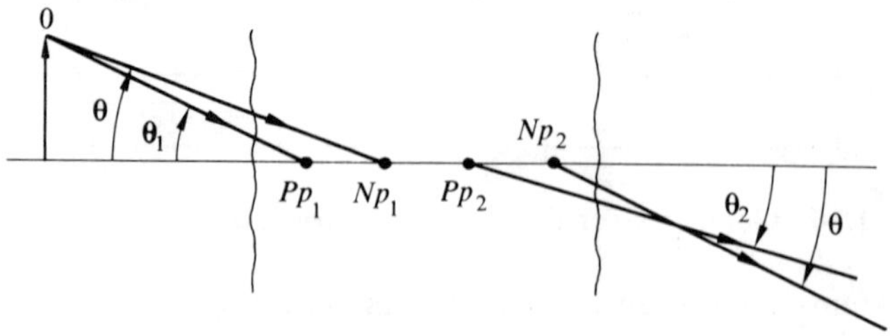

Fig. 1.17. Location of the nodal points of an optical system.

only if $D = 1$. This follows, since if $y = 0$, Eq. (1.39) implies $\theta_2 = D\theta_1$. This is illustrated in Fig. 1.17, where $\theta_2 \neq \theta_1$. Thus, the principal ray (the ray which strikes the axis at Pp_1 and leaves the axis at Pp_2) has unequal initial and final inclinations. The nodal ray is defined as that ray for which the initial and final inclinations are equal. As in Fig. 1.17, the intersections with the optic axis of the entrant and emergent nodal rays are called the first and second nodal points of the system, Np_1 and Np_2. The planes perpendicular to the axis which pass through the nodal points are called the first and second *nodal planes*, NP_1 and NP_2. The existence of the nodal points is verified by determining their location in terms of the matrix elements which determine the system.

Assuming that the nodal points exist, the matrix which connects them can be found. The matrix connecting the nodal points will be given the symbol N, with elements n_{ij}. It is seen that n_{12} must be zero, since y_2 is zero if y_1 is zero. Since θ_2 must equal θ_1 when $y_1 = y_2 = 0$, n_{22} must be unity. These conditions imply

$$\mathsf{N} = \begin{pmatrix} D & 0 \\ -1/f_2 & 1 \end{pmatrix}. \tag{1.47}$$

Equation (1.47) allows us to determine the locations of the nodal points for an arbitrary system. Assume that the system matrix S is given between two points 1 and 2. Let Np_1 be located at point 1′ a distance z to the left of 1 and Np_2 be located at 2′ a distance z' to the right of 2. Then, if the form of Eq. (1.47) is imposed on Eq. (1.20), we obtain

$$z' = \frac{D - s_{11}}{s_{21}} = Np_2, \tag{1.48}$$

$$z = \frac{1 - s_{22}}{s_{21}} = Np_1. \tag{1.49}$$

By comparing Eqs. (1.34), (1.35) and Eqs. (1.48), (1.49), it is seen that the nodal points coincide with the principal points if and only if $D = +1$.

It remains to be seen how the nodal points may be determined experimentally. In Fig. 1.18 a set of parallel rays enters the system and is brought

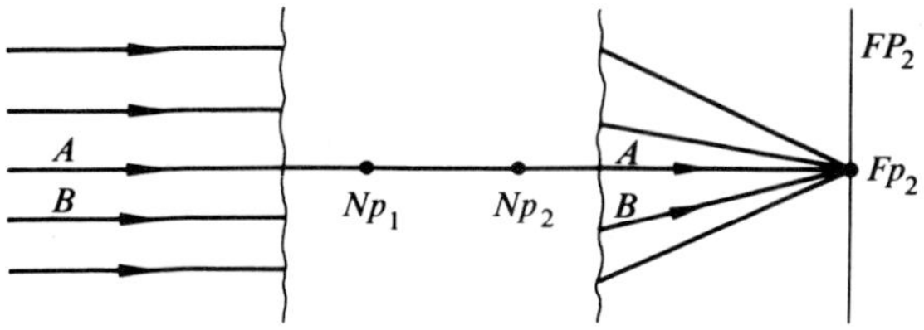

Fig. 1.18. System with nodal points on the optic axis.

to a focus at Fp_2. The axial ray, A, passes through both nodal points at zero inclination and reaches the image point on the axis. In Fig. 1.19 the entire system has been rotated through a small angle ϕ about the second nodal point. It is now ray B which enters the system headed toward Np_1 and which leaves the system parallel to its entering direction. The entering rays are still parallel and must be brought to a focus at the point where ray B crosses FP_2. Thus, by rotating the system about an axis through Np_2 the lateral position of the image of a parallel set of rays does not change. Thus, by examining the image of a distant object, while rotating the system about axes perpendicular to the optic axis, one may locate that axis which passes through Np_2. It should be obvious how one can locate Np_1 in a similar manner. The six points Fp_1 and Fp_2, Pp_1 and Pp_2, and Np_1 and Np_2 are called the *cardinal points* of an optical system. Any four noncoincident cardinal points completely determine the system matrix and the Gaussian behavior of the optical system to which they belong.

1.7 SUMMARY

In the matrix formulation of an optical system in the Gaussian approximation, such a system is determined when the locations of the cardinal points are found. Given an optical system, one must be able to form the system matrix as a product of reflection, refraction, and translation matrices, and from the system matrix obtain the location of the cardinal points.

The location of the cardinal points in terms of the matrix elements are summarized below:

$$f_2 = -1/s_{21}, \qquad f_1 = Df_2,$$
$$Fp_1 = -s_{22}/s_{21}, \quad Fp_2 = -s_{11}/s_{21},$$
$$Pp_1 = \frac{D - s_{22}}{s_{21}}, \quad Pp_2 = \frac{1 - s_{11}}{s_{21}},$$
$$Np_1 = \frac{1 - s_{22}}{s_{21}}, \quad Np_2 = \frac{D - s_{11}}{s_{21}}.$$

The use of these equations may be demonstrated by the following example.

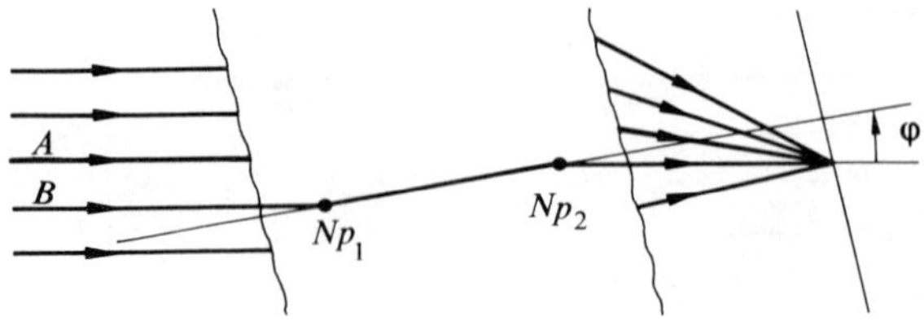

Fig. 1.19. System rotated about nodal point two.

Example 3. As in Fig. 1.20, a glass hemisphere has its flat face silvered. Light impinges on the spherical side from the left. The index of refraction of the glass is 1.5 and it is surrounded by air. Locate the cardinal points of the system. The radius of the spherical surface is taken as the positive quantity R.

The vertex matrix is given by

$$\mathsf{V}_{11} = \mathsf{R}_{12} \quad \mathsf{T}_{23} \quad \mathsf{M}_{33} \quad \mathsf{T}_{32} \quad \mathsf{R}_{21},$$

which, written out in terms of n and R, gives

$$\mathsf{V}_{11} = \begin{pmatrix} 1 & 0 \\ (n-1)\dfrac{1}{R} & n \end{pmatrix} \begin{pmatrix} 1 & -R \\ 0 & 1 \end{pmatrix} \begin{pmatrix} 1 & 0 \\ 0 & -1 \end{pmatrix} \begin{pmatrix} 1 & R \\ 0 & 1 \end{pmatrix} \begin{pmatrix} 1 & 0 \\ \left(\dfrac{1}{n}-1\right)\dfrac{1}{R} & \dfrac{1}{n} \end{pmatrix}$$

$$= \begin{pmatrix} \dfrac{2}{n}-1 & \dfrac{2R}{n} \\ \dfrac{2}{R}\left(1-\dfrac{1}{n}\right) & 1-\dfrac{2}{n} \end{pmatrix}.$$

It is important that the reader understand the origin of each term, thus checking his understanding of the conventions being used.

A useful check on the calculation is the determinant, which is found to be -1.

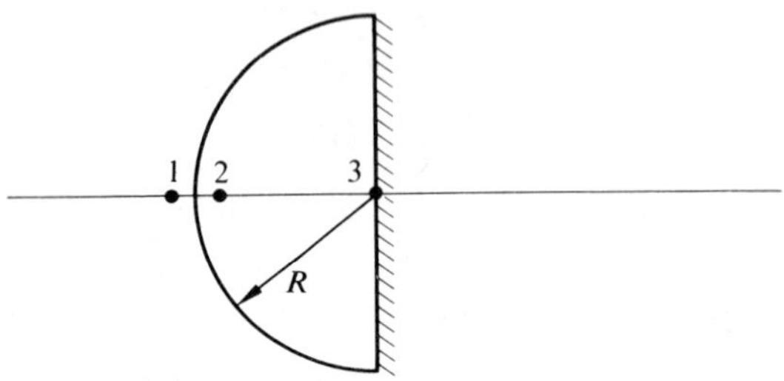

Fig. 1.20. Hemispherical thick mirror with the plane face silvered.

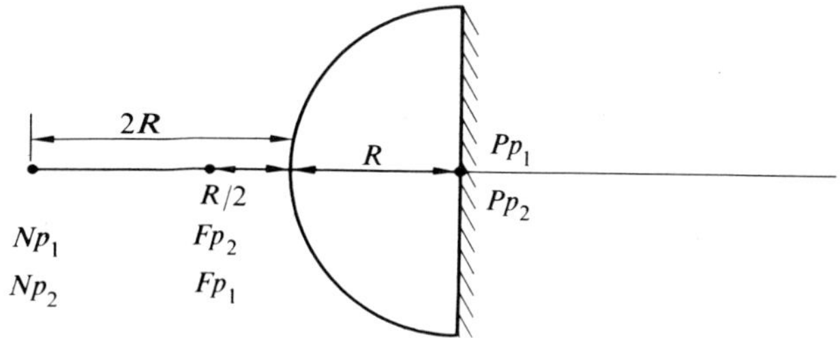

Fig. 1.21. Cardinal points for a hemispherical thick lens.

For $n = 1.5$, we have

$$V = \begin{pmatrix} \dfrac{1}{3} & \dfrac{4R}{3} \\ \dfrac{2}{3R} & -\dfrac{1}{3} \end{pmatrix}.$$

The cardinal points are thus found to be

$$\begin{aligned} Fp_1 &= +R/2, & Fp_2 &= -R/2, \\ Pp_1 &= -R, & Pp_2 &= +R, \\ Np_1 &= +2R, & Np_2 &= -2R. \end{aligned}$$

Since the reference points of the system are both at the spherical vertex, the cardinal points are as shown in Fig. 1.21.

In conclusion, it must be pointed out that we have not discussed the major practical problems of geometric optics, the aberrations of optical systems. For all but the most unsophisticated systems, designs based on the Gaussian approximation are insufficient for the production of quality images. In addition, for all real optical materials the index of refraction is not independent of wavelength and thus the locations of the cardinal points are functions of wavelength or color. Discussion of these difficulties belongs more properly to a textbook of optical design, and would take us too far afield to be included here. However, matrix methods exist to handle these problems also, and the interested reader is referred to the bibliography.

PROBLEMS

1.1 Locate the cardinal points of an optical system consisting of two thin lenses in air, separated by a distance t. If one thin lens has a positive focal length and the other a negative one, show that it is possible for the system to have a positive focal length with neither principal plane between the lenses.

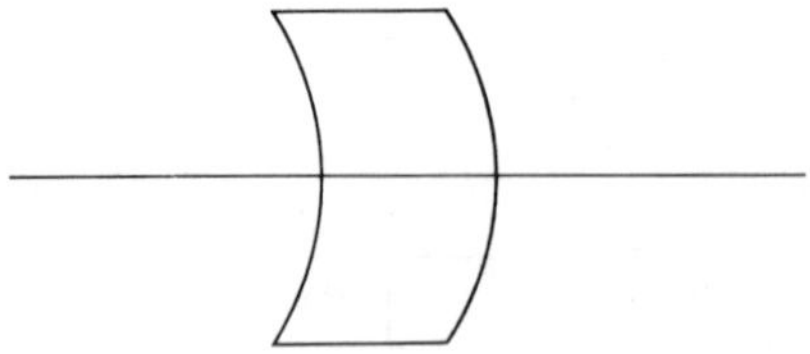

Figure P1.1

Answer:

$$V = \begin{bmatrix} 1 - t/f_1 & t \\ -1/f_1 - 1/f_2 + t/f_1 f_2 & 1 - t/f_2 \end{bmatrix}.$$

1.2 Locate the cardinal points of a sphere of radius R and index of refraction n, with air on both sides of the sphere. By considering the definition of the nodal points prove that they must be located at the center of the sphere without performing a calculation.

Answer: $f = nR/2(n - 1)$.

1.3 Locate the cardinal points of a glass hemisphere whose spherical side is silvered. Light enters from the left through the plane side.

Answer: Principal points, R/n to right of plane surface. Focal point, $R/2n$ to right of plane surface.

1.4 The object-image relation for a thin lens is given by $1/s + 1/s' = 1/f$, where $1/f = (n - 1)(K_1 - K_2)$. For such a thin lens a second, much dimmer, real image is often found to the right of the lens, whose location is given by $1/s + 1/s' = (3n - 1)(K_1 - K_2)$. Explain the origin of this image and derive the equation giving its location.

1.5 The lens shown in Fig. P1.1 is surrounded by air. The radii of curvature are equal and the thickness of the lens is t. Prove that the principal planes are separated by the same distance t and that the focal length is positive.

1.6 If the surfaces of the lens in Problem 1.5 are not of equal curvature but have a common center of curvature, prove that the principal planes coincide at the common center of curvature and that the focal length is negative.

1.7 In Fig. P1.2 a ray is shown proceeding toward the first nodal point of a

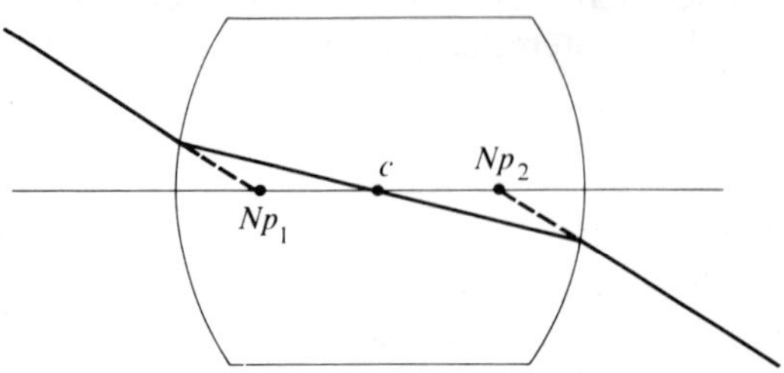

Figure P1.2

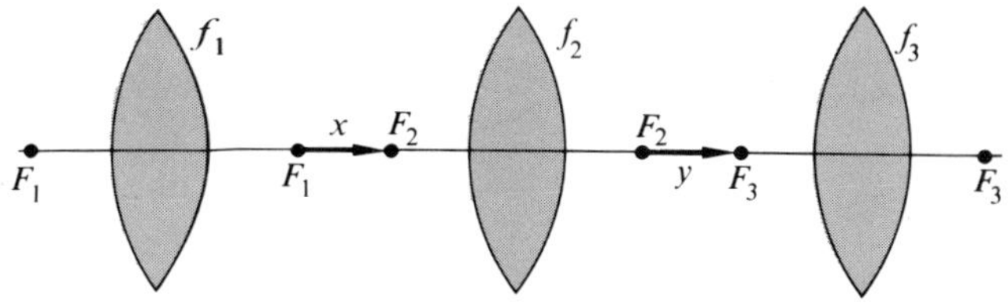

Figure P1.3

single thick lens in air. The point on the axis through which this ray passes is known as the optical center of the lens. Prove that the location of this point is independent of the index of refraction of the lens.

1.8 A doublet is constructed of two thin lenses separated by a distance t. The lenses are made of the same glass and have an index of refraction n_0 for light of wavelength λ_0. Prove that the focal length of the doublet is the same for all wavelengths near λ_0 if $f_1 + f_2 = 2t$.

1.9 In Fig. P1.3 a triple combination of thin lenses is shown. The positions of the first and last lenses are fixed but the middle lens is allowed to move. Prove that the distance between the focal points of the system is constant if $x + y = 0$ and $f_1 = f_3$. Under these conditions show that the focal length of the system is given by $f = -(f_1^2 f_2)/(x^2 + f_2^2)$.

1.10 The first focal point and the first principal point of an optical system are given as distances F_1 and P_1 to the left of some point A. The second focal point and second principal point are likewise distances F_2 and P_2 to the right of some point B. Show that the matrix S between these two points is given by

$$\mathsf{S}_{BA} = \begin{pmatrix} \dfrac{F_2}{F_2 - P_2} & \dfrac{P_1P_2 - F_1P_2 - F_2P_1}{F_2 - P_2} \\ \dfrac{1}{P_2 - F_2} & \dfrac{F_1}{F_2 - P_2} \end{pmatrix}.$$

1.11 Let the vertex matrix of an optical system be given as

$$\mathsf{V} = \begin{pmatrix} s_{11} & s_{12} \\ s_{21} & s_{22} \end{pmatrix}.$$

If the system is now reversed so that vertex a becomes vertex b and vice versa, the system will possess a new vertex matrix, V'. Under the condition that $|\mathsf{V}'| = |\mathsf{V}| = 1$, show that

$$\mathsf{V}' = \begin{pmatrix} s_{22} & s_{12} \\ s_{21} & s_{11} \end{pmatrix}.$$

1.12 In Fig. P1.4 an arbitrary ray is shown for a system with an odd number of reflecting surfaces. If S is the system matrix connecting the coincident points 1 and 2, show that $s_{11} = -s_{22}$.

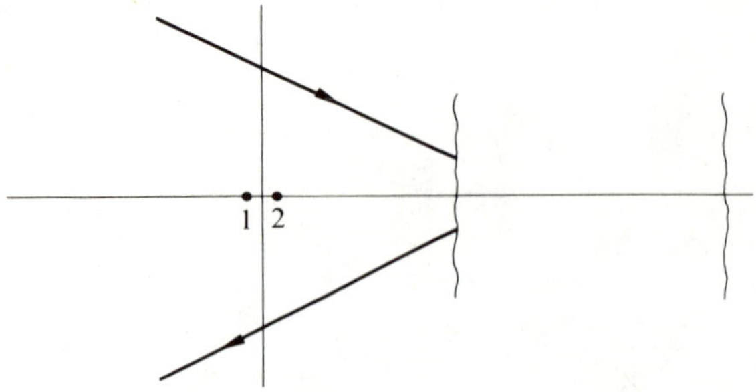

Figure P1.4

1.13 From the results of Problem 1.12 prove that the focal points of a system with an odd number of reflecting surfaces coincide, as do the principal points and the nodal points. Prove that the focal points are midway between the principal points and the nodal points.

1.14 Illustrate the results of the last problem for the particular case of a thin lens of focal length $+f$ placed a distance d to the left of a mirror of radius $-R$.

1.15 Negative principal points are defined as conjugate points of unit negative lateral magnification. For a single lens in air show that the negative principal points are two focal lengths from the principal points.

1.16 Let the matrix of a system be given between two reference points 1 and 2 with S_{21} having $D(S_{21}) = +1$. If points 1 and 2 are fixed but the system is moved a distance d to the right, show that the new matrix between 1 and 2 is

$$\mathsf{S}'_{21} = \mathsf{R}^{-1}\mathsf{S}_{21}\mathsf{R}, \qquad \mathsf{R} = \begin{pmatrix} 1 & d \\ 0 & 1 \end{pmatrix}.$$

1.17 In Fig. P1.5 a refracting system is shown. Assume that the system matrix S_{ba} is given with $D(S) = 1$. Point p is midway between points a and b and point p' is a distance d to the left of point p. Let the system now be rotated about the line OO' so that p' is fixed but the system is reversed. Show that the new matrix between a and b is the same as the old matrix if $d = (s_{11} - s_{22})/2s_{11}$.

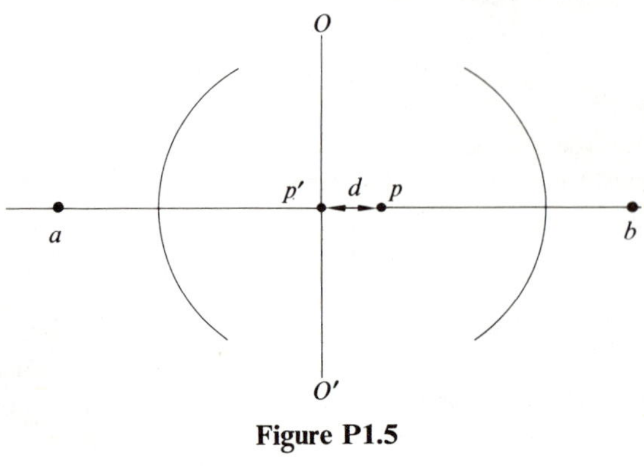

Figure P1.5

CHAPTER 2

THE EXPERIMENTAL EVIDENCE

Ordinarily, in introductory courses great ingenuity is exercised in making the concepts of physics as concrete as possible. Difficulty is frequently experienced by the student in later courses because formerly familiar concepts must be extended or radically revised when applied in a different domain. Such is usually the case with regard to the concepts of moment of inertia and angular momentum in the transition from particle dynamics to rigid body motion. The concept of angular momentum in quantum mechanics is significantly different from that in classical physics.

It is important to realize that our study of light will be from a perspective very much different from that in a traditional course. The principal emphasis in such a course, implicitly at least, is the consideration of light as an electromagnetic wave. This is possibly the easiest way, but often enough this results in over-dependence on the electromagnetic interpretation.

Our initial preference is for a description of light in which the emphasis is on the observable states of light known as polarization states. The intention of this chapter is to describe experiments which give data essential for our perspective, to directly correlate experiments with polarization states, and to indicate how these states in turn may be correlated with a mathematical formalism. The sections dealing with the different kinds of refracting media are important to an understanding of the devices constructed to produce and analyze polarization states.

2.1 THE QUANTITATIVE MEASUREMENT OF LIGHT AND QUANTA

To begin our study, we must have some way of detecting light. There are many possible detectors which could be employed, and normally the exact type will not be specified. For our present purpose, let the detector be a large hollow metal sphere with a tiny hole. After careful elimination of all other possible energy inputs, it is found that if light is allowed to enter the sphere, the temperature of the sphere increases, and we conclude that light must be a carrier of energy. By measuring the temperature change of the sphere, the time rate of change of the energy per unit area of the beam entering the hole can be deduced. In principle, we have an intensity meter which can be calibrated to read out in W/m^2 and which can be used to calibrate more sophisticated detectors. In order to make life easier, let us connect to an intensity meter an automatic recorder which will continuously plot the power input to the meter as a function of time.

We begin a second experiment by taking a source of light, say an electric lamp powered by a constant voltage source, and enclose it in a box with a small hole through which light can emerge. A detector and the source of light are placed at a fixed distance from each other, and enclosed in a room with no other sources of light. If a number of filters are placed successively in the path of the light between the source and the detector, the automatic recorder will plot a graph like that shown in Fig. 2.1(a) (the filters could be simply dark pieces of smoked glass), where each jump down represents the addition of another filter. If the amplification of the recorder is increased and more filters are placed in the beam, it is found that the plot is not so smooth as before and begins to appear as in Fig. 2.1(b). Continuing the process of filter addition eventually results in a graph such as that in Fig. 2.1(c).

Since the plots are of intensity *vs.* time, the area under each peak in Fig. 2.1(c) represents an energy absorption by the intensity meter, and we are led to conclude that the energy which the light carries is not being absorbed in a continuous manner, but in discrete bursts. This is truly a remarkable and very important result. Our thought experiment is simply one instance of what is always observed: *In every experiment performed to date in which the total absorption of light energy has been studied at sufficiently low intensities, this energy has always been observed in discrete amounts.* Furthermore, it can be shown by other experiments that this absorption takes place in a highly localized region of space, e.g., at a single silver atom inside one silver grain in a photographic emulsion—and during a very small interval of time. Since the absorption is discrete and highly localized in space and time, we associate with light the concept of a particle. This highly localized discrete energy or quantum of energy is called a *photon.*

The concept of a photon as a particle must, however, be handled with care. Later we shall see that the location of a photon usually has a completely definite meaning only after the fact. For example, we can say where a photon was absorbed in a photographic place; but once the energy is absorbed, the photon no longer exists. A word of caution, then, before we proceed. *We have not said, and will not at any time say, that light is discrete.* We will say only that its energy is absorbed in discrete amounts.

Returning then to the quanta of energy associated with a light beam, it is soon discovered that these quanta are not all of the same energy. However,

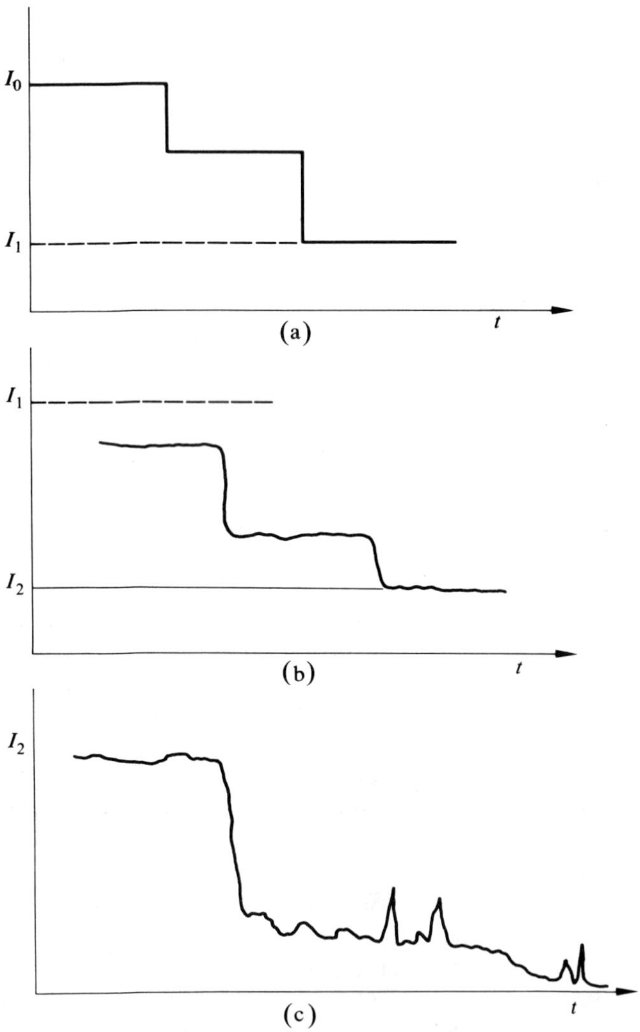

Fig. 2.1. Transmissions of light through a series of opaque filters.

it is possible to extract from a beam of light a sub-beam all of whose quanta are almost of equal energy. There are many ways to do this, but a discussion of these methods is not relevant at present. We will refer to *monoenergetic* beams and sometimes to *monochromatic* beams, since energy and color prove to be related.

2.2 POLARIZING FILTERS

There are commercially available a number of devices which can generally be grouped under the heading "polarizing filters". The nature of these devices can be discussed in terms of some theoretical description of light. Since we have no such description at present, we will examine these devices by a completely empirical investigation of their effect on a beam of light. Assume an unlimited supply of three types of filter, which will be labeled types R, L, and P. Later they will be called Right Circular (R), Left Circular (L) and Plane (P), but the applicability of these terms is not as yet evident.

Our first observation is that when only one of a given type of filter is placed in a beam of light, the intensity of the transmitted beam is one-half that of the incident beam. Moreover, it is found that the reduction of the intensity by one-half does not occur for all possible sources of light. For example, it does not occur for certain laser beams. In what follows in this and the next chapter, it will be assumed that the light source emits nearly monochromatic light and is such that each filter, upon encountering directly light from the source, transmits only one-half the intensity of the light incident on the filter. We will call such light *unpolarized* light. Many common light sources, though not monochromatic, produce unpolarized light.

Secondly, it is found that if either an R- or an L-filter is in the beam and an additional filter of the same type is placed in the beam, no further reduction of the beam intensity occurs. Rotation of the R- or L-types about the beam direction (which will be called the optic axis) does not affect the transmitted intensity, nor does rotation by 180° about an axis perpendicular to the optic axis have any effect. If an R-filter is already in the beam and it is followed by an L-filter, the transmission drops to zero. Interchanging the positions of the R- and L-filters still results in zero transmission. Rotation of either or both R and L about the optic axis has no effect on this result.

Now, what about the P-filter? Well, if two P-filters are placed in the beam, it is found that on rotating only one of the filters about the optic axis the intensity of the transmitted light varies between zero and 100% of $I_{\text{max.}}$, the intensity of the light which is transmitted by the filter which first intercepts the beam. Let us orient the two filters for maximum transmission and draw a line on the face of each filter such that the lines on the two filters are parallel. These lines have been chosen such that when they are parallel maximum transmission is obtained. This process could be repeated for all the other P-filters at our disposal. Note that the direction of the lines of the P-filters has not been chosen in a unique way. This arbitrariness will be removed in Section 2.8. If we now measure the intensity transmitted as a function of the angle between the lines of any two P-filters, we find the intensity to vary as

$$I = I_{\text{max.}} \cos^2 \theta. \tag{2.1}$$

This experimental relationship is known as Malus' law. Finally, if a P-filter is placed before or after an R- or L-filter, an additional decrease in intensity by one-half occurs and rotation of either filter has no effect.

2.3 STATES AND PREDICTION

The experimental results discussed in the preceding section and summarized in Fig. 2.2 indicate that the beams emerging from an R- or L-filter are different in some property which has not been discussed. That is, two light beams can both be monochromatic and still differ. Let us simply say that

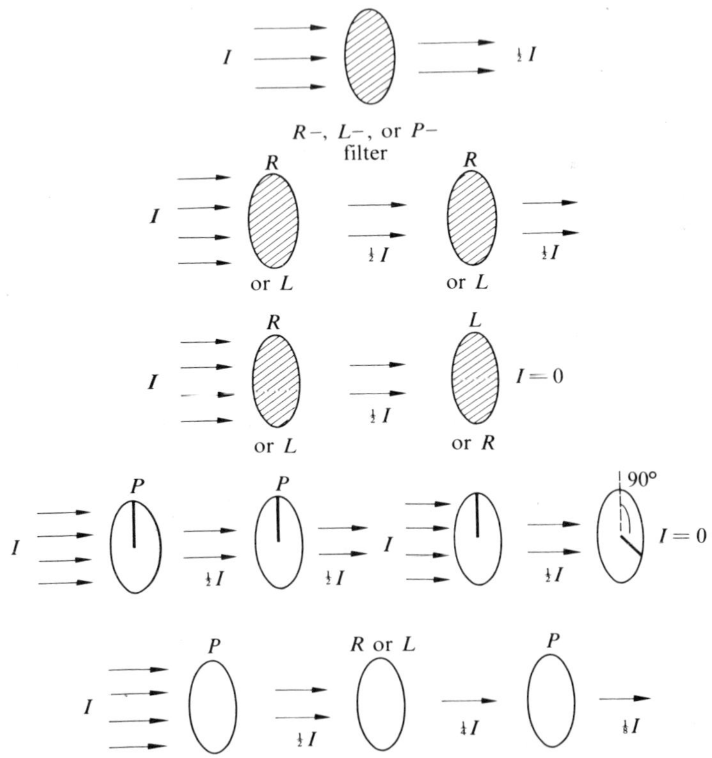

Fig. 2.2. Transmission through polarizing filters.

light which has traversed an L-filter is in a L-state; similarly, that which has traversed an R- or P-filter is in an R- or P-state, respectively. Light which has been prepared in an R-state will always traverse a following R-filter but never a following L-filter. Similar statements can be made concerning a prepared L-state. In other words, the outcome of an experiment involving light in either an R- or L-state incident on a series of R- and L-filters in any combination can be completely predicted. By complete prediction is meant that the outcome of an experiment can be stated with certainty (in terms of probability, the probability of an outcome is either zero or one). If complete prediction is possible, we will say that the state involved is *pure*. Note that the word "state" applies to the light but that the term "pure state" is defined in terms of the light *and* the type of experiment performed.

In the example involving R- and L-filters complete prediction was possible, but is this always the case? Can a state be pure in relation to one experiment and not pure in relation to another? Can there be experiments in relation to which there are no pure states?

To answer these questions let us prepare a monochromatic beam and place an R-filter followed by a P-filter in the path of the beam. We find that the light which has first traversed the R-filter comes through the P-filter with its intensity reduced by one-half. Consequently, light in an R-state is not pure in relation to the P-filter. Are there pure states relative to the RP-filter pair? No, because only light in an R-state is totally transmitted by the R-filter, and light in an R-state does not have 100% transmission through a P-filter. In terms of the photon concept, we can only give a 50/50 probability that the "next" photon which passes through the R-filter will either pass through or be absorbed in the P-filter.

2.4 SUPERPOSITION OF STATES

We now wish to describe an experiment which shows how two different states may be combined to produce yet another state. Consider the experimental arrangement shown in Fig. 2.3, which is known as a Mach-Zehnder interferometer. M_1 and M_2 are half-silvered mirrors and M and M' are full-silvered mirrors. A monochromatic incident beam striking M_1 is split and recombines at M_2. A and A' are variable absorbers whose sole function is to control the intensity of the light in each beam. The paths $M_1M'M_2$ and M_1MM_2 are chosen such that each beam within the apparatus travels the same distance. The filters placed in various beams are designated by the nomenclature already adopted. Two wedges of transparent glass form a variable retarder, F, whose thickness can be varied by simply moving the wedges relative to each other. The role of the retarder will become evident as the experiment is discussed.

Assume for the moment that the thickness of the retarder F is zero, and

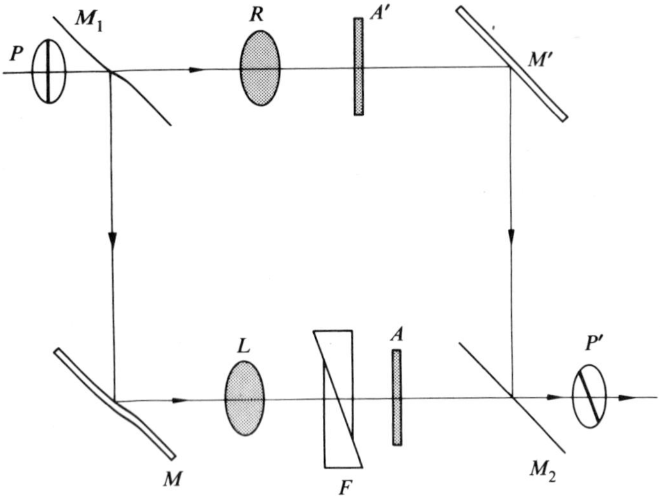

Fig. 2.3. The Mach-Zehnder interferometer.

bear in mind that the light striking M_1 is in a P-state since it has passed through a P-filter. If the absorbers are varied such that only the upper beam is transmitted by the apparatus, the outgoing light is observed to be in an R-state; if only the lower beam is transmitted, the outgoing light is in an L-state. Now suppose that each absorber is set to give complete transmission of each beam. An analysis of the emergent beam by P' shows that it is in a pure P-state whose line is the same as the incident P-state. This is true even in the case of low intensity, when individual photons can be counted.

Our experiment indicates that a P-state can be represented in some way as a superposition, or an adding together, of R- and L-states. The final P-state in our experiment has somehow been prepared by the entire apparatus. The individual R- and L-states observed when one path is closed are prepared by a different apparatus. The state obtained by the entire apparatus is not obtained by randomly mixing R- and L-states. We do not find that statistically half of the emerging quanta are R's and half L's. They are *all* P's.

Let us now examine the outcome of an experiment in which the thickness of the retarder F is varied. In Fig. 2.3 the polarizing and analyzing filters, P and P', respectively, have their "lines" initially parallel. While the lower beam traverses the retarder, it travels at a lower speed given by $v = c/n$, where c is the velocity of light *in vacuo* and n is the index of refraction of F. If τ is the time required for the lower beam to traverse a retarder of thickness t, then the *optical path length* is defined to be the distance in vacuum which the beam would traverse in the same time τ. It may be shown that the optical path length is given by nt. Since the beam travels a distance t through F and would have traversed the distance nt in the absence of F, the beam is retarded by the distance $t(n - 1)$, which will be called the *retardation*. As we continuously increase the retarder thickness from zero, the emerging light remains in a P-state, but the line of P' must be rotated about the emergent beam axis in order to maintain maximum transmission. After a certain thickness d has been inserted, the line of P' has rotated through 180° and we have recovered a pure P-state in relation to filter P. For each increase d in F, the optical path of the lower beam increases over that of the upper beam by an amount $(n - 1)d$, and a pure P-state in relation to filter P is recovered. We conclude that light somehow repeats itself in space; it is characterized by a spatial periodicity. Thus, we can assign a length to our light which is as typical of the light as is its energy. If this experiment is performed with various energies of light and the characteristic length for each energy is measured, we find that the energy $E = b/\lambda$, where b is a constant and λ is the characteristic length or *wavelength*. The wavelength is numerically equal to the increase in optical path length necessary in our experiment to cause the line of P' to complete one half-revolution of 180°. For visible light the common unit of wavelength is the *angstrom* (Å), which is a length of 10^{-8} cm, with visible light falling in the range 3000–8000 Å.

Since they are derived from the same source, there is a very definite relationship between the two beams in the apparatus. The emerging beam is always in a P-state when both arms of the apparatus are open. If one attempts to combine R- and L-states from different independent sources, the combined beam is not in general pure in relation to any of the filters and consists of a random mixture of R- and L-states. If a definite relationship exists, the resultant superposition is said to be *coherent*; otherwise, it is *incoherent*. The concept of coherence will be examined in more detail in later chapters.

2.5 *E*-STATES

At this point the student may be wondering whether we are finished with this business of describing experiments and giving their results. Not quite. The apparatus shown in Fig. 2.3 has possibilities which have not been explored. Recall that the separate beams combined to give a P-state when the absorbers A and A' were set to allow total transmission of each beam. Can other states be produced? Of course. Set A' to block the upper beam and the emerging light is in an L-state; block the lower beam to get an R-state. Now, for a given retarder thickness, let A' give total transmission and vary A continuously to give transmission from 0 to 100%. We observe a continuous transition of the emerging light from an R-state to a P-state. On interchanging A and A' a similar transition is observed from an L-state

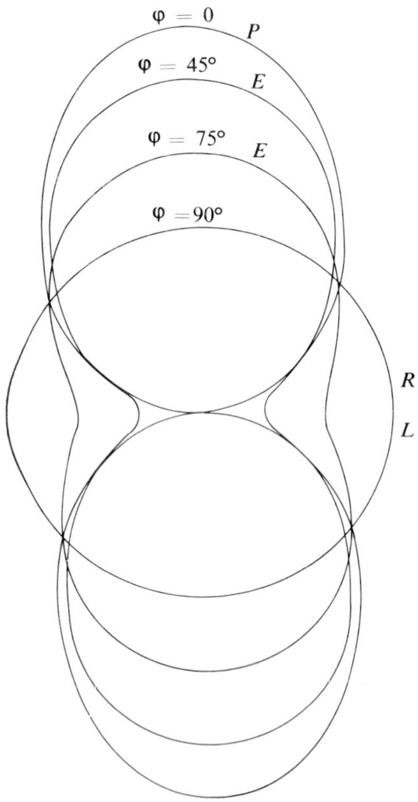

Fig. 2.4. The intensity transmitted through a *P*-filter for various incident states as a function of the orientation of the *P*-filter. The quantity ϕ is as used in Eq. (3.11).

to a *P*-state. The states observed "in between" the *R*- and *P*-, and *L*- and *P*-, are called *E*-states. Since any *E*-state is a superposition of *R*- and *L*-states, the *R*-, *L*-, and *P*-states are simply particular *E*-states.

If the light from the interferometer is analyzed with the filter P' in Fig. 2.3, the intensity of the light transmitted by P' generally varies with the angle which the line of P' makes with the filter *P*. In Fig. 2.4 is shown a polar plot of the intensity of the light transmitted by P' *vs.* the angle of rotation of P' for different emergent states from the interferometer. Two states are labeled *E*. *R* and *P* designate *R*- and *P*-states, respectively. The values of ϕ labeling the states will be explained in Chapter 3.

We have seen that the *R*-, *L*-, and *P*-states are pure in relation to the type of filter which produced these states, i.e., light in each of these states is transmitted without loss through a filter identical to that which produced the initial state. Furthermore, from the set of *R*-, *L*-, and *P*-filters we can extract two pairs of filters which possess a common property: the states of light produced by each member of a given pair are mutually exclusive in the sense that the state transmitted by one filter is not transmitted by its partner. Such mutually exclusive pure states are said to be *orthogonal.* Two P-filters, for example, whose lines are oriented at 90° relative to each other produce orthogonal states.

To each E-state produced by our interferometer a filter exists in relation to which it is pure. If this is so, do E-states come in orthogonal pairs? The answer is, "yes". To create an E-state orthogonal to another, the absorber in one arm of the interferometer is interchanged with that in the other arm and an additional retardation of one half-wavelength must be introduced in either arm.

One conclusion to be drawn from the interferometer experiments is that any polarization state can be obtained by the coherent superposition of an R-state and an L-state. It is pertinent to emphasize that this is not the only possibility. The assertion has been made that for each E-state there exists a corresponding filter. Now assume that we have two filters E_1 and E_2 which produce orthogonal states. What is observed if the R- and L-filters are replaced by the E_1- and E_2-filters? It is found again that any polarization state can be obtained by proper adjustment of the retarders and absorbers. Thus, the choice of representation of an arbitrary E-state as a superposition of R- and L-states is not unique; any pair of orthogonal states could be used.

A crucial experimental result is now given. As above, let E_1 and E_2 be orthogonal pure states. For any given beam of intensity I_0 which strikes an E_1-filter, let the transmitted intensity be I_1. If the same beam strikes an E_2-filter, let the transmitted intensity be I_2. Experimentally it is found that

$$I_0 = I_1 + I_2.$$

Since the probabilities, p_1 and p_2, that the given state will pass an E_1- or an E_2-filter are given by

$$p_1 = I_1/I_0. \qquad p_2 = I_2/I_0,$$

it follows that

$$p_1 + p_2 = 1. \tag{2.2}$$

2.6 KETS

In classical mechanics little progress can be made until vector calculus is developed and a consistent notation is used to represent vectors. The problem can be solved simply be printing a symbol in bold type to indicate a vector. We will indicate that our symbols R, L, etc., represent states by using a notation due to Dirac, $|R\rangle$, $|L\rangle$, $|E\rangle$, etc. Just as we say that $\boldsymbol{F}$ is a vector, we will say that $|E\rangle$ is a *ket*. Now, it appears natural to express the result of one of our experiments in Section 2.4 with an equation such as $|P\rangle = |R\rangle + |L\rangle$. This simply says that a P-state can be obtained by the superposition of an R- and an L-state. But we know that a simple superposition of an R- and an L-state will not necessarily give a P-state. In fact, a continuous succession of states from R- to P- to L- can be produced by varying the absorber in one arm of the apparatus (Fig. 2.3) from 0 to 100% transmission, while the absorber in the other arm is varied from 100 to 0%. Since we think of the absorbers as determining how much of each beam contributes to the superposition, our "addition" can be extended to $|E\rangle = a|R\rangle + b|L\rangle$, where the constants a and b give the proportional amount of $|R\rangle$ and $|L\rangle$ contributed to the final combination.

What does our experimental evidence tell us about these constants?

In the interferometer experiments the retarder F also plays a role in determining the state of the emergent light. In Fig. 2.3 the L-state is modified in the superposition by the presence of the retarder, and this modification is to be included in the multiplicative constant b. The retarder changes the state L to the state L' with $|L'\rangle = b|L\rangle$. If t is the thickness of the retarder and n its index of refraction, then the retardation (s) is given by $t(n - 1)$. Since it is the retarder that affects the state $|L\rangle$, b is a function of the retardation and we write $|L'\rangle = b(s)|L\rangle$. It will be assumed that $b(s)$ is a continuous function of s. If an additional retarder of thickness t' is inserted, the state L' is changed to the state L'' such that $|L''\rangle = b(s')|L'\rangle$. Now, the same state L'' must result if the total thickness $(t + t')$ is inserted at once. Consequently, we must have $|L''\rangle = b(s + s')|L\rangle$, and $b(s')b(s) = b(s + s')$. The only continuous function satisfying this condition is the exponential function. Thus, $b(s) = Be^{\alpha s}$, where B and α are constants. Recall that with a suitable continuous increase of retarder thickness the emergent state rotates through a series of states until it returns to its initial state. This demands that the exponential function be periodic, and this can be so only if the exponent is a pure imaginary. Therefore, we write $b(s) = Be^{iks}$, where B and k are real numbers.

From the above argument we can see that if retarders are in both arms of the interferometer in Fig. 2.3, then the superposition can be written as

$$|E\rangle = a|R\rangle + b|L\rangle, \tag{2.3}$$

where a and b may both be complex.

This equation bears an extraordinary resemblance to the vector addition of two orthogonal vectors in two-dimensional vector space. It suggests that we can think of a polarization state as a "vector sum" of two mutually exclusive states. This analogy will be developed at some length in the following chapter.

2.7 THE LINEAR AND ANGULAR MOMENTA OF LIGHT

There are very few basic quantities discussed in physics. In other words, there are few quantities which obey a conservation law. Other than energy, the remaining quantities which we will consider are linear and angular momenta. Let us imagine a torsional pendulum suspended from a very fine quartz fiber lying along the axis, as in Fig. 2.5, and enclosed in a vacuum chamber to reduce frictional effects. The pendulum consists of a horizontal cross-arm whose equilibrium position is along the y-axis, connected to the vertical quartz fiber. The oscillating mass can be a black surface at one end of the cross-arm with a similar mass at the opposite end to act as a counterweight. The plane of the black surface is parallel to the quartz fiber. We now shine an off-axis beam of light, traveling parallel to the x-axis on to our black surface, in short bursts whose frequency coincides with the natural frequency of the pendulum, and a low amplitude oscillation is observed. As the pendulum swings through its equilibrium position, it possesses a certain amount of linear momentum. By an appeal to conservation of linear momentum, we can only deduce that the light which has been absorbed in the black surface carried with it a certain amount of linear momentum. If we have arranged for our beam of light to be monoenergetic, or nearly so, we may compute the number of energy quanta which have been absorbed by the pendulum, knowing the intensity of the beam. By measuring the amplitude of the oscillations we may compute the linear momentum which has also been absorbed. If we divide the number of energy quanta into the linear momentum, we find that, regardless of the particular energy used, the linear momentum associated with the discrete energy bundles is given by

$$|\boldsymbol{p}| = E/c = h/\lambda, \tag{2.4}$$

where h is Planck's constant, which has the experimentally determined value $h = 6.63 \times 10^{-34}$ J sec, and c is the velocity of light *in vacuo*.

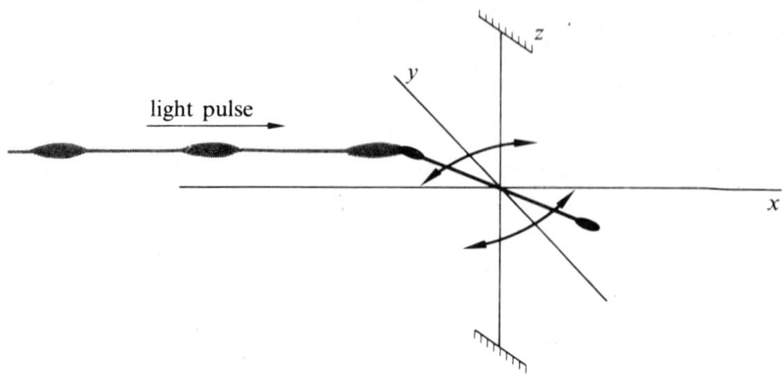

Fig. 2.5. Torsion pendulum for detecting the linear momentum associated with a light beam.

Not only does our pendulum possess linear momentum, but through its rotation it must also possess an angular momentum. This rotational momentum, however, arises from the particular moment arm of our black surface. If the beam and surface had been perfectly centered on the fiber instead of the cross-arm, no rotation would have taken place. In a given system angular momentum is computed from $\boldsymbol{r} \times \boldsymbol{p}$, where $\boldsymbol{r}$ is the vector from the origin about which the angular momentum is computed; and in a particular system a portion, if not all, of this can be transformed away by an appropriate choice of origin. This type of angular momentum is often called *orbital* angular momentum. As an example, consider the purely classical situation in Fig. 2.6. A ball of mass M travels in the xy-plane with velocity $\boldsymbol{v}$ parallel to the x-axis. It is also spinning about an axis through its center which is parallel to its velocity.

The total angular momentum can be considered as being made up of two parts. The orbital part $\boldsymbol{L}$ whose magnitude is given by

$$|\boldsymbol{L}| = |\boldsymbol{r}||\boldsymbol{v}|M \sin\theta = |\boldsymbol{v}|Md$$

can be transformed away by shifting the origin to the point P on the y-axis. The other angular momentum, $\boldsymbol{S}$, is that associated with the spinning of the ball about its center. This cannot be altered by shifting the origin.

The distinction between orbital and spin angular momentum is not unambiguous in classical mechanics. For example, the spin momentum of the ball may be thought of as being the sum of the orbital momenta of each particle of the ball about its center. In Chapter 9, however, we shall see that a clearer mathematical distinction may be made from the viewpoint of quantum mechanics.

The existence of orbital angular momentum in our experiment was unavoidable since light has linear momentum. We now ask whether light possesses angular momentum which is independent of a translational origin.

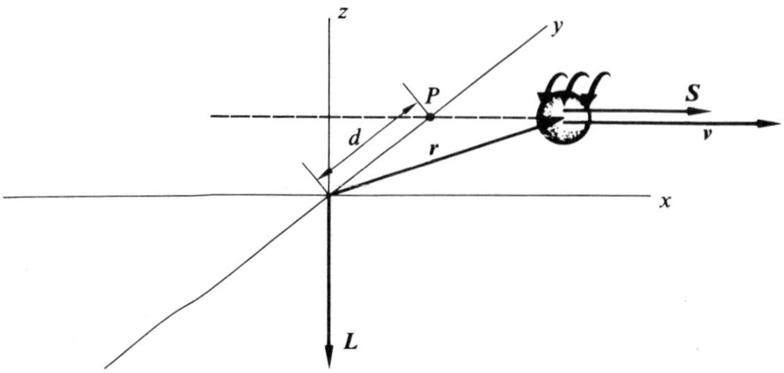

Fig. 2.6. The orbital angular momentum, **L**, and the spin angular momentum, **S**, of a classical particle.

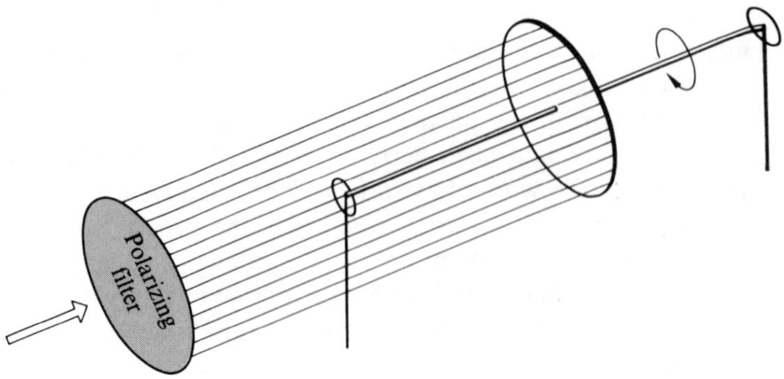

Fig. 2.7. Polarized light incident on an absorbing disk which is free to rotate about the axis of the beam.

If for the moment we adopt the dangerous viewpoint of regarding the photon as a particle, we ask whether this particle can be "spinning". We construct another experiment, shown in Fig. 2.7, in which the plane of our absorber is perpendicular to the axis of our support and we shine our beam along the axis of support. If we use our original raw beam as it comes from the source, we will find no detectable induced rotation of the disk. Even if there were a nonorbital or intrinsic angular momentum associated with the many photons striking the disk, this result is what we would expect. The odds are that the individual photons would have their spin angular momentum vectors oriented randomly to the axis, and, hence, have no net angular momentum along the direction of the axis or along the direction of their velocity, which is the same thing.

At our disposal, however, are our polarizing filters, which have the ability to reduce randomness in a beam and give us pure states. We thus repeat our experiment for the detection of intrinsic angular momentum using pure states of one type or another. Using P-states, we find no rotation. If we allow a beam which has been placed in an R-state to strike our apparatus, we find that the disk does indeed begin to rotate and in such a direction that its angular momentum vector points along the velocity of the incident beam. If we use L-light, we find a rotation in the opposite sense with the angular velocity vector pointing antiparallel to the velocity of the incident beam. We can measure the total angular momentum of the disk after a given amount of energy is absorbed. If we determine the number of quanta which have been absorbed and divide this into the total angular momenta, we find that with each quantum of energy absorbed an amount of angular momentum is absorbed equal to $\hbar = h/2\pi$.

If we now prepare our beam by using a variable E-filter, we find, as we change our beam in a continuous manner from R to L, that the apparent component of angular momentum in the direction of propagation absorbed with each photon varies from $+\hbar$ to $-\hbar$. One interpretation would be that each quantum carries an intrinsic angular momentum of magnitude $\hbar$ whose orientation with the velocity can vary from zero to 180°. Thus, *R*-states have the velocity and angular momentum parallel; *L*-states have them antiparallel; and the general *E*-state has some intermediate orientation. *P*-states would thus have their angular momentum vector perpendicular to the velocity and the plane determined by the angular momentum and the velocity could determine the plane of *P*-polarization. The only thing we shall presently find correct about this interpretation is the interpretation of *R*- and *L*-states.

Let us recall that if a *P*-state hits an *R*-filter followed by an *L*-filter, we will find half of the quanta being absorbed in the *R* and the remaining half absorbed in the final *L*. Yet we cannot think of the *P*-state itself as being a random combination of *R*'s and *L*'s, for then half of the *P*-states would be absorbed in a second *P*-filter. We have seen that we can write

$$|P\rangle = a|R\rangle + b|L\rangle,$$

and we must again attempt to understand the nature of this coherent superposition. Recalling our interferometer experiment, we interpreted the superposition of two states over the two possible paths not to mean so many photons coming along each path, but to indicate the number we would receive over each path if we forced the energy to appear in either one path or the other. We said that the photon observed with both paths open did not come by one or the other but came through the whole apparatus. The particle nature of light, which is sometimes implied by the use of the word "photon", is only valid if we force the light to appear in some particular place. If we ask the light where it is, it will answer us by appearing somewhere. Less anthropomorphically, light can obviously be detected only where we have placed a detector. The location of a photon is meaningful with any sort of rigor only during the process of absorption or emission. To apply this same interpretation to our polarization states and their interpretation in terms of angular momentum, we will say the following: To ask for the orientation of the spin angular momentum of a photon is meaningless unless we carry out an experiment to determine that orientation. When such experiments are performed, the spin angular momentum of an individual photon is always found to be either parallel or antiparallel to its velocity.

Thus, representing elliptically polarized light as a coherent superposition of *R*- and *L*-states as

$$|E\rangle = a|R\rangle + b|L\rangle$$

tells us through the constants a and b the relative probability of the photon's appearing with $\hbar$ units of angular momentum along the direction of its motion or with $\hbar$ units opposed to this direction.

Our interpretation of *R*- and *L*-states in terms of angular momentum allows us to represent these states by certain geometric representations which have basically the same spatial symmetry. Angular momentum is essentially a quantity which defines in space both a direction and a sense of handedness: that is, a rotation about a certain direction and a rotation in a certain sense, either clockwise or counterclockwise. We could thus represent these states as is shown in Fig. 2.8(a).

As for all entities which possess a "handedness", it is well known that reflection in a mirror changes the handedness from right to left or vice versa, as shown in Fig. 2.8(b). If we now shine a beam of light through an *R*-filter, allow it to reflect from a mirror, and back to an *R*-filter, we find that it is completely absorbed. If we let it reflect back to an *L*-filter, it is completely transmitted. Reflection interchanges *R*- and *L*-stages, as shown in Fig. 2.8(c).

2.8 ANISOTROPIC RETARDERS

Substances may be classified in a variety of ways: as conductors-insulators, gases-liquids-solids, etc. For a given macroscopic property materials may be classified as being homogeneous or inhomogeneous, and isotropic or anisotropic. If a given property is the same everywhere in the material, then the substance is said to be homogeneous relative to the given property. A substance is said to be isotropic relative to a given property if at any point within the substance the given property is independent of direction. As an example, consider a wooden board. If there are knot-holes, the board is inhomogeneous: its properties differ from one spot to another. On the other hand, even without knot-holes, the board is anisotropic. It is easier to break it along the direction of the wood grain than it is to break it across the grain.

Let us first consider an optical material which is both homogeneous and isotropic relative to the index of refraction. Ordinary glass is an example of such a material. Consider a plate of glass in air with a narrow beam of monochromatic light striking the glass at an angle of incidence, α. Since the refracted ray is rectilinear, the index of refraction for each point along the ray is the same. If the index of refraction were not the same for each point, the refracted ray would curve. By varying the angle of incidence and the horizontal position of the plate we see that there are an infinite number of possible refracted rays, and for all of these possibilities we get the same number for the index of refraction.

This means that the index of refraction is the same at each and every point in the glass and that the glass is homogeneous relative to the index of refraction. The glass is also isotropic relative to the index of refraction, since the index of refraction is independent of the direction of the refracted ray which passes through. Since the index of refraction is $n = c/v$, where v is the

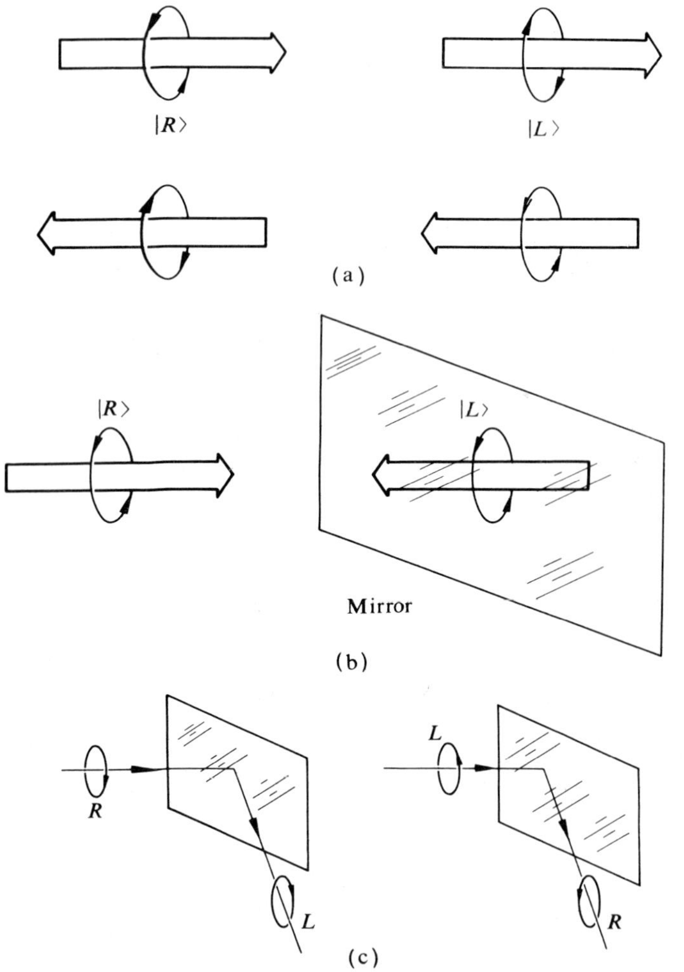

Fig. 2.8. Geometric symmetries of *R*- and *L*-states.

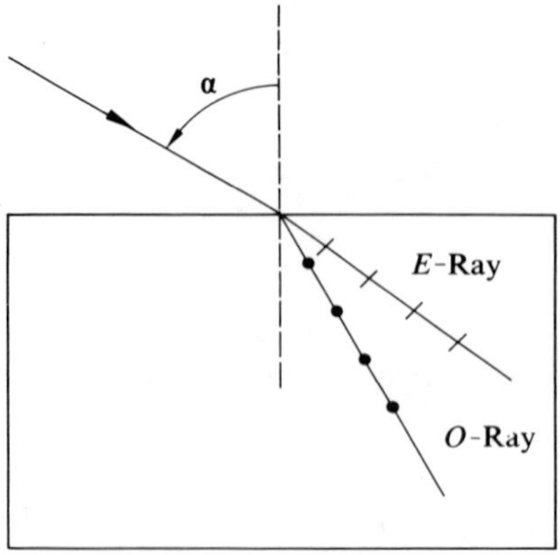

Fig. 2.9. Light incident on a birefringent crystal. The optic axis of the crystal is parallel to the plane of the paper. Hatch marks indicate that the *E*-ray is a *P*-state whose plane of polarization is in the plane of the paper. The dots on the *O*-ray indicate a *P*-state whose line is perpendicular to the paper.

speed of light in the refracting medium, the speed of light in the glass is the same everywhere and is independent of the direction of the refracted ray.

There do exist materials which are homogeneous, but anisotropic, relative to the index of refraction. Such materials are said to be *birefringent.* The index of refraction is not a function of position within the substance but is determined by the direction and polarization state of the refracted light. A typical example of such a material is calcite, and in Fig. 2.9 is shown a narrow beam of unpolarized light striking one face of a calcite crystal. In this case the phenomenon of double refraction occurs. The original beam splits into two beams, which are labeled *O* and *E*. If analyzed with a *P*-filter, these beams are found to be in orthogonal *P*-states. In crystal optics the *O*-ray (ordinary ray) is defined as the ray which obeys Snell's law, and the *E*-ray (extraordinary ray) as that which does not. Thus, at normal incidence the *O*-ray will not be deviated and the *E*-ray in general will. Now, it is possible to cut the crystal in such a way that neither the *O*- or *E*-ray is deviated at normal incidence and thereafter travels through the crystal at the same speed. This direction defines the *optic axis* of the crystal. If the crystal has only one such axis, it is said to be uniaxial or monoaxial. Calcite is a uniaxial crystal. If the crystal is cut so that at normal incidence the light strikes at right angles to the optic axis, neither the *O*- or *E*-ray is deviated, but in this case the speeds of the rays are different.

In defining the lines of *P*-states, we have until now been somewhat arbitrary. A line was placed on one *P*-filter and the lines of all the others were drawn in such a way that any pair of filters would obey Malus' law. We now redraw all such lines on our *P*-filters in such a way that the line of the P-state which is the O-ray is always perpendicular to the optic axis of a calcite crystal. This standardizes our P-filters with all other P-filters. For any angle of incidence, the polarization of the O-ray is always such that the line of its P-state is perpendicular to the optic axis.

With the remarks above in mind let us now see how anisotropic retarders may be constructed from uniaxial crystals. If a flat plate is cut from a uniaxial crystal in such a way that the optic axis is parallel to the flat faces, and light in a P-state impinges on this plate normal to the faces, the P-state's line can make any angle with the optic axis from 0 to 90°. Since the transmission through the plate is not along the optic axis, the speed is dependent on what P-state is used. However, if the line of the P-state is either parallel or perpendicular to the optic axis, the state is transmitted unchanged but the speeds of transmission are different in the two cases. Let $n_\parallel$ and $n_\perp$ be the indices of refraction for the P-states whose lines are parallel and perpendicular to the optic axis. Uniaxial crystals are classified as positive or negative according to whether $n_\perp < n_\parallel$ or $n_\perp > n_\parallel$. Calcite, for example, is a negative crystal with $n_\parallel = 1.486$ and $n_\perp = 1.658$ for monochromatic light of wavelength $\lambda = 5893$ Å. The important point is that the P-state parallel to the optic axis will have a retardation different from that of a P-state perpendicular to the optic axis. In either case the index of refraction for a P-state whose line is along one of these two directions will be less than that along the other. The line of this P-state will be called hereafter the fast axis and the other the slow axis of the plate.

2.9 POLARIZATION BY ABSORPTION

The birefringent materials discussed in the previous section were all assumed to be transparent. Although a polarization state may be modified by an anisotropic substance, the total incident energy flux was assumed to be transmitted. A birefringent crystal which has anisotropic absorption properties is tourmaline. If a sufficiently thick plate is cut from a tourmaline crystal such that its broad faces are parallel to the optic axis, the E-ray is transmitted but the O-ray is partially absorbed. This phenomenon is known as dichroism.

A common and inexpensive material which polarizes by absorption is known by the name of Polaroid. Sheets of Polaroid can be used to construct a device which approximates the P-filter. If light in a P-state is incident on a Polaroid sheet, rotation of the sheet about the incident beam direction gives a maximum and minimum transmission.

In an absorption polarizer the absorption of light other than the desired P-state is never perfect, and some of the desired P-state will be absorbed. In a later chapter the generalization of Malus' law for a pair of identical dichroic polarizers will be found to be

$$I = H_0 \cos^2 \theta + H_{90} \sin^2 \theta. \tag{2.5}$$

H_0 and H_{90} may be written as

$$H_0 = \tfrac{1}{2}(k_1^2 + k_2^2),$$
$$H_{90} = k_1 k_2.$$

Here k_1 and k_2 are known as the major and minor principal transmittances, respectively. Values of these transmittances are given in Table 2.1 as a function of wavelength for three commercial types of Polaroid.

2.10 OPTICAL ACTIVITY

In addition to birefringence there is the interesting phenomenon known as optical activity. Materials exhibiting this property may be isotropic and homogeneous, but as light in a P-state travels through the material the line of the P-state will rotate about the beam direction. In a later chapter optical activity will be seen to result from different velocities for R- and L-states. The amount of rotation depends on the thickness of the material and, in the case of solutions, also on the concentration. Common table sugar is optically active. In fact, a common test for the purity of sugar is based on the amount of rotation produced by a standard length of sugar solution.

Table 2.1

Values of the Principal Transmittances k_1 and k_2 of Polaroid Type HN Polarizers

Wave-length	HN22		HN32		HN38	
	k_1	k_2	k_1	k_2	k_1	k_2
4000	0.21	0.00001	0.47	0.003	0.67	0.04
5000	0.55	0.000002	0.75	0.00005	0.86	0.005
6000	0.43	0.000002	0.67	0.00002	0.79	0.0003
7000	0.59	0.000003	0.77	0.00003	0.86	0.0007

The values express transmittance as a *pure number*, not a percentage.

PROBLEMS

2.1 A beam of light in an R-state is incident on a totally absorbing disk which is free to rotate about an axis parallel to the beam. The mass of the disk is m, its specific heat is C, and its moment of inertia about the axis of rotation is I. Starting from rest the disk acquires an angular velocity Ω. Show that the temperature increase ΔT of the disk in this time is given by

$$\Delta T = \frac{I\Omega}{mC}\left[\frac{2\pi c}{\lambda} - \frac{\Omega}{2}\right],$$

where λ is the wavelength of the incident light, and c is the velocity of light.

2.2 A beam of light in an L-state passes through a P-filter which is free to rotate about the axis of the beam. Why would you expect the P-filter to rotate? In what direction will it rotate?

Answer: Counter-clockwise when viewed from the source.

2.3 A calcite plate is cut with the optic axis parallel to the plane of the plate. How thick must the plate be to retard the O-ray one quarter-wavelength in respect to the E-ray if the wavelength of the light is 5893 Å?

Answer: 8.57×10^{-4} mm.

2.4 The average number of photons absorbed in 1 sec by a counter placed in a monoenergetic beam is N. If the counter now moves with velocity V in the direction of the beam's propagation, what average number will be counted? Where did the extra photons "go"?

Answer: $N' = (1 - V/c)N$.

2.5 A photon of wavelength λ is absorbed in a dectector of mass m, which is at rest but free to move. Show that the fraction of the incident photon's energy which is not available for increasing the dectector's temperature is $\frac{1}{2}h/\lambda mc$. Compute this fraction for $\lambda = 5000$ Å and $m = 1$ kg.

Answer: 2.21×10^{-36}.

2.6 Repeat Problem 2.5 for a detector moving with velocity $\pm V$ in the direction of the beam's propagation. If the increase in temperature of the detector is used to measure the energy of the photons and, hence, their wavelength, show that approximately

$$\frac{\lambda}{\lambda'} = \left(1 \pm \frac{V}{c}\right),$$

where λ is the wavelength measured when the detector is at rest and λ' is the measured wavelength when it is moving. The observer moving with the detector assumes that all of the photon's energy is available for increasing the temperature.

2.7 As shown in Fig. 2.3, what pure state would be detected in each segment of path along both arms of the interferometer? What would be the consequence of moving the L-filter in the interferometer from the segment MM_2 to the segment M_1M? If this were done, where would the R-filter have to be placed to restore the original results?

2.8 Prove that the optical path, nd, is the distance light would have traveled *in vacuo* during the time in which it travels the distance d in a medium of refractive index n.

2.9 Design an interferometer which could be used in place of the Mach-Zehnder interferometer, using two full-silvered mirrors but only one half-silvered mirror. Utilize orthogonal P-filters instead of R- and L-filters in the arms of the interferometer.

2.10 When the index of refraction is not constant over the path length of a ray, the optical path length is defined by

$$L = \int_A^B n\, dl,$$

where A and B represent two points on the ray. Geometrical optics may be derived from Fermat's principle, which states that of all possible paths from A to B a light ray will travel that path for which L is an extremum. For our purposes this means that L is either larger or smaller on this path than it is on all nearby paths. Derive Snell's law and the law of reflection from Fermat's principle.

CHAPTER 3

THE REPRESENTATION AND PRODUCTION OF STATES

Once it has been observed that the expression $|E\rangle = a|R\rangle + b|L\rangle$ is similar to that encountered in the addition of vectors in a two-dimensional vector space, it may be desirable to set down a set of axioms which the elements of any vector space must satisfy. If we did this, the physical interpretation of many of the axioms in terms of polarized states would be obvious. However, since the interpretation of others would be impossible at this point, we will not take this approach. Instead, we will assume that polarization states can be represented by two-dimensional complex vectors and proceed to develop a self-consistent calculus of polarization states.

3.1 THE VECTOR AND MATRIX REPRESENTATION

Vectors are written in many texts as

$$\boldsymbol{V} = A\boldsymbol{e}_1 + B\boldsymbol{e}_2 + C\boldsymbol{e}_3,$$

where A, B and C are the vector components and the $\boldsymbol{e}_i$ are orthogonal unit vectors. There is no special reason, aside from uniformity, why vectors must be written in this fashion. So long as the order of the components is specified and maintained, the vector $\boldsymbol{V}$ can be written as

$$\begin{pmatrix} A \\ B \\ C \end{pmatrix} \qquad \text{or} \qquad (A, B, C).$$

These representations are, respectively, called column (3×1-matrix) and row (1×3-matrix) vectors. We will remove the restrictions imposed by our initial choice of a vector in three-dimensional Cartesian space. First, we can define an n-dimensional vector space, where n is any integer. Ordinarily, this means that a vector in this space can be expressed as a linear combination of n unit orthogonal vectors. Second, the components of a vector will not be limited to the domain of real numbers; they now may be complex.

Since any two orthogonal states in superposition give another E-state, we will represent polarization states by two-dimensional complex vectors. We may choose either row or column vectors and Dirac's notation allows the use of both. With each state we associate not only a ket symbol $|E\rangle$ but also a symbol $\langle E|$ called a bra. Ket symbols will be represented by column vectors and bras will be represented by row vectors. How are the elements in the bra to be related to those of the ket? Assume that a state E is represented by a ket as

$$|E\rangle = \begin{pmatrix} A \\ B \end{pmatrix}.$$

Then the bra representing the same state is defined by

$$\langle E| = (A^\star, B^\star).$$

In other words, the bra vector is the complex conjugate of the transpose of the ket vector or its *adjoint*. We will see the usefulness of this definition shortly.

Suppose E_1 and E_2 are two different E-states with

$$|E_1\rangle = \begin{pmatrix} A_1 \\ B_1 \end{pmatrix}, \qquad |E_2\rangle = \begin{pmatrix} A_2 \\ B_2 \end{pmatrix},$$

and let us consider the meaning of the product $\langle E_1|E_2\rangle$. Since the bra vector is a row vector and the ket is a column vector, the symbol $\langle E_1|E_2\rangle$ is the product of two matrices. The first matrix is 1×2, the second is 2×1, and the product will be 1×1 or a scalar. Multiplication yields

$$\langle E_1|E_2\rangle = A_1^\star A_2 + B_1^\star B_2,$$

and we note that this implies

$$\langle E_1|E_2\rangle = \langle E_2|E_1\rangle^\star.$$

The product of a bra on the left with a ket on the right is called a *bracket* product, and is a generalization of the familiar scalar product of vectors whose components are real. If $|E_1\rangle = |E_2\rangle$, then the bracket product is a real number and is a generalization to complex vector space of the square of the length of a vector. If any two kets $|E_1\rangle$ and $|E_2\rangle$ satisfy the relation $\langle E_1|E_2\rangle = 0$, they are said to be *orthogonal*. If they also satisfy the relation

$$\langle E_1|E_1\rangle = \langle E_2|E_2\rangle = 1,$$

they are said to be *orthonormal*. Another possible product is formed when the bra $\langle E_1|$ is multiplied from the left by the ket $|E_2\rangle$. This is written as $|E_2\rangle\langle E_1|$. Interpreting this as a matrix multiplication, we have a

2×1-matrix on the left and a 1×2-matrix on the right. This product yields the 2×2-matrix

$$|E_2\rangle\langle E_1| = \begin{pmatrix} A_2A_1^\star & A_2B_1^\star \\ B_2A_1^\star & B_2B_1^\star \end{pmatrix}.$$

For future use we observe that this matrix satisfies the relation

$$|E_2\rangle\langle E_1| = \{|E_1\rangle\langle E_2|\}^\dagger,$$

where the dagger symbol indicates the adjoint of $|E_1\rangle\langle E_2|$.

Since the bracket product, $\langle E_1|E_2\rangle$, is a scalar, it may be freely moved through a matrix expression. Thus, for example, for any three kets $|E_1\rangle$, $|E_2\rangle$, and $|E_3\rangle$ we may write

$$\langle E_1|E_2\rangle|E_3\rangle = |E_3\rangle\langle E_1|E_2\rangle.$$

But, since matrix multiplication is associative, this yields

$$\langle E_1|E_2\rangle|E_3\rangle = |E_3\rangle\langle E_1||E_2\rangle.$$

This manipulative trick will be frequently used.

3.2 THE BRACKET PRODUCT AND PROBABILITY AMPLITUDES

We have seen in Chapter 2 that a correspondence can be made between kets and states of light, and the problem arises as to how the bracket product can be interpreted in view of this correspondence. The correspondence between kets and states of light was inferred from the interferometer experiments, which indicated that any E-state can be expressed as

$$|E\rangle = a|R\rangle + b|L\rangle, \tag{3.1}$$

where a and b may both be complex. Since the kets $|R\rangle$ and $|L\rangle$ are column vectors, we are free to choose any pair of column vectors to represent the R- and L-states, provided that our choice reflects the fact that these states are mutually exclusive. This situation is the same as that for real vectors, where any vector with real components can be expressed as a sum of three orthogonal vectors. In real space unit vectors along the x-, y-, and z-axes are linearly independent; not one of them can be expressed as a vector sum of the other two. In this sense the unit vectors are mutually exclusive. But the ket vectors have this property; there is no way of obtaining an R-state through a superposition of L-states.

Of the many column vectors which satisfy the above requirements, the simplest we can think of is the choice

$$|R\rangle = \begin{pmatrix} 1 \\ 0 \end{pmatrix}, \qquad |L\rangle = \begin{pmatrix} 0 \\ 1 \end{pmatrix},$$

which allows us to write the ket $|E\rangle = a|R\rangle + b|L\rangle$ as

$$|E\rangle = \begin{pmatrix} a \\ b \end{pmatrix}.$$

We emphasize that this particular choice of a vector representation is not unique or forced on us by physical requirements. It is merely the simplest choice. Later we shall find other representations for $|R\rangle$ and $|L\rangle$ which are convenient for certain problems. To assign the unit vectors $\begin{pmatrix} 1 \\ 0 \end{pmatrix}$ and $\begin{pmatrix} 0 \\ 1 \end{pmatrix}$ to two particular mutually exclusive states is known as picking a *basis* for representing polarization states.

Four possible bracket products can be written with the pair of kets $|R\rangle$ and $|L\rangle$, namely,

$$\langle R|R\rangle = \langle L|L\rangle = 1,$$
$$\langle R|L\rangle = \langle L|R\rangle = 0.$$

What operational significance can these products have in terms of our experiments in the preceding chapter?

The following interpretation is made. The ket involved in the product is a statement that light has been prepared in either a pure R- or a pure L-state. The bra in the product indicates the filter used to analyze the prepared state. For example, the product $\langle R|R\rangle$ indicates that light in an R-state is analyzed by an R-filter. Light in an R-state analyzed by an R-filter is completely transmitted as an R-state, and $\langle R|R\rangle = 1$ means that the probability of transmission of light in an R-state through an R-filter is unity; the outcome of the experiment is completely predictable. Similarly, $\langle L|R\rangle = 0$ says that the probability of transmission of an R-state through an L-filter is zero. Another way this may be phrased is to say that if light is known to be in an R-state, the probability that it is in an R-state is unity; the probability that the R-state is in its orthogonal state is zero. It is important to recognize that the bracket products listed above possess a common essential feature, i.e., if a beam of light is known to be in a pure polarization state, then analysis of this beam has a completely predictable outcome only with filters which are identical to that which produced the state or to its mutually exclusive partner. *Only when this condition is fulfilled is the interpretation of the bracket product as a probability consistent with experiment.* Since the probability that any E-state will pass through the E-filter that creates it is unity, we are consistent in placing $\langle E|E\rangle = 1$ for any pure state $|E\rangle$.

Let us investigate the implications of this interpretation of the bracket product for the superposition in Eq. (3.1). The light emerging from the interferometer in Fig. 2.3 is in the state $|E\rangle$, and analysis of this state by a filter in relation to which it is pure has a completely predictable outcome.

By taking the bracket product of $|E\rangle$ in Eq. (3.1) with $|R\rangle$ and $|L\rangle$, we obtain

$$\langle R|E\rangle = a, \qquad \langle L|E\rangle = b.$$

Thus,

$$|E\rangle = \langle R|E\rangle|R\rangle + \langle L|E\rangle|L\rangle,$$

and

$$\begin{aligned}\langle E|E\rangle &= \langle R|E\rangle\langle E|R\rangle + \langle L|E\rangle\langle E|L\rangle \\ &= |\langle R|E\rangle|^2 + |\langle L|E\rangle|^2 \\ &= |a|^2 + |b|^2.\end{aligned}$$

If p_1 and p_2 are the probabilities of transmission of the state $|E\rangle$ through an R- and an L-filter, respectively, we know from Eq. (2.2) that $p_1 + p_2 = 1$ and we may make the correspondence

$$p_1 = |\langle R|E\rangle|^2, \quad p_2 = |\langle L|E\rangle|^2,$$

since $\langle E|E\rangle = 1$. In other words, the probability of transmission of the state $|E\rangle$ through an R-filter is *not* the bracket product $\langle R|E\rangle$ but is $|\langle R|E\rangle|^2$. Since bracket products do not generally give probabilities, they are called *probability amplitudes* or simply amplitudes. The product $\langle R|E\rangle$, for example, is the amplitude for an E-state to be transmitted through an R-filter.

A method of measurement for these probabilities is suggested by the very meaning of the amplitudes. Let the intensity of the light emergent from the interferometer be measured as I_0. By placing an R-filter in the emergent beam, a measurement of the intensity of the light transmitted by the R-filter can be made, and a similar measurement is possible with an L-filter. Since the amplitudes for the transmission of an E-state through an R- or an L-filter are, respectively, $\langle R|E\rangle$ and $\langle L|E\rangle$, then the probabilities for such transmission are

$$I_1/I_0 = |\langle R|E\rangle|^2, \qquad I_2/I_0 = |\langle L|E\rangle|^2,$$

where I_1 and I_2 are the light intensities transmitted by the R- and L-filters.

Since the intensity measurements correspond to the squares of the absolute values of the amplitudes, such measurements cannot distinguish the amplitudes $\langle R|E\rangle$ and $e^{i\phi}\langle R|E\rangle$. In other words, the phases, ϕ, of these amplitudes cannot be determined. It is possible, however, to have apparatus in which amplitudes may interfere such that *phase differences* between amplitudes become very important. Interfering amplitudes will be of importance in Chapter 6.

A very important amplitude is that given by the bracket product $\langle P'|P\rangle$. From Eq. (2.1) we know that the probability that a P-state will pass through a P'-filter is given by

$$p = I/I_{\text{max.}} \cos^2\theta,$$

where θ is the angle between the line of the P-filter which prepared the P-state and the line of the P'-filter. From the previous discussion we may set

$$\langle P'|P\rangle = e^{i\phi}\cos\theta.$$

But we have no way of evaluating the phase ϕ. However, as will become apparent, if we arbitrarily fix this phase, the phases in all other bracket products will be determined. Without loss of generality we will therefore set $\phi = 0$ and write

$$\langle P'|P\rangle = \langle P|P'\rangle = \cos\theta. \tag{3.2}$$

In an effort to correlate the ket vector representation and bracket products with the observable polarization states, we have relied almost entirely on the interferometer. We will now discuss a device which has just the opposite effect to that of an interferometer. Instead of combining light beams to produce a polarization state, this device resolves or analyzes light in a given polarization state into its components. Our device splits the incident beam into R- and L-states. We want to utilize this apparatus shown in Fig. 3.1 to give another illustration of how amplitudes are correlated with experiment. It will be assumed that the device which resolves light into its components completely transmits light, and it will simply be called a "resolver".

Light coming from the left strikes the E-filter and is transmitted as a prepared E-state, which traverses the resolver and is split into two beams labeled 1 and 2 which are in states $|R\rangle$ and $|L\rangle$, respectively. D_1 and D_2 are identical photon counters which are polarization-insensitive.

When the incident light is very weak, both counters never click simultaneously; the light coming out of the resolver is detected in either an R- or an L-state. As before, we can express the light prepared in the E-state as a superposition

$$|E\rangle = a|R\rangle + b|L\rangle,$$

where the bracket products are related by

$$\langle E|E\rangle = |\langle R|E\rangle|^2 + |\langle L|E\rangle|^2.$$

The light incident on the resolver is known to be in the state $|E\rangle$ and the probability that the light is detected in an R-state is $|\langle R|E\rangle|^2$. The probability

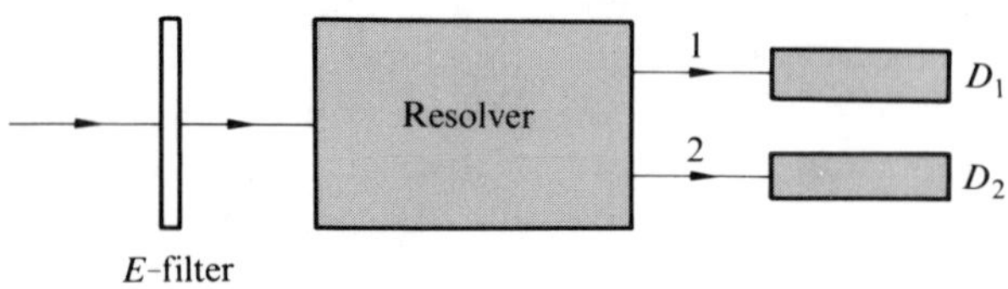

Fig. 3.1. Schematic representation of a resolver.

of detecting the light in an L-state is $|\langle L|E\rangle|^2$. If counts are taken for a sufficiently long time, the number of counts registered by D_1 is proportional to $|\langle R|E\rangle|^2$ and those recorded by D_2 over the same interval are proportional to $|\langle L|E\rangle|^2$. Since the light is prepared in an E-state, a measure of $\langle E|E\rangle$ can be obtained by removing the resolver and using either counter to register the number of counts in the same interval. This last step would not be necessary if we continued to use the convention $\langle E|E\rangle = 1$. However, this much rigidity is not always desirable.

There are a number of alternative interpretations of the bracket product which are intimately related. While keeping in mind our remarks about the dangers of regarding light as discrete, we can think of $|\langle R|E\rangle|^2$ as giving the probability of a single photon in an E-state traversing an R-filter. We can also think of a beam passing an initial E-filter such that N photons per second pass the filter. Then we could write $\sqrt{N}|E\rangle = |E'\rangle$, so that $\langle E'|E'\rangle = N$. A similar interpretation is likewise possible with the intensity of a beam I, such that $\langle E''|E''\rangle = I$, where $|E''\rangle = \sqrt{I}|E\rangle$. In general, the bracket product of a state with itself can have many related meanings Exactly which meaning is intended will either be explicitly stated or evident from the context. This procedure of multiplying a bra or ket by a real constant to make $\langle E|E\rangle$ fit a particular interpretation is known as *normalization*.

3.3 THE SUPERPOSITION OF *P*-STATES

Until now we have written E-states as a coherent superposition of the orthogonal kets $|R\rangle$ and $|L\rangle$ only. That this is not the only possibility was discussed in Section 2.5, and in what follows we derive the general standard form for a state expressed as the superposition of any pair of orthogonal states. The interpretation of the bracket product in the preceding section will then permit an analysis of a state which is expressed as the superposition of orthogonal P-states.

When a polarization state is written as a superposition of P-states with the basis

$$|P_x\rangle = \begin{pmatrix}1\\0\end{pmatrix} \qquad |P_y\rangle = \begin{pmatrix}0\\1\end{pmatrix},$$

the vector representing the state is called the Jones vector and the manipulation of these vectors is called the Jones calculus. However, we shall refer to any two-dimensional complex representation of a polarization state as a Jones vector and to the manipulation of Jones vectors as the Jones calculus.

Now, let $|E_1\rangle$ and $|E_2\rangle$ be any two orthogonal states and superimpose them as

$$|E\rangle = A|E_1\rangle + B|E_2\rangle,$$

where the constants A and B may be complex and can be written in the form

$$A = ae^{i\psi} \qquad B = be^{i\phi},$$

with a, b, ψ, and ϕ being real. This superposition can be rewritten as

$$|E\rangle = e^{i\alpha}\{ae^{-i\beta/2}|E_1\rangle + be^{+i\beta/2}|E_2\rangle\}, \tag{3.3}$$

where

$$\alpha = \tfrac{1}{2}(\psi + \phi), \qquad \beta = \phi - \psi.$$

In any experiment in which the state $|E\rangle$ is to be detected in another pure state $|E'\rangle$, the phase factor $e^{i\alpha}$ will vanish in $|\langle E|E'\rangle|^2$. For this reason the factor $e^{i\alpha}$ is often neglected and the standard form for the coherent superposition of two orthogonal states is that of Eq. (3.3) with $e^{i\alpha}$ neglected.

Let $|P_x\rangle$ and $|P_y\rangle$ designate pure states which have been prepared by P-filters whose lines are along the x- and y-axes. When light in the state $|P_x\rangle$ is analyzed by a P-filter, it is completely transmitted when the line of the filter is along the x-axis. Similarly, the state $|P_y\rangle$ is pure in respect to a P-filter whose line is along the y-axis. Using these two states as a basis, we may express any E-state as

$$|E\rangle = a_x e^{-i\phi/2}|P_x\rangle + a_y e^{+i/\phi 2}|P_y\rangle.$$

Using an intensity normalization for this state, we obtain

$$I_0 = \langle E|E\rangle = a_x^2 + a_y^2.$$

In what follows we will identify particular E-states, e.g., R- and L-states, with specific choices of a_x, a_y, and ϕ.

We now analyze the state $|E\rangle$ by using a P-filter whose line is oriented at angle θ with the x-axis and, hence, at angle $(\pi/2 - \theta)$ with the y-axis. We obtain

$$\langle P|E\rangle = a_x e^{-i\phi/2}\langle P|P_x\rangle + a_y e^{+i\phi/2}\langle P|P_y\rangle, \tag{3.4}$$

and, using Eq. (3.2),

$$\begin{aligned}\langle P|E\rangle &= a_x e^{-i\phi/2}\cos\theta + a_y e^{+i\phi/2}\cos(\pi/2 - \theta)\\ &= a_x e^{-i\phi/2}\cos\theta + a_y e^{+i\phi/2}\sin\theta.\end{aligned}$$

First consider the special case $\phi = 0$ and compute the intensity transmitted by the P-filter:

$$I = |\langle P|E\rangle|^2 = (a_x\cos\theta + a_y\sin\theta)^2. \tag{3.5}$$

As we would expect, this is highly dependent on θ. An insight can be obtained by looking for maxima and minima of $|\langle P|E\rangle|^2$. Setting $dI/d\theta = 0$ yields the solutions

$$\tan\theta_1 = -a_x/a_y, \tag{3.6}$$

$$\cot\theta_2 = a_x/a_y. \tag{3.7}$$

Investigation of $d^2I/d\theta^2$ shows that Eq. (3.6) yields a minimum and Eq. (3.7) a maximum. To find the maximum and minimum transmissions, we place the values given by these equations into Eq. (3.5) and obtain

$$(a_x\cos\theta_1 + a_y\sin\theta_1)^2 = 0,$$

$$(a_x\cos\theta_2 + a_y\sin\theta_2)^2 = a_x^2 + a_y^2 = I_0.$$

Since $\tan\theta_1 = -\cot\theta_2$, θ_1 and θ_2 must define two mutually perpendicular directions. Along the θ_1-direction the transmission through a P-filter is zero and along θ_2 it is equal to the incident intensity. We thus conclude that $|E\rangle$ is another P-state with its line oriented at angle θ_2 with the x-axis, whenever in standard form $\phi = 0$.

To see this more clearly, let us transform our reference system to an x'-axis in the θ_2-direction, as shown in Fig. 3.2, where θ' and θ are the angles which the P-filter makes with the x'- and x-axes, respectively. Equation (3.5) becomes

$$I = [a_x\cos(\theta' + \theta_2) + a_y\sin(\theta' + \theta_2)]^2,$$

which reduces to

$$I = (a_x^2 + a_y^2)\cos^2\theta'$$
$$= I_0\cos^2\theta'.$$

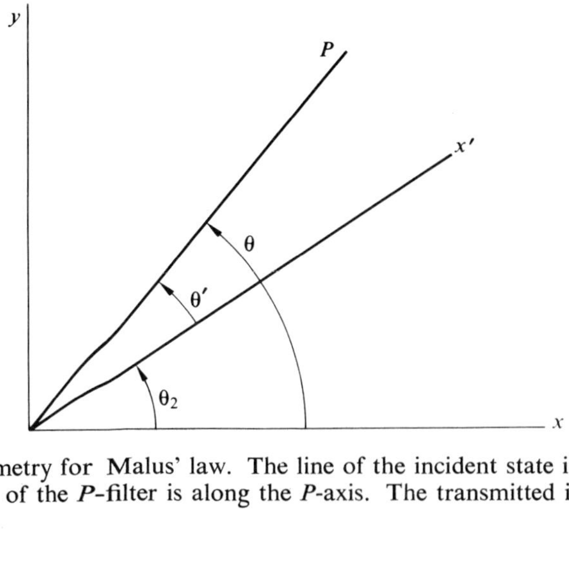

Fig. 3.2. Geometry for Malus' law. The line of the incident state is along the x'-axis. The line of the P-filter is along the P-axis. The transmitted intensity obeys $I = I_0\cos^2\theta'$.

Since the transmitted intensity obeys Malus' law, this completely identifies $|E\rangle$ as a pure P-state whose line is oriented along θ_2. Consequently, any P-state can be written as a superposition

$$|P\rangle = a_x|P_x\rangle + a_y|P_y\rangle, \tag{3.8}$$

where a_x and a_y are real. The line of this P-state makes an angle $\theta = \tan^{-1}(a_y/a_x)$ with the x-axis.

Let us now consider the general case with $\phi \neq 0$. From Eqs. (3.4) and (3.5) we have

$$\langle P|E\rangle = a_x e^{-i\phi/2} \cos\theta + a_y e^{+i\phi/2} \sin\theta,$$
$$|\langle P|E\rangle|^2 = a_x^2 \cos^2\theta + a_y^2 \sin^2\theta + a_x a_y \cos\theta \sin\theta(e^{i\phi} + e^{-i\phi}).$$

But $e^{i\phi} + e^{-i\phi} = 2\cos\phi$; thus,

$$|\langle P|E\rangle|^2 = a_x^2 \cos^2\theta + a_y^2 \sin^2\theta + 2a_x a_y \cos\theta \sin\theta \cos\phi,$$

which, of course, agrees with the previous result when $\phi = 0$.

Following the development we have given, this equation should indicate the transmission of a general elliptical state $|E\rangle$ through a P-filter. Particular cases of E-states are R- and L-states.

Thus, some particular choices of a_x, a_y, and ϕ should yield the representations for R- and L-states. The test for circular polarization is that transmission through a P-filter must be independent of the orientation of the P-filter, θ. We proceed as follows: first we find the maximum and minimum of $|\langle P|E\rangle|^2$ as a function of θ, then demand that the values of a_x, a_y, and ϕ which yield maximum and minimum values be independent of θ; that is, we find those values for which every value of θ yields both a maximum and a minimum, or for which $|\langle P|E\rangle|^2$ as a function of θ is constant. Taking the derivative, we find

$$\frac{\partial}{\partial\theta}|\langle P|E\rangle|^2 = -2a_x^2 \cos\theta \sin\theta + 2a_y^2 \sin\theta\cos\theta + 2a_x a_y \cos^2\theta \cos\phi - 2a_x a_y \sin^2\theta \cos\phi.$$

Setting this equal to zero to find the maxima and minima yields

$$\left(\frac{a_y}{a_x} - \frac{a_x}{a_y}\right) + (\cot\theta - \tan\theta)\cos\phi = 0.$$

For this to be independent of θ, we must have $\cos\phi = 0$, with $\phi = \pm\pi/2$. If $\cos\phi = 0$, then a_x must equal a_y, and in order to normalize the state to unity if a probability interpretation is desired, we place $a_x = a_y = 1/\sqrt{2}$. We shall pick a convention in which the choice $\phi = +\pi/2$ corresponds to

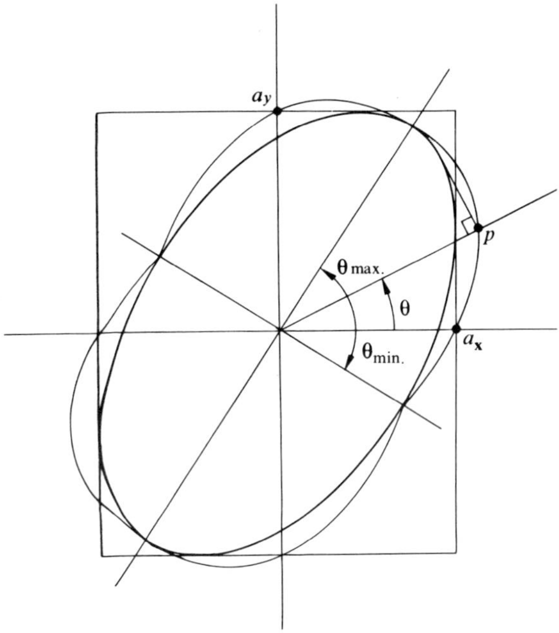

Fig. 3.3. The probability amplitude $|\langle P|E\rangle|$ as a function of the orientation, θ, of a *P*-filter.

a right circular state and $\phi = -\pi/2$ corresponds to a left circular state. Thus,

$$|R\rangle = \frac{1}{\sqrt{2}}\{e^{-i\pi/4}|P_x\rangle + e^{+i\pi/4}|P_y\rangle\},$$
$$|L\rangle = \frac{1}{\sqrt{2}}\{e^{+i\pi/4}|P_x\rangle + e^{-i\pi/4}|P_y\rangle\}. \tag{3.9}$$

The general case of elliptical polarization is not quite so easy when $a_x \neq a_y$ and ϕ may take on any value in the range $0 \leqslant \phi \leqslant \pi/2$. For a typical *E*-state a plot of $|\langle P|E\rangle|$ as a function of θ is given in Fig. 3.3. It is to be emphasized that $|\langle P|E\rangle|$ is *not* the ellipse which is shown but is the rather strange curve which touches the ellipse at the four points of contact. The relation between the ellipse and the curve is illustrated at the point p. The value of $|\langle P|E\rangle|$ in a direction θ is the projection of the ellipse along that direction. Thus, calling *E*-states elliptical is not an obvious choice.

Why $|\langle P|E\rangle|$ should be obtainable from an ellipse in this fashion and the physical interpretation of the ellipse itself will be delayed until the electromagnetic or wave interpretation is discussed in a later chapter. At the moment the problem is to relate the values of θ at which $|\langle P|E\rangle|$ takes on its maximum and minimum values to a_x, a_y, and ϕ. Returning to $(\partial/\partial\theta)|\langle P|E\rangle|^2$, we have

$$(a_y/a_x - a_x/a_y) + (\cot\theta - \tan\theta)\cos\phi = 0.$$

Setting $\tan R = a_y/a_x$ gives

$$-(\cot R - \tan R) + (\cot\theta - \tan\theta)\cos\phi = 0.$$

But by a trigonometric identity

$$\cot\alpha - \tan\alpha = 2\cot 2\alpha,$$

so that

$$-2\cot 2R + 2\cot 2\theta\cos\phi = 0,$$

or

$$\theta = \tfrac{1}{2}\tan^{-1}[\tan 2R\cos\phi],$$

which has two solutions separated by 90°. By considering the sign of $(\partial^2/\partial\theta^2)|\langle P|E\rangle|^2$ the maximum transmission direction is found to be that direction which lies closest to the x- or y-axis, according as $a_x > a_y$ or $a_x < a_y$. If $\phi = \pm\pi/2$, $\cos\phi = 0$ and

$$\theta_{\text{max.,min.}} = 1/2\tan^{-1}(0) = 0, \pi/2,$$

so that the maximum and minimum transmissions occur at 0 and $\pi/2$. Thus, an E-state with maximum and minimum directions along the x- and y-axes can always be written as

$$|E\rangle = a_x e^{\mp i\pi/4}|P_x\rangle + a_y e^{\pm i\pi/4}|P_y\rangle.$$

We have shown that the directions of the maximum and minimum transmission of an E-state through a P-filter are always separated by 90°. Thus, different E-states having the same maximum and minimum transmissions differ (apart from an arbitrary phase factor) only in the orientation of these directions and in the handedness of the state. Without a formal trigonometric proof we can therefore say that for any arbitrary E-state

$$|E\rangle = a_x e^{-i\psi/2}|P_x\rangle + a_y e^{+i\psi/2}|P_y\rangle,$$

expressed in a particular coordinate system x, y, there also exists a coordinate system x', y' in which $|E\rangle$ can be written

$$|E\rangle = a_{x'} e^{\mp i\pi/4}|P_{x'}\rangle + a'_y e^{\pm i\pi/4}|P_{y'}\rangle. \tag{3.10}$$

A simple proof of this statement will be given after the Poincaré sphere has been discussed in Chapter 4. It will also be shown there that for any state $|E\rangle$ there exists a coordinate system x'', y'' in which $|E\rangle$ may be written

$$|E\rangle = \frac{1}{\sqrt{2}}\{e^{\mp i\phi/2}|P_{x''}\rangle + e^{\pm i\phi/2}|P_{y''}\rangle\}. \tag{3.11}$$

It is the ϕ of Eq. (3.11) which is used to label the states in Fig. 2.4. By analogy with the circular case the plus and minus signs refer to right and left elliptical polarizations.

3.4 RETARDATION PLATES AND THEIR MATRIX REPRESENTATION

The derivation in the preceding section of the representation for *R*- and *L*-states as a superposition of orthogonal *P*-states is based on the requirement that the intensity of the light transmitted by the *P*-filter be independent of the orientation of the filter. This requirement leads to the particular exponential phase factor in Eq. (3.9). In this section the physical significance of phase factors will be examined in terms of retarders, and it will be shown how retarders may be described by matrices. The primary motivation in what follows below and in the next section is to show how any polarization state can be produced by a device called an elliptical polarizer. We shall find for any state $|E\rangle$ the filter in respect to which it is pure.

It was seen in Chapter 2 that the effect of a retarder on a pure state $|E\rangle$ is mathematically represented by a multiplicative phase factor. Thus, if the state $|E\rangle$ passes through a glass plate, the emergent state is given by $|E'\rangle = e^{iks}|E\rangle$, where $s = t(n - 1)$ is the retardation of the plate; n and t are the index of refraction and thickness of the plate. If the retardation is increased by an amount Δs, the emergent state is $|E''\rangle = e^{ik(s+\Delta s)}|E\rangle$.

From Section 2.4 we know that when the increase in retardation is equal to the wavelength λ, the ket $|E''\rangle$ will be the same as the ket $|E'\rangle$. Since the exponential phase factor is periodic with period 2π, we must have $k = 2\pi/\lambda$, where k is now defined as the *wave number*. As a result we can express the light emerging from a retarder in terms of the wavelength of the light, and the thickness and index of refraction of the plate in the form

$$|E'\rangle = e^{(2\pi i/\lambda)t(n-1)}|E\rangle.$$

Let us now insert a birefringent plate in a beam which is written as a coherent superposition of two *P*-states $|P_x\rangle$ and $|P_y\rangle$ as

$$|E\rangle = A|P_x\rangle + B|P_y\rangle,$$

where A and B may be complex. We insert the plate so that its fast axis is parallel to the x-axis and its slow axis is along the y-axis. We call n_s and n_f the indices of refraction along the slow and fast directions. After passing through the plate each component has undergone a retardation. If the incident beam is a pure $|P_x\rangle$ state, then the transmitted state is $|P_{x'}\rangle$, given by

$$|P_{x'}\rangle = e^{ikt(n_f-1)}|P_x\rangle.$$

If the incident state is a pure $|P_y\rangle$ state, then the emergent state is

$$|P_{y'}\rangle = e^{ikt(n_s-1)}|P_y\rangle.$$

If the incident state is the superposition given above, then the emergent state must be given by

$$\begin{aligned} |E'\rangle &= Ae^{ikt(n_f-1)}|P_x\rangle + Be^{ikt(n_s-1)}|P_y\rangle \\ &= e^{-ikt}\{Ae^{iktn_f}|P_x\rangle + Be^{iktn_s}|P_y\rangle\}, \end{aligned} \tag{3.12}$$

which may be written in standard form

$$|E'\rangle = e^{-i\psi}[Ae^{-i\delta/2}|P_x\rangle + Be^{+i\delta/2}|P_y\rangle], \tag{3.13}$$

where

$$\psi = kt\left[1 - \frac{n_s + n_f}{2}\right],$$

$$\delta = kt[n_s - n_f].$$

The quantity δ in Eq. (3.13) we call the *retardance* of the birefringent plate. From Eq. (3.12) we see that the components of $|E'\rangle$ have different retardations, and we will call the component whose line is along the fast axis of the plate the fast component and the other the slow component. The difference in retardations, $t(n_s - n_f)$, is often used to classify birefringent retardation plates by expressing this retardation difference as a fraction of a wavelength. Thus, we have a half-wave plate when $t(n_s - n_f) = \lambda/2$ and $\delta = \pi$, or a quarter-wave plate when $t(n_s - n_f) = \lambda/4$ and $\delta = \pi/2$.

A matrix which describes a retarder can now be written without much effort. Let the state $|E\rangle$ be written in column vector representation as

$$|E\rangle = A|P_x\rangle + B|P_y\rangle = \begin{pmatrix} A \\ B \end{pmatrix}.$$

If the light in this state is incident on a retarder plate, the emergent state is given by

$$|E'\rangle = \begin{pmatrix} Ae^{-i\delta/2} \\ Be^{+i\delta/2} \end{pmatrix}.$$

We now wish to find a matrix operator to represent the effect of the plate. That is, we wish to find a matrix M such that

$$|E'\rangle = \mathsf{M}|E\rangle.$$

It is obvious from Eq. (3.13) that, ignoring the factor $e^{-i\psi}$,

$$\mathsf{M} = \begin{pmatrix} e^{-i\delta/2} & 0 \\ 0 & e^{+i\delta/2} \end{pmatrix}. \tag{3.14}$$

The matrix M given above is a valid representation of the retarder only if the fast and slow axes of the plate are along the x- and y-axes, respectively.

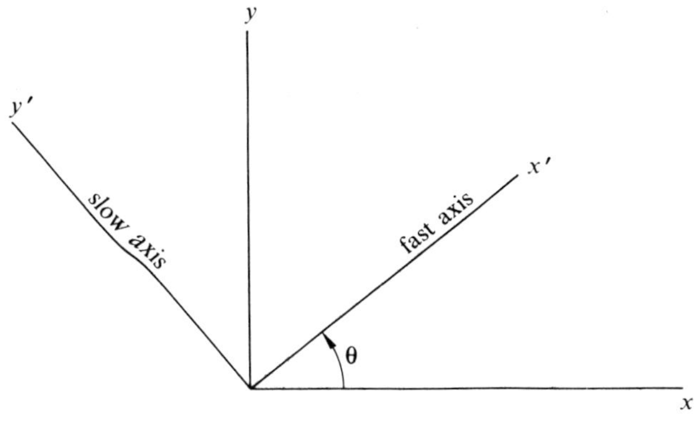

Fig. 3.4. The orientation of the fast and slow axes, x' and y', of a retarder plate.

If the fast and slow axes of the plate are interchanged, it is seen that the correct matrix is

$$\mathsf{M}^\star = \begin{pmatrix} e^{i\delta/2} & 0 \\ 0 & e^{-i\delta/2} \end{pmatrix}. \tag{3.15}$$

What is the representation of the retarder when its fast axis makes an angle θ with the x-axis? This representation can be obtained in the following way. We will rotate the retarder plate until its fast axis makes an angle θ with the x-axis. Let x' and y' be the new directions of the fast and slow axes of the plate. We will take an arbitrary state $|E\rangle$ expressed in terms of $|P_x\rangle$ and $|P_y\rangle$ and express it in terms of $|P_{x'}\rangle$ and $|P_{y'}\rangle$. In the primed system the effect of the retarder, after rotation, is given by M, since the axes of the retarder are along the axes of the primed coordinate system. After applying the retarder matrix in the primed system we will then transform back to the original system. Looking down the beam toward the retarder, the geometry is as shown in Fig. 3.4. The beam direction is perpendicular to the page. Expressing the same state E in both systems,

$$|E\rangle = A|P_x\rangle + B|P_y\rangle,$$
$$|E'\rangle = A'|P_{x'}\rangle + B'|P_{y'}\rangle,$$

and, taking the bracket product of both equations with $\langle P_{x'}|$ and $\langle P_{y'}|$, we obtain

$$A' = A\langle P_{x'}|P_x\rangle + B\langle P_{y'}|P_y\rangle,$$
$$B' = A\langle P_{y'}|P_y\rangle + B\langle P_{y'}|P_y\rangle.$$

From the interpretation of the bracket product we know that the amplitude for the detection of an unprimed P-state in a primed P-state is the cosine of the angle between the lines of these states. Therefore, we may write

$$A' = A\cos\theta + B\sin\theta,$$
$$B' = -A\sin\theta + B\cos\theta.$$

It is readily seen that the above equations may be written in matrix form as

$$|E'\rangle = \mathsf{R}|E\rangle,$$

where the rotation matrix R (by which the components of the ket vector in the primed system are expressed in terms of the components of the ket vector in the unprimed system) is given by

$$\mathsf{R} = \begin{pmatrix} \cos\theta & \sin\theta \\ -\sin\theta & \cos\theta \end{pmatrix}. \tag{3.16}$$

The state vector in the primed system after passing through the retarder plate is then given by $\mathsf{M}|E'\rangle = \mathsf{MR}|E\rangle$. To express this in the original system we must transform back to that system by using R^{-1}. This matrix, R^{-1}, represents a rotation through an angle $-\theta$ and, hence, is equal to R with θ replaced by $-\theta$. Finally, the state $|E''\rangle$ obtained after transmitting a state $|E\rangle$ through a retarder plate whose fast axis makes an angle θ with the x-axis is given by $|E''\rangle = \mathsf{M}'|E\rangle$, where $\mathsf{M}' = \mathsf{R}^{-1}\mathsf{MR}$. The explicit result is

$$\mathsf{M}' = \begin{pmatrix} (e^{-i\delta/2}\cos^2\theta + e^{+i\delta/2}\sin^2\theta) & -2i\sin\delta/2\cos\theta\sin\theta \\ -2i\sin\delta/2\cos\theta\sin\theta & (e^{-i\delta/2}\sin^2\theta + e^{+i\delta/2}\cos^2\theta) \end{pmatrix}. \tag{3.17}$$

As an example, consider a state $|P_y\rangle$ incident on a quarter-wave plate whose fast axis is midway between the x- and y-axes. For this case $\delta = \pi/2$ and $\theta = \pi/4$; therefore,

$$\mathsf{M}' = \frac{1}{\sqrt{2}}\begin{pmatrix} 1 & -i \\ -i & 1 \end{pmatrix}.$$

The emergent state is obtained as

$$|E'\rangle = \mathsf{M}'\begin{pmatrix} 0 \\ 1 \end{pmatrix} = \frac{e^{-i\pi/4}}{\sqrt{2}}\begin{pmatrix} e^{-i\pi/4} \\ e^{+i\pi/4} \end{pmatrix},$$

which is identified as an R-state.

3.5 AN ELLIPTICAL POLARIZER

We are now in a position to describe a device consisting of three elements which can produce any E-state. Our elliptical polarizer or E-filter simply consists of a P-filter placed between two retarder plates. In Fig. 3.5 is shown such an E-filter with the light incident from the left. The retarder plates are

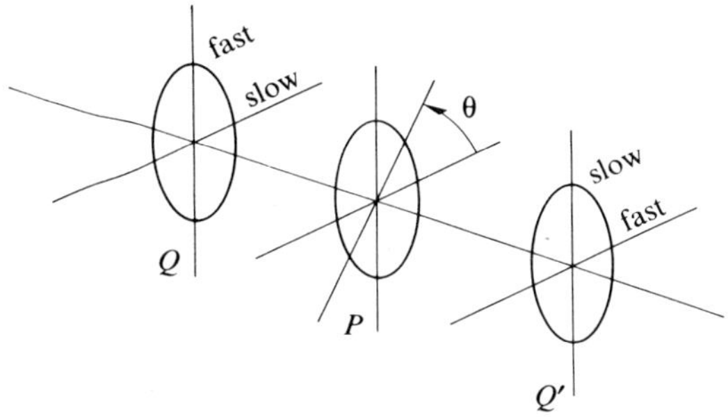

Fig. 3.5. An elliptical polarizer consisting of two retarder plates, Q and Q', and a P-filter.

indicated by Q and Q', and the P-filter by P. The retarder Q is oriented such that its slow and fast axes are parallel to the x- and y-axes, respectively. The line of the P-filter makes an angle θ with the x-axis, and the fast axis of Q' is parallel to the slow axis of Q.

No matter what the state of the incident light (polarized or unpolarized), only light in a P-state will be transmitted by the P-filter, and we write for this transmitted state

$$|P\rangle = A_x|P_x\rangle + A_y|P_y\rangle,$$

where A_x and A_y are real, and the angle the P-filter makes with the x-axis is given by $\tan\theta = A_y/A_x$. The retardance of the plate Q' is δ, and since the fast axis of Q' is along the x-axis, we find by using Eq. (3.14) that the light emerges from Q' in the state

$$\begin{aligned} |E\rangle &= \mathsf{M}|P\rangle \\ &= A_x e^{-i\delta/2}|P_x\rangle + A_y e^{+i\delta/2}|P_y\rangle, \end{aligned} \tag{3.18}$$

where M is given by Eq. (3.14) and $|E\rangle$ is a right elliptical state. If left states are desired, Q and Q' are interchanged. If the emergent state is incident on an E-filter identical to that which we have described, the retarder Q of the second E-filter whose matrix is given by Eq. (3.15) produces the state

$$|P\rangle = \mathsf{M}^\star|E\rangle = A_x|P_x\rangle + A_y|P_y\rangle,$$

which is transmitted completely by the P-filter. The final retarder Q', represented by M, gives the emergent state $\mathsf{M}|P\rangle$, which is identical to that given by Eq. (3.18). For the particular state $|E\rangle$, the effect of the two retarders effectively cancel each other, and since the P-filter has no effect, the emergent state is simply the incident state. It should be evident that no E-state other than that given by Eq. (3.18) will be transmitted unaltered, since no other

E-state will emerge from the first retarder as a P-state whose line is exactly $\theta = \tan^{-1} A_y/A_x$.

We may now write the matrix for the elliptical polarizer T_e as a product of the matrices for the retarders and the P-filter:

$$\mathsf{T}_e = \mathsf{MTM}^\star = \begin{pmatrix} e^{-i\delta/2} & 0 \\ 0 & e^{i\delta/2} \end{pmatrix} \begin{pmatrix} \cos^2\theta & \cos\theta\sin\theta \\ \cos\theta\sin\theta & \sin^2\theta \end{pmatrix} \begin{pmatrix} e^{i\delta/2} & 0 \\ 0 & e^{-i\delta/2} \end{pmatrix}$$

$$= \begin{pmatrix} \cos^2\theta & e^{-i\delta}\cos\theta\sin\theta \\ e^{+i\delta}\cos\theta\sin\theta & \sin^2\theta \end{pmatrix},$$

where M and $\mathsf{M}^\star$ are given by Eqs. (3.14) and (3.15) and T is the transmission matrix of the P-filter. It is left as an exercise to show that the transmission matrix for a P-filter whose line makes an angle θ with the x-axis is given by T.

In practice it is possible to construct a general elliptical filter using quarter-wave plates rather than general retarder plates. Recall that in Section 3.4 it was stated that any state such as that given by Eq. (3.18) could be written

$$|E\rangle = A_{x'}e^{-i\pi/4}|P_{x'}\rangle + A_{y'}e^{+i\pi/4}|P_{y'}\rangle$$

in some coordinate system x', y'.

Now, it should be apparent that in this coordinate system the matrix for the elliptical polarizer will have the same form as it had in the old system if it is rewritten in terms of $\theta' = \tan^{-1} A_{y'}/A_{x'}$ and $\delta'/2 = 1/2(\pi/2)$. The matrix in the (x', y')-system is thus

$$\mathsf{T}_{e'} = \begin{pmatrix} \cos^2\theta' & -i\cos\theta'\sin\theta' \\ i\cos\theta'\sin\theta' & \sin^2\theta' \end{pmatrix}.$$

From the forms of the matrices T_e and $\mathsf{T}_{e'}$ it is seen that $\mathsf{T}_{e'}$ can be taken to represent two quarter-wave plates separated by a P-filter. The slow axis of the first plate and the fast axis of the second plate would lie along the x'-axis. The line of the P-filter would make angle θ' with the x'-axis.

Thus, a general elliptical polärizer may be constructed using quarter-wave plates and a P-filter if we allow the entire filter to be rotated about the axis of the beam.

3.6 TRANSMISSION MATRICES IN OTHER BASES

Some examples of the type of problem which will be of concern in this section have already been given, but these were specific instances of a rather broad class of transformations. The transformations in the last section could be visualized in terms of an actual rotation of an E-filter about the optic axis. In many cases the relation of the transformations with a physical rotation is not clear, if indeed there is such a relation. We now pause and solve the problem in more general terms. If the transmission matrix of a given device is known when the state of light entering the device is expressed as a superposition of a given pair of orthonormal kets, what is the corresponding transmission matrix when the entrant state is expressed in terms of a different set of orthonormal kets?

In what follows, sets of orthonormal kets such as $|P_x\rangle$, $|P_y\rangle$ or $|R\rangle$, $|L\rangle$ will be referred to as basic kets because they provide a basis in which any polarization state may be represented. Also, in order to minimize notation difficulties, any two different sets of basic kets will be labeled $|\phi_i\rangle$ and $|\psi_i\rangle$, with $i = 1,2$, and the kets $|\psi\rangle$ and $|\phi\rangle$ which are superpositions of these basic kets will be said to be in the ψ- and ϕ-bases, respectively. For example, if the E-state in question is an R-state, and $|\psi_i\rangle = |R\rangle, |L\rangle$; $|\phi_i\rangle = |P_x\rangle, |P_y\rangle$, then

$$|R\rangle = |\psi\rangle = \begin{pmatrix} 1 \\ 0 \end{pmatrix},$$

$$|R\rangle = |\phi\rangle = \frac{1}{\sqrt{2}} \begin{pmatrix} e^{-i\pi/4} \\ e^{+i\pi/4} \end{pmatrix}.$$

The matrix which transforms a ket in the ψ-basis to one in the ϕ-basis is taken as U. Let $|\psi\rangle$ and $|\phi\rangle$ be two different expansions for the *same* polarization state of light incident on an optical device, and assume that the transmission matrix M corresponding to the ψ-expansion is known. We can then write

$$|\psi'\rangle = \mathsf{M}|\psi\rangle, \tag{3.19}$$

$$|\phi\rangle = \mathsf{U}|\psi\rangle, \tag{3.20}$$

where a primed symbol inside a ket indicates the emergent state.

To determine the transmission matrix M' in the ϕ-representation, we multiply Eq. (3.19) by U and obtain

$$|\phi'\rangle = \mathsf{U}|\psi'\rangle = \mathsf{UM}|\psi\rangle, \tag{3.21}$$

and from Eq. (3.20) we have $|\psi\rangle = \mathsf{U}^{-1}|\phi\rangle$, which is substituted in Eq. (3.21) to give

$$|\phi'\rangle = \mathsf{UMU}^{-1}|\phi\rangle.$$

We are assuming for the moment that U is unitary, i.e. $\mathsf{U}^{-1} = \mathsf{U}^{\dagger}$; this will be explicitly demonstrated later. As a result, the transmission matrix M' in the ϕ-basis is

$$\mathsf{M}' = \mathsf{UMU}^{\dagger}. \tag{3.22}$$

Before proceeding to some examples of this matrix transformation, it is convenient to have available a powerful identity which is useful in transforming kets from one basis to another. Let us expand any ket $|\phi\rangle$ in terms of the basic kets $|\phi_i\rangle$:

$$|\phi\rangle = \langle\phi_1|\phi\rangle|\phi_1\rangle + \langle\phi_2|\phi\rangle|\phi_2\rangle$$
$$= |\phi_1\rangle\langle\phi_1|\phi\rangle + |\phi_2\rangle\langle\phi_2|\phi\rangle,$$

which can be written

$$|\phi\rangle = \{\sum_i |\phi_i\rangle\langle\phi_i|\}\,|\phi\rangle. \tag{3.23}$$

It is seen that the operator in the brackets leaves $|\phi\rangle$ unchanged and acts as an identity operator or matrix. In the above manipulation we expanded in the $|\phi_i\rangle$ but we did not have to give explicit column vectors for these kets. For example, if $|\phi_1\rangle$ is $|R\rangle$, we have not said whether we wish to write

$$|\phi_1\rangle = \begin{pmatrix}1\\0\end{pmatrix}$$

or

$$|\phi_1\rangle = \frac{1}{\sqrt{2}}\begin{pmatrix}e^{-i\pi/4}\\ e^{+i\pi/4}\end{pmatrix}.$$

Thus, Eq. (3.23) is actually independent of which basis or representation for that basis we use. Thus, we can write

$$|E\rangle = \sum_i |\phi_i\rangle\langle\phi_i||E\rangle, \tag{3.24}$$

where no particular basis is explicitly indicated for the ket $|E\rangle$. If actual column vectors are used in Eq. (3.24), the same representation must, of course, be used throughout the equation. Finally, we can identify

$$\mathsf{I} = \sum_i |\phi_i\rangle\langle\phi_i|, \tag{3.25}$$

and if $|\psi_i\rangle$ are any other set of orthonormal kets,

$$\mathsf{I} = \sum_i |\psi_i\rangle\langle\psi_i|,$$

where I is taken to be an identity operator or identity matrix.

We may now find the matrix U which transforms the ket $|E\rangle = |\phi\rangle$ in the ϕ-basis to the ket $|E\rangle = |\psi\rangle$ of the same state written in the ψ-basis. Using the identity operator, we explicitly write

$$|\phi\rangle = \sum_i |\phi_i\rangle\langle\phi_i|E\rangle, \tag{3.26}$$

$$|\psi\rangle = \sum_j |\psi_j\rangle\langle\psi_j|E\rangle. \tag{3.27}$$

Placing Eq. (3.27) into Eq. (3.26), we obtain

$$\sum_i |\phi_i\rangle\langle\phi_i|E\rangle = \sum_{ij} |\phi_i\rangle\langle\phi_i|\psi_j\rangle\langle\psi_j|E\rangle$$
$$= \sum_{ij} \langle\phi_i|\psi_j\rangle\langle\psi_j|E\rangle|\phi_i\rangle, \tag{3.28}$$

giving

$$\langle\phi_i|E\rangle = \sum_j \langle\phi_i|\psi_j\rangle\langle\psi_j|E\rangle. \tag{3.29}$$

But $\langle\phi_i|E\rangle$ is the i-th component of $|E\rangle$ in the ϕ-basis and $\langle\psi_j|E\rangle$ is the j-th component of $|E\rangle$ in the ψ-basis. Using explicit vector representations for these states Eq. (3.29) allows us to write

$$\begin{pmatrix}\langle\phi_1|E\rangle\\ \langle\phi_2|E\rangle\end{pmatrix} = \begin{pmatrix}\langle\phi_1|\psi_1\rangle & \langle\phi_1|\psi_1\rangle\\ \langle\phi_2|\psi_1\rangle & \langle\phi_2|\psi_2\rangle\end{pmatrix}\begin{pmatrix}\langle\psi_1|E\rangle\\ \langle\psi_2|E\rangle\end{pmatrix}.$$

Thus, the element U_{ij} of the transformation matrix is

$$\mathsf{U}_{ij} = \langle\phi_i|\psi_j\rangle. \tag{3.30}$$

The explicit form of U having been obtained, it is left as an exercise to prove that U is unitary, $\mathsf{U}^+ = \mathsf{U}^{-1}$, as was assumed.

As an example, consider the retarder matrix

$$\mathsf{M} = \begin{pmatrix}e^{-i\delta/2} & 0\\ 0 & e^{i\delta/2}\end{pmatrix},$$

which corresponds to a retarder when the entrant state of light is expressed as a superposition of $|P_x\rangle = \begin{pmatrix}1\\0\end{pmatrix}$ and $|P_y\rangle = \begin{pmatrix}0\\1\end{pmatrix}$. What is the retarder matrix which must be used when the entrant light is expressed as a superposition of $|R\rangle = \begin{pmatrix}1\\0\end{pmatrix}$ and $|L\rangle = \begin{pmatrix}0\\1\end{pmatrix}$? Let the basic kets of our ψ-representation be $|P_x\rangle$, $|P_y\rangle$ and those of the ϕ-representation be $|R\rangle$ and $|L\rangle$. The transformation matrix U becomes

$$\mathsf{U} = \begin{pmatrix}\langle R|P_x\rangle & \langle R|P_y\rangle\\ \langle L|P_x\rangle & \langle L|P_y\rangle\end{pmatrix} = \frac{1}{\sqrt{2}}\begin{pmatrix}e^{i\pi/4} & e^{-i\pi/4}\\ e^{-i\pi/4} & e^{i\pi/4}\end{pmatrix}, \tag{3.31}$$

where the bracket products are evaluated by use of Eq. (3.9). The calculation of M' in Eq. (3.22) gives the result

$$\mathsf{M}' = \mathsf{U}\mathsf{M}\mathsf{U}^\dagger = \begin{pmatrix}\cos\delta/2 & \sin\delta/2\\ -\sin\delta/2 & \cos\delta/2\end{pmatrix}.$$

3.7 OPTICAL ACTIVITY

The retarders which have been used until now possess the property of resolving a light beam into two orthogonal P-states and transmitting these orthogonal states at different speeds. Such retarders are said to be linear retarders. Just as linear retarders resolve beams into P-states, there exist circular retarders which resolve beams into R- and L-states, which are transmitted with different speeds. Such materials, as we have mentioned before, are said to be optically active. We show in what follows that different transmission velocities for R- and L-states result in the rotation of the line of a P-state transmitted through the material.

It has been shown in Section 3.4 that the linear retarder introduces a phase difference between the components of a state expressed as a superposition of P-states. Similarly, if a state is expressed as a superposition of $|R\rangle$ and $|L\rangle$,

$$|E\rangle = A|R\rangle + B|L\rangle,$$

a phase difference between the components of the state is introduced when light in this state passes through an optically active substance. If this phase difference is β, the emergent state can be expressed as

$$|E'\rangle = Ae^{-i\beta/2}|R\rangle + Be^{i\beta/2}|L\rangle.$$

The retarder matrix for the optically active material expressed in an (R, L)-basis is therefore

$$\mathsf{N} = \begin{pmatrix} e^{-i\beta/2} & 0 \\ 0 & e^{i\beta/2} \end{pmatrix}. \tag{3.32}$$

By analogy with linear polarizers as discussed in Section 3.4, it can be seen that in the case of optical activity

$$\beta = kt(n_L - n_R), \tag{3.33}$$

where k is the wave number, t is the thickness of the material, and the indices of refraction for the R- and L-states are n_R and n_E, respectively.

Since the matrix U of Eq. (3.31) transforms from the (P_x, P_y)-basis to the (R, L)-basis, $\mathsf{U}^\dagger = \mathsf{U}^{-1}$ transforms from the (R, L)-basis to the (P_x, P_y)-basis.

Therefore, the matrix N expressed in the (P_x, P_y)-basis is given by

$$\begin{aligned} \mathsf{N}' &= \mathsf{U}^\dagger \mathsf{N} \mathsf{U} \\ &= \begin{pmatrix} \cos\beta/2 & -\sin\beta/2 \\ \sin\beta/2 & \cos\beta/2 \end{pmatrix}. \end{aligned}$$

As a special case let an incident beam be the P-state,

$$|P\rangle = \cos\theta|P_x\rangle + \sin\theta|P_y\rangle,$$

where the line of the $\boldsymbol{P}$-state is at angle θ with the x-axis. After traversing the optically active material the emergent state is given by

$$\mathsf{N}'|P\rangle = \begin{pmatrix} \cos(\theta + \beta/2) \\ \sin(\theta + \beta/2) \end{pmatrix},$$

which is simply a P-state whose line has rotated through an angle $\beta/2$ relative to the line of the entrant P-state. Thus, if the phase difference is positive, the rotation is in a counterclockwise sense with the beam coming toward an observer. If β is negative, the rotation would be in a clockwise sense.

3.8 DICHROIC POLARIZERS

The birefringent materials which we have been discussing so far have all been assumed to be transparent. Although the polarization state may have been modified during transmission, the total incident energy flux was transmitted and not absorbed. We will now examine the cases of isotropic and anisotropic absorption, and derive the equation, given in Section 2.9, which is a generalization of Malus' law for imperfect polarizers.

The simplest nontransparent case is that of an isotropic uniform absorber such as a piece of evenly smoked glass or a partially silvered mirror. In such cases a certain fraction of the light incident on the plate will be transmitted independent of the polarization state. The remaining fraction will either be absorbed within the material, thus raising its temperature, or reflected into another beam. It is fairly obvious that if k is the fraction of the incident energy which is transmitted, then the matrix which represents transmission through the plate is given by

$$|E'\rangle = \sqrt{k}|E\rangle = \begin{pmatrix} \sqrt{k} & 0 \\ 0 & \sqrt{k} \end{pmatrix} \begin{pmatrix} A \\ B \end{pmatrix}.$$

The interpretation of the bracket product $\langle E'|E'\rangle$ will clearly depend on the normalization of $|E\rangle$, since $\langle E'|E'\rangle = k\langle E|E\rangle$. If $\langle E|E\rangle = 1$, then k is the probability that a photon will be transmitted. If the normalization is such that the ket representing the incident beam is $\sqrt{N}|E\rangle$ or $\sqrt{I_0}|E\rangle$, where N is the number of incident photons per unit time and I_0 is the intensity of the incident beam, then the bracket product $\langle E'|E'\rangle$ is interpreted as the number of photons per unit time that are transmitted or the intensity of the transmitted beam. In the following we adopt a probability normalization.

Anisotropy in materials is exhibited in absorption phenomena as well as in transmission. As in the case of birefringence there were fast and slow axes, so for linear dichroic materials there are two principal absorption directions. A P-state whose line is parallel to one of these directions is found to undergo a maximum absorption; when its line is perpendicular to this

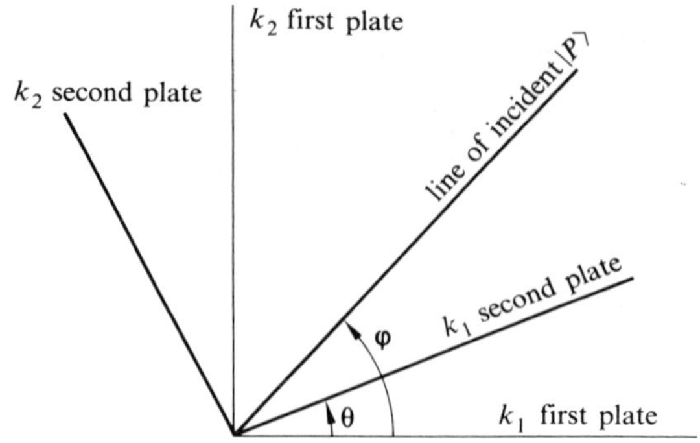

Fig. 3.6. Illustrating the geometry of the principal axes of a pair of imperfect polarizers.

direction, the state suffers a minimum absorption. If we express an arbitrary state $|E\rangle$ in terms of two P-states whose directions are along the maximum and minimum transmission directions of a dichroic plate, the transmission matrix is given by

$$|E'\rangle = \begin{pmatrix} \sqrt{k_1} & 0 \\ 0 & \sqrt{k_2} \end{pmatrix} \begin{pmatrix} A \\ B \end{pmatrix} = \begin{pmatrix} \sqrt{k_1}A \\ \sqrt{k_2}B \end{pmatrix}.$$

A dichroic sheet would be a perfect linear polarizer if either k_1 or k_2 were to be zero and the other equal to unity.

As an example, consider the transmission of a pure P-state through a pair of identical partial polarizers whose absorption axes make an angle θ with each other. The geometry is as shown in Fig. 3.6, and the incident light is perpendicular to the planes of the parallel polarizers. Let us first express the incident P-state in terms of $|P_x\rangle$ and $|P_y\rangle$ along the axis of the first plate. This is given by

$$|P\rangle = \cos\phi|P_x\rangle + \sin\phi|P_y\rangle.$$

Applying the transmission matrix of the first plate to this beam gives the state of the transmitted beam as

$$|P'\rangle = \sqrt{k_1}\cos\phi|P_x\rangle + \sqrt{k_2}\sin\phi|P_y\rangle.$$

We can now rewrite this in terms of $|P_{x'}\rangle$ and $|P_{y'}\rangle$ along the axes of the second plate by using the rotation matrix for angle θ:

$$\begin{pmatrix} \cos\theta & \sin\theta \\ -\sin\theta & \cos\theta \end{pmatrix} \begin{pmatrix} \sqrt{k_1}\cos\phi \\ \sqrt{k_2}\sin\phi \end{pmatrix} = \begin{pmatrix} \sqrt{k_1}\cos\theta\cos\phi + \sqrt{k_2}\sin\theta\sin\phi \\ -\sqrt{k_1}\sin\theta\cos\phi + \sqrt{k_2}\cos\theta\sin\phi \end{pmatrix}.$$

Applying the transmission matrix of the second plate to this result yields the state of the light transmitted by both polarizers expressed in terms of P-states along the second plate's axes as

$$|P''\rangle = \begin{pmatrix} k_1 \cos\theta \cos\phi + \sqrt{k_1 k_2} \sin\theta \sin\phi \\ -\sqrt{k_1 k_2} \sin\theta \cos\phi + k_2 \cos\theta \sin\phi \end{pmatrix}.$$

We now compute $\langle P''|P''\rangle$:

$$\langle P''|P''\rangle = k_1^2 \cos^2\theta \cos^2\phi + k_2^2 \cos^2\theta \sin^2\phi + 2[k_1\sqrt{k_1k_2} - k_2\sqrt{k_1k_2}] \sin\theta \cos\theta \sin\phi \cos\phi + k_1k_2 \sin^2\theta.$$

If the incident beam is a completely random incoherent superposition of P-states equally distributed angularly from $\phi = 0$ to $\phi = \pi$, the transmission is given by the average of the above expression over ϕ from 0 to π. Since

$$\frac{1}{\pi}\int_0^\pi \cos\phi \sin\phi \, d\phi = 0, \qquad \frac{1}{\pi}\int_0^\pi \sin^2\phi \, d\phi = \frac{1}{\pi}\int_0^\pi \cos^2\phi \, d\phi = \frac{1}{2},$$

we obtain

$$\langle P''|P''\rangle_{\text{average}} = \tfrac{1}{2}(k_1^2 + k_2^2)\cos^2\theta + (k_1k_2)\sin^2\theta,$$

which may be considered as a generalization of Malus' law for the transmission of an unpolarized beam through a pair of imperfect polarizers. Comparing this result with Eq. (2.5), one may obtain the major and minor transmittances in terms of k_1 and k_2.

PROBLEMS

3.1 Given the states below, with $\langle R|R\rangle = \langle L|L\rangle = 1$,

$$|E_1\rangle = 3e^{i\pi/3}\,|R\rangle + 2e^{-i\pi/3}\,|L\rangle,$$
$$|E_2\rangle = 2e^{-i\pi/2}\,|R\rangle + 3e^{i\pi/2}\,|L\rangle,$$

a) normalize each state so that $\langle E_1|E_1\rangle = \langle E_2|E_2\rangle = 1$;
b) evaluate the products $\langle E_1|E_2\rangle$ and $|E_1 \times E_2|$;
c) what is the probability that a photon in the state $|E_1\rangle$ will be transmitted by the filter which produces state $|E_2\rangle$?

Answer: $\dfrac{1}{\sqrt{13}}, \; -\dfrac{3\sqrt{3}}{\sqrt{13}}, \; \dfrac{1}{13}\begin{pmatrix} 6e^{i5\pi/6} & 9e^{-i\pi/6} \\ 4e^{i\pi/6} & 6e^{-i\pi5/6} \end{pmatrix}, \; \dfrac{27}{169}.$

3.2 The states $|E_1\rangle$ and $|E_2\rangle$ given below form an orthonormal pair:

$$|E_1\rangle = \frac{1}{\sqrt{2}}\,[e^{-i\pi/8}\,|P_x\rangle + e^{+i\pi/8}\,|P_y\rangle],$$

$$|E_2\rangle = \frac{1}{\sqrt{2}}\,[e^{3\pi/8i}\,|P_x\rangle + e^{-3\pi/8i}\,|P_y\rangle].$$

If the right circular state $|R\rangle$ is expanded as

$$|R\rangle = A|E_1\rangle + B|E_2\rangle,$$

find the constants A and B.

Answer: $\cos \pi/8,\ -\sin \pi/8$.

3.3 What is the probability that the left circular state $|L\rangle$ will be transmitted by a filter which always transmits state $|E_1\rangle$ of Problem 3.2 and never transmits state $|E_2\rangle$ of that problem.

Answer: $\cos^2 3\pi/8$.

3.4 If the states $|P_x\rangle$ and $|P_y\rangle$ are used as a basis, show that the transmission matrix for a P-filter whose line makes an angle θ with the x-axis is given by

$$\begin{pmatrix} \cos^2\theta & \cos\theta\sin\theta \\ \cos\theta\sin\theta & \sin^2\theta \end{pmatrix}.$$

3.5 A matrix which is often useful is the operator $|E\rangle\langle E|/\langle E|E\rangle$. Show that when it operates on the ket $|E\rangle$, the result is the identical ket $|E\rangle$. If the states $|E_1\rangle$ and $|E_2\rangle$ are an orthonormal pair, show that the matrix of the filter which produces $|E_1\rangle$ is given by $|E_1\rangle\langle E_1|$. If $|E\rangle$ is any polarization state incident on the filter which produces state $|E_1\rangle$ show that the probability that $|E\rangle$ will be transmitted by the filter is given by

$$|\langle E_1|E\rangle|^2 = \langle E|\mathsf{M}|E\rangle,$$

where $\mathsf{M} = |E_1\rangle\langle E_1|$.

3.6 If the matrix N corresponding to a polarizing device is unitary, $\mathsf{N}^\dagger\mathsf{N} = 1$, what does this imply about the transmission properties of N? Is the matrix for a polarizing filter unitary? Is the matrix of a retarder plate unitary?

3.7 The intensity of a state

$$|E\rangle = A_x e^{-i\phi/2}\,|P_x\rangle + A_y e^{\,i\phi/2}|P_y\rangle$$

is given by $I_0 = \langle E|E\rangle = A_x^2 + A_y^2$. What is the intensity of this state after transmission through a half-wave plate whose slow axis is along the x-axis. What is the intensity transmitted by a P-filter whose line is at angle θ with the x-axis? What is the intensity of the beam after it goes through the retarder and then through the P-filter. What is its intensity after going through the P-filter and then through the retarder?

Answer: I_0,

$$A_x^2 \cos^2 \theta + A_y^2 \sin^2 \theta + 2A_xA_y \cos \theta \sin \theta \cos \phi,$$
$$A_x^2 \cos^2 \theta + A_y^2 \sin^2 \theta - 2A_xA_y \cos \theta \sin \theta \cos \phi,$$
$$A_x^2 \cos^2 \theta + A_y^2 \sin^2 \theta + 2A_xA_y \cos \theta \sin \theta \cos \phi.$$

3.8 Light in an *R*-state strikes a 3/4-wave plate and the transmitted light encounters a *P*-filter followed by a half-wave plate. Determine the intensity and state of the final beam. Let both plates have their slow axes in the *x*-direction and consider the cases where the line of the *P*-filter is at $\pm 45°$ to the *x*-axis.

Answer: 0, I_0.

3.9 Determine the emergent state in each of the following cases.

a) A *P*-state incident on a quarter-wave plate with the line of the *P*-state midway between the fast and slow axes of the plate.

b) A *P*-state incident on a half-wave plate with the line of the *P*-state midway between the fast and slow axes of the plate.

c) *R*- and *L*-states incident on a quarter-wave plate.

d) *R*- and *L*-states incident on a half-wave plate.

Answer: *R* or *L*, the orthogonal *P* state, *P* states, *L* or *R*.

3.10 Show that the transmission matrix for the *P*-filter of Problem 3.4 expressed in the basis $|R\rangle$, $|L\rangle$ is as given in Appendix II.

3.11 Show that the matrix for an optically active circular retarder expressed in the basis $|P_x\rangle$, $|P_y\rangle$ is as given in Appendix II.

3.12 Show that the matrix operator for a half-wave plate whose fast axis makes an angle of 45° with the *x*-axis is given by

$$\begin{pmatrix} 0 & -i \\ -i & 0 \end{pmatrix}.$$

3.13 If a general polarization state is given by

$$|E\rangle = A_x e^{-i\phi/2} |P_x\rangle + A_y e^{+i\phi/2} |P_y\rangle,$$

show that the orthogonal state $|E'\rangle$, $\langle E'|E\rangle = 0$ is given by

$$|E'\rangle = A_y e^{-i\phi/2} |P_x\rangle - A_x e^{i\phi/2} |P_y\rangle.$$

In the case $|E\rangle = |R\rangle$, $|E'\rangle = |L\rangle$ show that the above forms are consistent with those given by Eq. (3.9).

3.14 Following Eq. (3.16), the transformation for a matrix from the basis $|P_x\rangle$, $|P_y\rangle$ to the basis $|P_{x'}\rangle$, $|P_{y'}\rangle$ was given as

$$\mathsf{M}' = \mathsf{R}^{-1}\mathsf{M}\mathsf{R} = \mathsf{R}^{\dagger}\mathsf{M}\mathsf{R}.$$

Why does this not have the form of Eq. (3.22),

$$\mathsf{M}' = \mathsf{U}\mathsf{M}\mathsf{U}^{\dagger}?$$

3.15 Prove that the matrix U whose elements are given by Eq. (3.30) is unitary.

CHAPTER 4

THE STOKES PARAMETERS

Development of another description of polarization states and optical systems might well raise the question, "Why bother?" One reason we proceed to do this is that it will lead to a different insight into polarization states. Furthermore, the formalism in the preceding chapter is incapable of handling an important class of problems dealing with the incoherent superposition of light beams. In other words, how can states of light be described which are not pure in relation to *any* E-filter? What perhaps may be surprising is that this new formalism—the Stokes vectors and Mueller matrices—which is the offspring of that in the preceding chapter—the Jones calculus—gives a description of these "nonpure states" which its parent is incapable of grasping. What is more important, however, is that the description of both pure and nonpure states will open a path which can lead to the formulation and methods of quantum mechanics.

4.1 THE PROJECTION OPERATOR

Consider the pure state $|E\rangle$, normalized such that $\langle E|E\rangle = 1$, and let us construct the 2×2-matrix $|E\rangle\langle E|$, known as the projection operator. Clearly, the matrix elements of the projection operator depend on the particular ket from which it is formed. To distinguish projection operators, we shall write the symbol $\mathsf{M}(E)$, where the letter in parentheses refers to the ket which determines the matrix elements of a particular projection operator. There are three observations about $\mathsf{M}(E)$ we wish to make immediately. First, it is a Hermitian operator; that is,

$$\mathsf{M}(E) = |E\rangle\langle E| = \{|E\rangle\langle E|\}^\dagger = \mathsf{M}^\dagger(E). \tag{4.1}$$

Second, if $|E'\rangle$ is any other pure state, then

$$\begin{aligned} \mathsf{M}(E)|E'\rangle &= |E\rangle\langle E|E'\rangle = \langle E|E'\rangle|E\rangle, \\ \mathsf{M}^2(E)|E'\rangle &= |E\rangle\langle E|E\rangle\langle E|E'\rangle = \langle E|E'\rangle|E\rangle. \end{aligned} \tag{4.2}$$

Consequently, operating with $\mathsf{M}(E)$ any number of times on a ket yields the same result. Any operator such that, as in this case, $\mathsf{M}^2(E) = \mathsf{M}(E)$ is said to be *idempotent*. Finally, each member of the set of all possible polarization states corresponds to a definite projection operator. If $|E\rangle$ and $|E'\rangle$ are different polarization states, then $\mathsf{M}(E) \neq \mathsf{M}(E')$, and there exists a one-to-one correspondence between pure polarization states and projection operators. This correspondence is truly one-to-one, however, only if arbitrary phase factors are ignored; for if $|E'\rangle = e^{i\phi}|E\rangle$, then

$$|E'\rangle\langle E'| = |E\rangle\langle E|.$$

We wish now to prove a relationship which will be of considerable use in the following sections. Assume a pure state $|E\rangle$ and a matrix operator A which represents an optical device. We want to prove that

$$\langle E|\mathsf{A}|E\rangle = \mathrm{Tr}\,[\mathsf{M}(E)\mathsf{A}.] \tag{4.3}$$

In terms of the familiar matrix operators of the last chapter, this means that if A is a matrix representing a retarder, linear polarizer, etc., then $\mathsf{A}|E\rangle$ generally results in a polarization state different from $|E\rangle$, say $|E'\rangle$. The amplitude that the state $|E'\rangle$ will be transmitted through a filter which completely transmits the state $|E\rangle$ is given by the trace (Tr) of the product of the projection operator and the matrix operator for the device. In order to prove Eq. (4.3), we will evaluate directly both members of the equation.

Let us first expand $|E\rangle$ in terms of a basic set of kets $|\phi_i\rangle$. Let

$$\begin{aligned} |E\rangle &= \sum_i \langle\phi_i|E\rangle|\phi_i\rangle, \\ \langle E| &= \sum_j \langle\phi_j|\langle E|\phi_j\rangle, \end{aligned} \tag{4.4}$$

yielding

$$\langle E|\mathsf{A}|E\rangle = \sum_{ij} \langle E|\phi_j\rangle\langle\phi_j|\mathsf{A}|\phi_i\rangle\langle\phi_i|E\rangle. \tag{4.5}$$

To evaluate the right-hand member of Eq. (4.3), we proceed as follows:

$$\mathsf{M}(E) = |E\rangle\langle E| = \sum_{ij} \langle\phi_i|E\rangle\langle E|\phi_j\rangle|\phi_i\rangle\langle\phi_j|.$$

Multiplying on the right by the identity $\sum_k |\phi_k\rangle\langle\phi_k|$ and A,

$$\begin{aligned} \mathsf{M}(E)\mathsf{A} &= \sum_{ijk} \langle\phi_i|E\rangle\langle E|\phi_j\rangle|\phi_i\rangle\langle\phi_j|\mathsf{A}|\phi_k\rangle\langle\phi_k|, \\ \mathsf{M}(E)\mathsf{A} &= \sum_{ijk} \langle\phi_i|E\rangle\langle E|\phi_j\rangle\langle\phi_j|\mathsf{A}|\phi_k\rangle|\phi_i\rangle\langle\phi_k|. \end{aligned} \tag{4.6}$$

All the bracket products in Eq. (4.6) are scalars, and the coefficient of the term $|\phi_i\rangle\langle\phi_k|$ is simply the element in the i-th row and k-th column in the matrix $\mathsf{M}(E)\mathsf{A}$. Since the $|\phi_i\rangle$ are orthonormal, taking the trace of Eq. (4.6) involves summing only the diagonal elements of $\mathsf{M}(E)\mathsf{A}$ for which $i = k$; therefore

$$\mathrm{Tr}\,[\mathsf{M}(E)\mathsf{A}] = \sum_{ij} \langle\phi_i|E\rangle\langle E|\phi_j\rangle\langle\phi_j|\mathsf{A}|\phi_i\rangle, \tag{4.7}$$

and comparison with Eq. (4.5) gives Eq. (4.3).

4.2 THE STOKES PARAMETERS

The ket vectors and 2 × 2-matrix operators which we have been manipulating were chosen such that they correspond to experimental data. Evidently, if we are to develop a different mathematical representation, it likewise must correspond to such data. How can we ensure this? In principle, we simply need to find a set of quantities which are in one-to-one correspondence with the ket vectors. The projection operators fulfil this requirement and the implications of this correspondence will now be examined.

The projection operator is a 2 × 2-matrix, and can be expressed as a linear combination of other 2 × 2-matrices as

$$\mathsf{M}(E) = \tfrac{1}{2}[P_0\boldsymbol{\sigma}_0 + P_1\boldsymbol{\sigma}_1 + P_2\boldsymbol{\sigma}_2 + P_3\boldsymbol{\sigma}_3], \tag{4.8}$$

where P_0 and P_i ($i = 1, 2, 3$) are scalars, and the $\boldsymbol{\sigma}_i$ are linearly independent, Hermitian, and unitary matrices given by

$$\boldsymbol{\sigma}_0 = \begin{pmatrix} 1 & 0 \\ 0 & 1 \end{pmatrix} \quad \boldsymbol{\sigma}_1 = \begin{pmatrix} 1 & 0 \\ 0 & -1 \end{pmatrix} \quad \boldsymbol{\sigma}_2 = \begin{pmatrix} 0 & 1 \\ 1 & 0 \end{pmatrix} \quad \boldsymbol{\sigma}_3 = \begin{pmatrix} 0 & -i \\ i & 0 \end{pmatrix}. \tag{4.9}$$

The factor of 1/2 in Eq. (4.8) is included so that later results will conform to standard notation. Except for a permutation of indices, these matrices are most often referred to as the Pauli spin matrices, which are frequently used in atomic and nuclear physics. For the subscript order as given above, we shall simply refer to them as *sigma matrices*. The sigma matrices have the following properties, which are easily verified:

$$\boldsymbol{\sigma}_i^\dagger = \boldsymbol{\sigma}_i, \qquad \boldsymbol{\sigma}_i^2 = \boldsymbol{\sigma}_0, \qquad \boldsymbol{\sigma}_i\boldsymbol{\sigma}_j = i\boldsymbol{\sigma}_k,$$

$$\mathrm{Tr}\,(\boldsymbol{\sigma}_i\boldsymbol{\sigma}_j) = 2\delta_{ij}, \qquad \boldsymbol{\sigma}_i\boldsymbol{\sigma}_j = -\boldsymbol{\sigma}_j\boldsymbol{\sigma}_i.$$

The subscripts i, j, k take on the values 1, 2, 3 in cyclic order and the Kronecker delta symbol is defined by

$$\delta_{ij} = \begin{cases} 1, & i = j \\ 0, & i \neq j. \end{cases}$$

Since $\mathsf{M}(E)$ and the sigma matrices are Hermitian, the scalar coefficients of the sigma matrices must be real for

$$\mathsf{M}(E) = \tfrac{1}{2}\sum_i P_i\boldsymbol{\sigma}_i, \qquad \mathsf{M}^\dagger(E) = \tfrac{1}{2}\sum_i P^\star{}_i\boldsymbol{\sigma}_i \quad (i = 0, 1, 2, 3),$$

so we must have, since $\mathsf{M}(E)$ is Hermitian,

$$\sum_i P_i \boldsymbol{\sigma}_i = \sum_i P_i^{\star} \boldsymbol{\sigma}_i.$$

Multiplying both members of the preceding equation by $\boldsymbol{\sigma}_j$ and taking the trace of both members yields $P_j = P_j^{\star}$. Consequently, when $\mathsf{M}(E)$ is expressed as a linear combination of the sigma matrices, the projection operator is uniquely determined by the four real numbers (P_0, P_1, P_2, P_3) which determine the polarization state. When the projection operator $\mathsf{M}(E)$ is expressed in terms of $|P_x\rangle$ and $|P_y\rangle$ as a basis, these real parameters are called the *Stokes parameters* and they allow a description of light beams in terms of 4×1-column vectors.

At this point it is easy to show how the Stokes parameters can be calculated for any pure polarization state. Assume the pure state

$$|E\rangle = A|P_x\rangle + B|P_y\rangle; \qquad A = ae^{-i\phi/2},\ B = be^{i\phi/2}, \tag{4.10}$$

which has the projection operator

$$\mathsf{M}(E) = \begin{pmatrix} AA^{\star} & AB^{\star} \\ BA^{\star} & BB^{\star} \end{pmatrix} = \tfrac{1}{2}\sum_i \boldsymbol{\sigma}_i P_i.$$

By using Eq. (4.3) and the relation $\mathrm{Tr}\,(\boldsymbol{\sigma}_i\boldsymbol{\sigma}_j) = 2\delta_{ij}$, we obtain the results

$$\begin{aligned} \mathrm{Tr}\,[\mathsf{M}(E)\boldsymbol{\sigma}_j] &= \tfrac{1}{2}\,\mathrm{Tr} \sum_i \boldsymbol{\sigma}_i\boldsymbol{\sigma}_j P_i = P_j, \\ \langle E|\boldsymbol{\sigma}_j|E\rangle &= P_j. \end{aligned} \tag{4.11}$$

This not only makes the calculation of the Stokes parameters rather easy, but, used in conjunction with Eq. (4.3), enables us to obtain a very useful relation between the Stokes parameters. By Eqs. (4.3) and (4.8) we can write

$$\begin{aligned} |E\rangle\langle E| &= \tfrac{1}{2}\sum_i P_i\boldsymbol{\sigma}_i, \\ \langle E|E\rangle\langle E|E\rangle &= \tfrac{1}{2}\sum_i P_i\langle E|\boldsymbol{\sigma}_i|E\rangle, \\ \langle E|\boldsymbol{\sigma}_0|E\rangle\langle E|\boldsymbol{\sigma}_0|E\rangle &= \tfrac{1}{2}\sum_i P_i^2, \\ P_0^2 &= P_1^2 + P_2^2 + P_3^2, \end{aligned} \tag{4.12}$$

which is a result which will form the basis of our later discussion of the Poincaré sphere and is satisfied only for *pure* states.

The operational significance of these parameters becomes apparent when they are calculated directly from the 2×2-matrix $M(E)$. By use of Eq. (4.10) and Eq. (4.11) the Stokes parameters can be shown to be

$$\begin{aligned} P_0 &= a^2 + b^2, \qquad P_1 = a^2 - b^2, \\ P_2 &= 2ab\cos\phi, \qquad P_3 = 2ab\sin\phi. \end{aligned} \tag{4.13}$$

The intensity of the beam in the state $|E\rangle$ is simply P_0, and by adding the first two relations above we find that $P_1 = 2a^2 - P_0$, where a^2 is the intensity of the light transmitted by a P-filter whose line is along the x-axis. Thus, the Stokes parameter, P_1, of a beam is twice the intensity transmitted by a P_x-filter minus the intensity of the beam. If the beam is analyzed by a P-filter whose line is oriented at 45° relative to the x-axis, then we find that

$$|\langle P|E\rangle|^2 = \tfrac{1}{2}(P_0 + P_2),$$

and P_2 is twice the intensity of the transmitted beam minus the total intensity. Similarly, it can be shown that P_3 is twice the intensity transmitted by an R-filter minus the total intensity.

Since the Stokes parameters characterize polarization states uniquely, it is convenient to write the Stokes parameters as the elements of a column matrix (4 × 1) or a row matrix (1 × 4) where the order in which the parameters appear in the matrix will always be maintained as follows:

$$\begin{pmatrix} P_0 \\ P_1 \\ P_2 \\ P_3 \end{pmatrix}, \quad (P_0, P_1, P_2, P_3).$$

These row and column vectors will hereafter be called *Stokes vectors*. Let us determine the Stokes vectors for some of the familiar polarization states. We assume here that $\langle E|E\rangle = P_0 = 1$. Since R- and L-states are of the form

$$|E\rangle = \frac{1}{\sqrt{2}}(e^{\mp i\pi/4}|P_x\rangle + e^{\pm i\pi/4}|P_y\rangle),$$

we find

$$|R\rangle \Rightarrow \begin{pmatrix} 1 \\ 0 \\ 0 \\ +1 \end{pmatrix} \qquad |L\rangle \Rightarrow \begin{pmatrix} 1 \\ 0 \\ 0 \\ -1 \end{pmatrix}. \tag{4.14}$$

A P-state whose line is at an angle θ with the x-axis has a ket

$$|P\rangle = \cos\theta\,|P_x\rangle + \sin\theta\,|P_y\rangle,$$

and its Stokes vector is

$$|P\rangle \Rightarrow \begin{pmatrix} 1 \\ \cos^2\theta - \sin^2\theta \\ 2\cos\theta\sin\theta \\ 0 \end{pmatrix}. \tag{4.15}$$

In particular,

$$|P_x\rangle \Rightarrow \begin{pmatrix} 1 \\ 1 \\ 0 \\ 0 \end{pmatrix} \qquad |P_y\rangle \Rightarrow \begin{pmatrix} 1 \\ -1 \\ 0 \\ 0 \end{pmatrix}, \tag{4.16}$$

and, for $\theta = \pm 45°$,

$$|P\rangle_\pm \Rightarrow \begin{pmatrix} 1 \\ 0 \\ \pm 1 \\ 0 \end{pmatrix}.$$

4.3 THE MUELLER MATRICES

It has been observed in the preceding section that with each pure state $|E\rangle$, there is associated a four-component vector whose components are the four real Stokes parameters of the state. We now face the following problem. Let $|E\rangle$ be a polarization state of light before it passes through an optical device represented by a 2×2-matrix M. The transmitted state will be $|E'\rangle = \mathsf{M}|E\rangle$. If the Stokes vector for $|E\rangle$ has components P_j ($j = 0, 1, 2, 3$), then the Stokes vector for $|E'\rangle$ will have components P_j', and there must be a 4×4-matrix such that

$$\begin{pmatrix} P_0' \\ P_1' \\ P_2' \\ P_3' \end{pmatrix} = \mathscr{M} \begin{pmatrix} P_0 \\ P_1 \\ P_2 \\ P_3 \end{pmatrix}.$$

How can a 4×4-matrix representation of M be found?

We proceed by using the properties of the trace of a matrix and the fact that if $|E'\rangle = \mathsf{M}|E\rangle$, then $|E'\rangle\langle E'| = \mathsf{M}|E\rangle\langle E|\mathsf{M}^\dagger$. From Eq. (4.11),

$$\begin{aligned} P_i' &= \operatorname{Tr}\,[\sigma_i \mathsf{M}|E\rangle\langle E|\mathsf{M}^\dagger] \\ &= \tfrac{1}{2}\operatorname{Tr}\,[\sigma_i \mathsf{M} \sum_j \sigma_j P_j \mathsf{M}^\dagger] \\ &= \tfrac{1}{2}\operatorname{Tr} \sum_j (\sigma_i \mathsf{M} \sigma_j \mathsf{M}^\dagger)\, P_j \\ &= \sum_j \mathscr{M}_{ij} P_j, \end{aligned}$$

where

$$\mathscr{M}_{ij} = \tfrac{1}{2}\operatorname{Tr}\,(\boldsymbol{\sigma}_i \mathsf{M} \boldsymbol{\sigma}_j \mathsf{M}^\dagger). \tag{4.17}$$

In computing the $\mathscr{M}_{ij}$ a minor but important point must be made. Because the subscripts on the sigma matrices run 0, 1, 2, 3, rather than from 1 through

4, the matrix elements of $\mathscr{M}_{ij}$ are not numbered according to the usual convention. The element in the first row and first column, for example, is $\mathscr{M}_{00}$ not $\mathscr{M}_{11}$.

As a simple illustration of the application of the Mueller matrices and the Stokes vectors, let us consider the case where right circularly polarized light strikes a quarter-wave plate whose fast axis is along the x-axis. In the Jones calculus we have

$$\begin{aligned}|E'\rangle &= \mathsf{M}|R\rangle \\ &= \frac{1}{\sqrt{2}}\begin{pmatrix} e^{-i\pi/4} & 0 \\ 0 & e^{i\pi/4}\end{pmatrix}\begin{pmatrix} e^{-i\pi/4} \\ e^{+i\pi/4}\end{pmatrix} \\ &= \frac{i}{\sqrt{2}}[-|P_x\rangle + |P_y\rangle],\end{aligned}$$

which is a P-state which makes an angle of $-45°$ relative to the fast axis of the retarder. From the preceding section and Appendix II, the corresponding expression in the Mueller calculus is

$$\begin{pmatrix} 1 \\ 0 \\ -1 \\ 0 \end{pmatrix} = \begin{pmatrix} 1 & 0 & 0 & 0 \\ 0 & 1 & 0 & 0 \\ 0 & 0 & 0 & -1 \\ 0 & 0 & 1 & 0 \end{pmatrix}\begin{pmatrix} 1 \\ 0 \\ 0 \\ +1 \end{pmatrix},$$

and the components of the Stokes vector for the emergent light are also those for a P-state oriented at $-45°$ relative to the fast axis.

The construction of the Mueller matrix for a sequence of retarders and polarizers is the same as that in the Jones calculus. The matrices for the individual elements are multiplied from right to left in the order in which the beam encounters the elements. A tabulation of the Mueller matrices for a number of different elements is given in Appendix II.

4.4 PROBABILITY AMPLITUDES IN THE STOKES FORMALISM

In Section 3.3 it was shown that the calculation of probabilities involved the bracket product. In view of the correspondence between kets and the Stokes parameters, we expect that this will allow us to calculate probabilities using the Stokes vector rather than kets.

Suppose the $|E\rangle$ and $|E'\rangle$ are normalized so that $\langle E|E\rangle = \boldsymbol{P}_0$ and $\langle E'|E'\rangle = 1$. At the beginning of the chapter we proved that if $\mathbf{A}$ is a matrix operator and $|E\rangle$ is a pure state, then, by Eq. 4.3,

$$\langle E|\mathsf{A}|E\rangle = \operatorname{Tr}[\mathsf{M}(E)\mathsf{A}].$$

Now, let $\mathsf{A} = \mathsf{M}(E') = |E'\rangle\langle E'|$. Eq. (4.3) becomes

$$\langle E|E'\rangle\langle E'|E\rangle = \mathrm{Tr}\,[\mathsf{M}(E)\mathsf{M}(E')],$$
$$|\langle E'|E\rangle|^2 = \tfrac{1}{4}\,\mathrm{Tr}\,\{\sum_i P_i\boldsymbol{\sigma}_i \sum_j P'_j\boldsymbol{\sigma}_j\}.$$

The product of the two sums results in 16 terms, 12 of which involve products of the form $\boldsymbol{\sigma}_i\boldsymbol{\sigma}_j$ $(i \neq j)$, and since the trace of each of these terms vanishes, we have the result

$$|\langle E'|E\rangle|^2 = \tfrac{1}{2}\sum_i P_i P'_i$$
$$= \tfrac{1}{2}(1, P'_1, P'_2, P'_3)\begin{pmatrix} P_0 \\ P_1 \\ P_2 \\ P_3 \end{pmatrix}. \tag{4.18}$$

Very often the above equation is written more succinctly as

$$|\langle E'|E\rangle|^2 = \tfrac{1}{2}(1,\boldsymbol{P}')\begin{pmatrix} P_0 \\ \boldsymbol{P} \end{pmatrix}.$$

From the normalization of $|E\rangle$ and $|E'\rangle$ the left-hand side of Eq. (4.18) means that if a light beam with intensity P_0 is in a pure state $|E\rangle$, then the intensity of the light transmitted by an E'-filter is $|\langle E'|E\rangle|^2$. If the normalization were such that $\langle E|E\rangle = \langle E'|E'\rangle = 1$, then $|\langle E'|E\rangle|^2$ would be the probability that a photon in an E-state would traverse an E'-filter. The right-hand side of Eq. (4.18) shows how the transmitted intensity or transmission probability may be expressed in terms of the scalar product of Stokes vectors.

Instead of analyzing a given state $|E\rangle$ by an E'-filter, let us allow the state $|E\rangle$ to first pass through an optical device represented by the 2×2-matrix T and then analyze the emergent light by an E'-filter. The probability in this case is

$$|\langle E'|\mathsf{T}|E\rangle|^2,$$

and an explicit derivation of what this corresponds to in Stokes vector formalism is given below:

$$|\langle E'|\mathsf{T}|E\rangle|^2 = \langle E'|\mathsf{T}|E\rangle\langle E|\mathsf{T}^\dagger|E'\rangle$$
$$= \mathrm{Tr}\,[|E'\rangle\langle E'|\mathsf{T}|E\rangle\langle E|\mathsf{T}^\dagger].$$

We may expand the projection operators in terms of Stokes parameters

$$|\langle E'|\mathsf{T}|E\rangle|^2 = \mathrm{Tr}\,[\tfrac{1}{2}\sum_j P'_j\boldsymbol{\sigma}_j\mathsf{T}\tfrac{1}{2}\sum_i P_i\boldsymbol{\sigma}_i\mathsf{T}^\dagger],$$

and, since the trace of a sum is the sum of the traces,

$$|\langle E'|\mathsf{T}|E\rangle|^2 = \tfrac{1}{2}\sum_{ij} P_j'[\tfrac{1}{2}\,\mathrm{Tr}\,\boldsymbol{\sigma}_j\mathsf{T}\boldsymbol{\sigma}_i\mathsf{T}^\dagger]P_i$$

$$= \tfrac{1}{2}(P_0', P_1', P_2', P_3')\mathscr{M}\begin{pmatrix}P_0\\P_1\\P_2\\P_3\end{pmatrix}, \qquad (4.19)$$

where, by Eq. (4.17), $\mathscr{M}$ is the Mueller matrix corresponding to the Jones matrix T.

4.5 THE POINCARÉ SPHERE

In Section 4.2 it was indicated how the Stokes parameters can be obtained for pure states. However, there is one relation between the Stokes parameters characterizing a pure state which has not yet been exploited. If we take the entire ensemble of pure states, each of which satisfies the normalization condition $\langle E|E\rangle = 1$, then, according to Eq. (4.12), every such state satisfies the relation

$$1 = P_1^2 + P_2^2 + P_3^2,$$

which should remind us of the equation for a sphere of unit radius. This sphere, called the Poincaré sphere, is defined by a set of coordinate axes P_1, P_2, and P_3. Since each pure polarization state has a unique set of Stokes parameters, this implies that each point on the surface of the sphere corresponds to a definite pure polarization state. We shall call the vector from the origin to the point P_1, P_2, P_3 the Poincaré vector of the polarization state.

If a pure state $|E\rangle$ can be represented by a point on the Poincaré sphere, what point on the sphere represents the pure state orthogonal to $|E\rangle$? We will now show that mutually orthogonal pure states correspond to antipodal points on the sphere (i.e., opposite ends of a diameter). For any state

$$|E\rangle = ae^{-i\phi/2}|P_x\rangle + be^{i\phi/2}|P_y\rangle, \qquad (4.20)$$

where $\langle E|E\rangle = 1$, the corresponding orthogonal state is

$$|E'\rangle = be^{-i\phi/2}|P_x\rangle - ae^{i\phi/2}|P_y\rangle,$$

and, using Eq. (4.13), we see that

$$P_i' = -P_i \qquad (i = 1, 2, 3),$$

and that mutually orthogonal states are represented by antipodal points on the sphere. As particular examples we cite the special cases whose Stokes vectors have been evaluated: $|R\rangle$ and $|L\rangle$, $|P_x\rangle$ and $|P_y\rangle$, and the P-states

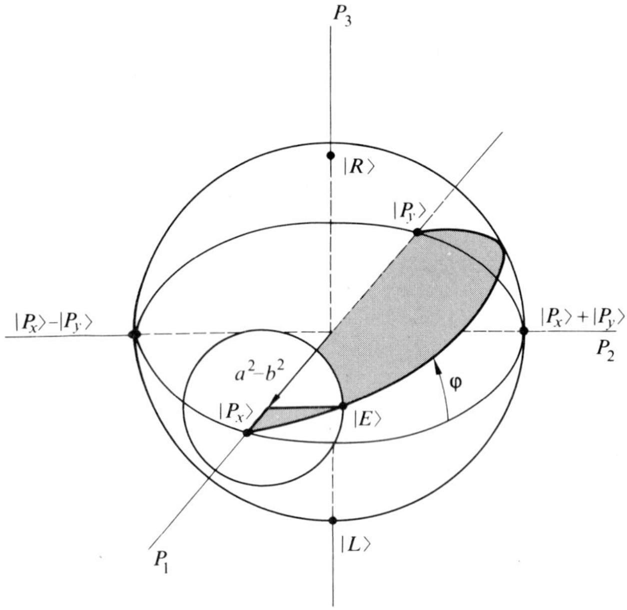

Fig. 4.1. The Poincaré sphere.

whose lines are at $\pm 45°$ relative to the x-axis. These states lie at opposite ends of the coordinate axis of the sphere, as shown in Fig. 4.1. It now remains to locate a general elliptical state on the sphere in terms of a, b, and ϕ in the expansion given by Eq. (4.20).

From Eq. (4.13) we have

$$a^2 - b^2 = P_1, \qquad P_3/P_2 = \tan \phi,$$

and we see that the coefficients a and b determine a plane in the Poincaré sphere which is perpendicular to the P_1-axis and intersects the sphere in a circle. The point on the sphere which represents $|E\rangle$ must lie somewhere on this circle. Likewise, the phase angle ϕ must determine a plane which contains the P_1-axis. This plane makes an angle ϕ with the P_1P_2-plane and intersects the sphere in a great circle. The point on the sphere which represents the state $|E\rangle$ is thus the intersection of these two circles.

The Poincaré sphere has utility in visualizing the outcome of sending a polarized beam through an optical device. For example, if a beam in a state given by Eq. (4.20) traverses a retarder plate whose fast and slow axes are along the x- and y-axes, respectively, the effect of the plate is to produce the emergent state

$$|E\rangle = ae^{-i(\phi+\delta)/2}|P_x\rangle + be^{i(\phi+\delta)/2}|P_y\rangle,$$

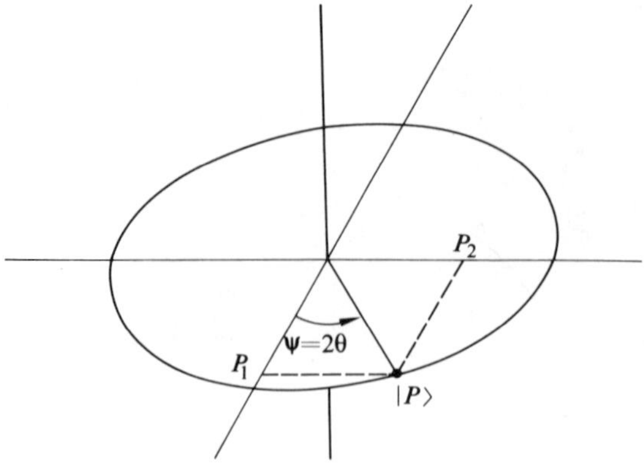

Fig. 4.2. Location of a general P-state on the equator of the Poincaré sphere.

where $\delta = kt(n_s - n_f)$. The effect of the retarder is simply to increase the phase angle. In terms of the Poincaré sphere, this is seen to be a rotation about the P_1-axis through an angle δ. If the retarder is a quarter-wave plate, then light in the state $|P_x\rangle + |P_y\rangle$ traversing the retarder will emerge as an R-state, which is obtained from the Poincaré sphere by rotation about the P_1-axis through the angle $\delta = \pi/2$. Note that in our convention all points on the upper hemisphere correspond to right elliptical states; those on the bottom hemisphere, to left elliptical states.

We have obtained the Poincaré sphere description of polarization states by expressing all such states as the superposition of the orthogonal states $|P_x\rangle, |P_y\rangle$. However, nowhere did we say exactly which axes were being used for x and y. We did not give an absolute orientation for the x- and y-axes in space. Thus, the Poincaré sphere description is valid so long as the P_1-axis passes through those points on the sphere which we have chosen to be $|P_x\rangle$ and $|P_y\rangle$. Since ϕ is zero for all P-states, such states must lie on the great circle in the P_1P_2-plane. If we wish to express our states in terms of axes x', y' rotated through angle θ from the original axes x, y, we need only rotate the P_1P_2-axis about the P_3-direction through an angle 2θ to find the appropriate new $P_1'P_2'$-axis on the Poincaré sphere. Let us prove this.

In Fig. 4.2 we show the equator of the Poincaré sphere with a general P-state indicated on the equator. In real space assume the line of this P-state is at angle θ with the x-axis. Thus,

$$|P\rangle = \cos\theta|P_x\rangle + \sin\theta|P_y\rangle.$$

The Stokes parameters for the state are thus

$$P_0 = 1, \qquad P_1 = \cos^2\theta - \sin^2\theta,$$
$$P_2 = 2\cos\theta\sin\theta = \sin 2\theta, \qquad P_3 = 0.$$

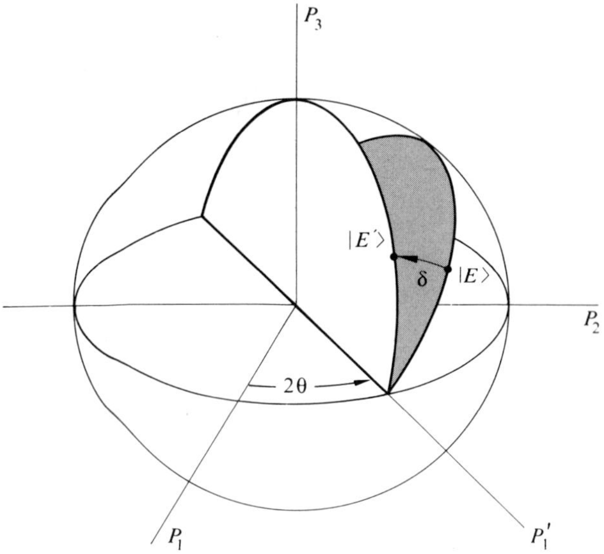

Fig. 4.3. Representation on the Poincaré sphere of the transmission of the state $|E\rangle$ through an anisotropic retarder whose fast axis makes an angle θ with the x-axis, and which introduces a retardation δ.

The tangent of ψ is given by

$$\tan \psi = P_2/P_1 = (\sin 2\theta)/(\cos 2\theta) = \tan 2\theta.$$

Thus,

$$\psi = 2\theta.$$

We can now easily see how a retarder plate whose axes are not along the original x- and y-axes affect the location of states on the Poincaré sphere. Let the retarder impose a phase difference δ and have its fast axis along x' making an angle θ with x. We find the axis in the P_1P_2-plane which makes an angle 2θ with P_1. Wherever the point for the incident state happens to be on the sphere, the new state is found by rotating the old point through an angle δ about the line P_1' in the plane. This is illustrated in Fig. 4.3. The state $|E\rangle$ is transformed into the state $|E'\rangle$. For example, if an R-state passes through a quarter-wave plate whose fast axis makes an angle of 67.5° with the x-axis, the emergent state is seen to be a P-state whose line is at 22.5° relative to the x-axis.

As another example of visualizing the effect of an optical device on polarization states, consider optically active materials. Placing such a device in a beam rotates the plane of P-states and in particular the states $|P_x\rangle$ and

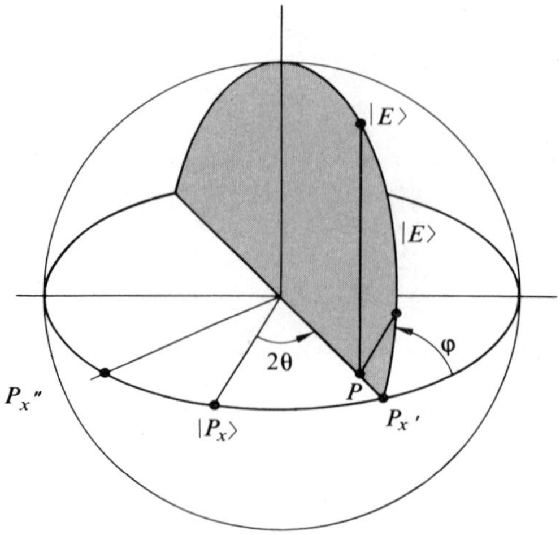

Fig. 4.4. Location on the Poincaré sphere of the state $|E\rangle$ when expressed in the basis $|P_{x'}\rangle$, $|P_{y'}\rangle$.

$|P_y\rangle$. It is readily seen that on the Poincaré sphere this amounts to a rotation about the P_3-axis. Thus, if a given slab of optically active material rotates $|P_x\rangle$ through an angle θ with the x-axis, then when an arbitrary state $|E\rangle$ passes through the slab, the new state can be found on the Poincaré sphere by rotating the old state through an angle 2θ about P_3.

Consider now a series of retarders and slabs of optically active materials. Each of these in order produces a rotation on the Poincaré sphere. Although it is not easy analytically to find the total rotation, any sequence of rotations about various axes is equivalent to a single rotation about some single axis. In Poincaré space, then, each pure polarization state corresponds to a point on the sphere and every nonabsorbing device corresponds to a rotation about some axis.

A simple proof is now easily given which was promised in Chapter 3. Let the P_1-axis pass through the Poincaré sphere at the states $|P_x\rangle$ and $|P_y\rangle$. Let $|P_{x'}\rangle$ and $|P_{y'}\rangle$ be orthonormal P-states such that $|P_{x'}\rangle$ makes an angle θ with $|P_x\rangle$, as shown in Fig. 4.4. Let the arbitrary state $|E\rangle$ be represented as

$$|E\rangle = ae^{-i\phi/2}|P_{x'}\rangle + be^{i\phi/2}|P_{y'}\rangle.$$

Then, ϕ is obtained from the Poincaré sphere as shown in Fig. 4.4. It is obvious that for every $|E\rangle$-state, axes x' and y' exist such that $\phi = \pi/2$. Since the distance from the center of the sphere to point P is $a^2 - b^2$, it is also clear that axes x'' and y'' exist such that $a = b$. Finally, the angle in real space between the axes x' and x'' must be 45°.

4.6 THE SIGMA MATRICES AS INSTRUMENT OPERATORS

There are a number of ways of grasping the physical significance of the sigma matrices which are intimately related. For the present we will concern ourselves with the interpretation of the sigma matrices in terms of actual physical devices. Since all of our devices are represented by 2×2-matrix operators which can be expressed as a linear combination of the sigma matrices, as in Eq. (4.8), the sigma matrices themselves should represent optical devices. The basic devices at our disposal are the linear and circular retarders, P-filters, and optically active materials. The linear and circular retarder matrices will give us access to the physical interpretation of the sigma matrices. For any pure state which is expressed as a coherent superposition of the states $|P_x\rangle$ and $|P_y\rangle$, the matrix representation for a circular retarder, according to Appendix II, is

$$\mathsf{R} = \begin{pmatrix} \cos \beta/2 & -\sin \beta/2 \\ +\sin \beta/2 & \cos \beta/2 \end{pmatrix},$$

where $\beta/2$ is the angle through which a P-state is rotated by an optically active material. When the angle of rotation is such that $\beta = \pi$, then

$$\mathsf{R} = \begin{pmatrix} 0 & -1 \\ +1 & 0 \end{pmatrix} = +i\boldsymbol{\sigma}_3.$$

The only difference between $\mathsf{R}|E\rangle$ and $\boldsymbol{\sigma}_3|E\rangle$ is the phase factor i; consequently, $\boldsymbol{\sigma}_3$ is an operator which represents a circular retarder that rotates a P-state through 90°. In terms of the Poincaré sphere, $\boldsymbol{\sigma}_3$ represents a rotation about the P_3-axis through an angle $\beta = \pi$, which implies that the only states unchanged by the instrument operator $\boldsymbol{\sigma}_3$ are the states $|R\rangle$ and $|L\rangle$.

In the case of the linear retarder whose fast axis is along the x-axis, we have, from Eq. (3.14),

$$\mathsf{M} = \begin{pmatrix} e^{-i\delta/2} & 0 \\ 0 & e^{i\delta/2} \end{pmatrix},$$

where, for a half-wave plate, $\delta = \pi$. Assume that a half-wave plate is rotated so that its fast axis is oriented at an angle of +45° relative to the x-axis. The matrix representation of the rotated half-wave plate for states expressed as a coherent superposition of the states $|P_x\rangle$ and $|P_y\rangle$ is obtained from Eq. (3.17):

$$\mathsf{M}' = -i\begin{pmatrix} 0 & 1 \\ 1 & 0 \end{pmatrix} = -i\boldsymbol{\sigma}_2,$$

and $-i\boldsymbol{\sigma}_2$ is seen to represent a half-wave plate whose fast axis makes an angle of +45° with the x-axis and is pure in relation to the state $|P_x\rangle + |P_y\rangle$. It is easy to verify that $(-i\boldsymbol{\sigma}_2)^\star = i\boldsymbol{\sigma}_2$ represents a half-wave plate whose *slow* axis makes an angle of $+45°$ with the x-axis; the operation of complex conjugation simply interchanges the fast and slow axes.

The matrix M' representing a half-wave plate whose slow axis is along the x-axis is

$$\mathsf{M}' = i\begin{pmatrix} 1 & 0 \\ 0 & -1 \end{pmatrix} = i\sigma_1,$$

and $i\sigma_1$ is seen to represent such a retarder which leaves unchanged the states $|P_x\rangle$ and $|P_y\rangle$.

The multiplication relation for the sigma matrices

$$\sigma_i\sigma_j = i\sigma_k \qquad (i, j, k = 1, 2, 3)$$

in cyclic order now has an operational significance. For example,

$$(i\sigma_1)(-i\sigma_2) = i\sigma_3$$

means that the effect on light in a pure state intercepted by the retarders $-i\sigma_2$ and $i\sigma_1$ in succession is equivalent to the effect when the light passes through an optically active retarder, $i\sigma_3$.

4.7 PARTIAL POLARIZATION

In our discussion so far, emphasis has been placed on the concept of pure states, and this provided the basis for the creation of the Stokes vector space. Since the projection operators were constructed from the kets which specify pure polarization states, it was assumed that each such state corresponded to a definite projection operator and to a definite Stokes vector. In Section 4.1 the statement was made that there exists a one-to-one correspondence between pure polarization states and projection operators. However, this is true only if we are talking about *pure* states. For every pure state there exists a corresponding Stokes vector, but the converse proposition, that for every Stokes vector there exists a pure state, is not true. In other words, in busily constructing the Stokes vector formalism, we have added something which is not a member of the set of pure polarization states!

An example of a Stokes vector which does not correspond to any pure polarization state is one in which all the Stokes parameters except P_0 are zero. It is important to bear in mind that the Stokes parameters can be obtained by measurement, as outlined in Section 4.2. Furthermore, we know that for *pure* states, by Eq. (4.12),

$$1 = P_1^2 + P_2^2 + P_3^2.$$

Any beam of light which does not satisfy Eq. (4.12) is said to be in a *nonpure*, or *mixed*, state. Specifically, a beam of light which has the measured Stokes parameters $(I, 0, 0, 0)$, where I is the intensity of the beam, is said to be

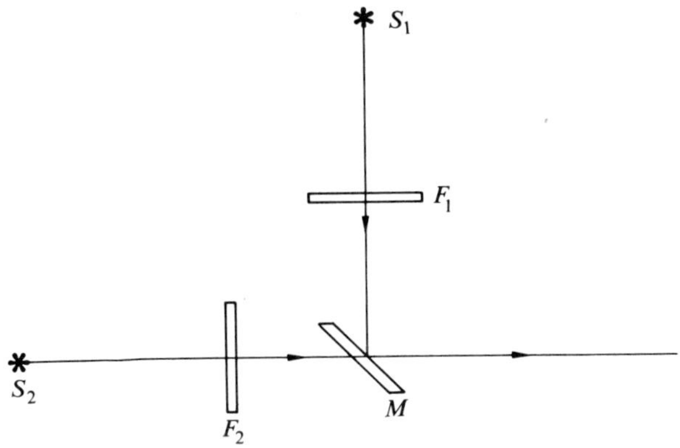

Fig. 4.5. The incoherent superposition of two light beams.

unpolarized. Since the Stokes parameters necessarily involve measurement, it is reasonable to describe the condition of any light beam in terms of a degree of polarization, $\mathscr{P}$, as

$$\mathscr{P} = \left(\frac{P_1 + P_2^2 + P_3^2}{P_0^2}\right)^{1/2}. \tag{4.21}$$

Any beam whose degree of polarization is less than 1 but greater than 0 is said to be *partially* polarized. Note that the degree of polarization does not depend on the particular orientation of the x- and y-axes. Rotating these axes will alter the direction of the P_1- and P_2-axes on the Poincaré sphere, but not change the intensity of the beam, P_0, or the length of the Poincaré vector, $\sqrt{P_1^2 + P_2^2 + P_3^2}$.

The problem remains as to how these partially polarized states may be encompassed in terms of Stokes vectors. This problem will be examined more thoroughly in Chapter 7, but some aspects of it are pertinent here.

Consider the situation shown in Fig. 4.5. The sources S_1 and S_2 are two ordinary independent sources; for example, light bulbs. If one wished, they could be made nearly monochromatic by the addition of filters, although this would not affect what is to follow. F_1 and F_2 are any two polarizing filters. M is a half-silvered mirror. At the mirror we are thus combining two beams each of which will be partially polarized. We may now block each beam in turn and measure the Stokes parameters of the emergent beams in each case, which we call P_i^1 and P_i^2. Now, with neither beam blocked, we may measure the Stokes parameters of the combined beams, which we will call P^{1+2}. If this experiment is performed, one finds the experimental result to be

$$P_i^{1+2} = P_i^1 + P_i^2. \tag{4.22}$$

This is not always so. If the two beams which are combined had come from the two arms of an interferometer, Eq. (4.22) would not be obeyed. For the moment let us assume that the two combining beams can be written as $e^{i\phi}|E_1\rangle$ and $e^{i\psi}|E_2\rangle$. We have explicitly written phase factors before these states, since in a superposition phase differences are of great importance. Now write the combined beam as

$$|E\rangle = e^{i\phi}|E_1\rangle + e^{i\psi}|E_2\rangle. \tag{4.23}$$

The matrix $\mathsf{M}(E)$ is then given by

$$\mathsf{M}(E) = \mathsf{M}(E_1) + \mathsf{M}(E_2) + e^{i(\phi-\psi)}|E_1\rangle\langle E_2| + e^{-i(\phi-\psi)}|E_2\rangle\langle E_1|.$$

Using Eq. (4.11) and the relation

$$\mathrm{Tr}\,(\mathsf{A} + \mathsf{B}) = \mathrm{Tr}\,(\mathsf{A}) + \mathrm{Tr}\,(\mathsf{B}),$$

we see that the Stokes parameters of the superposition would be the sum of the Stokes parameters of the individual states if the terms $|E_1\rangle\langle E_2|$ and $|E_2\rangle\langle E_1|$ could be ignored.

In Section 4.2 the interpretation of the Stokes parameters in terms of intensity measurements was discussed. Since an intensity measurement must be carried out over a finite length of time, in practice we are interested in the time average of $\mathsf{M}(E)$, which we indicate by $\overline{\mathsf{M}(E)}$. The experimental fact that the Stokes parameters of beams from independent sources add implies therefore that

$$\overline{e^{i(\phi-\psi)}} = \overline{e^{-i(\phi-\psi)}} = 0.$$

Thus, between independent sources there is a high instability in relative phase. For independent sources, a superposition in which the relative phase averages to zero and for which Eq. (4.22) is valid is called a completely incoherent superposition.

Since the Stokes parameters add for incoherent superposition, it is easy to see that a partially polarized beam may be represented as the incoherent sum of a pure polarization state and a state of zero polarization. Let the given state have parameters I, P_1, P_2, P_3. This vector may be written

$$\begin{pmatrix} I \\ P_1 \\ P_2 \\ P_3 \end{pmatrix} = \begin{pmatrix} I_p \\ P_1 \\ P_2 \\ P_3 \end{pmatrix} + \begin{pmatrix} I_0 \\ 0 \\ 0 \\ 0 \end{pmatrix},$$

where I_p is given by

$$I_p = \sqrt{P_1^2 + P_2^2 + P_3^2}$$

and

$$I_0 = I - I_p.$$

A useful representation of a completely unpolarized beam is that of an incoherent superposition of any of two pure orthogonal states. Since orthogonal states have antipodal Stokes vectors, we may write

$$\begin{pmatrix} I \\ 0 \\ 0 \\ 0 \end{pmatrix} = \begin{pmatrix} I/2 \\ P_1 \\ P_2 \\ P_3 \end{pmatrix} + \begin{pmatrix} I/2 \\ -P_1 \\ -P_2 \\ -P_3 \end{pmatrix},$$

with

$$I/2 = \sqrt{P_1^2 + P_2^2 + P_3^2}.$$

The relative phase instability which is assumed in incoherent superposition casts doubt on the validity of the superposition in Eq. (4.23). It is the beauty of the Stokes formalism that we may bypass this equation.

The path we have taken to arrive at the Stokes vector formulation for describing light, in both pure states and mixed states, is based only on the use of pure states. This is seen in the mathematical definition of the Stokes parameters as the expansion coefficients of a state which is the coherent superposition of two pure P-states. Thus, we appear to have reached an impasse; for although we can now describe nonpure states by Stokes parameters, we cannot define the parameters simply because the state is not pure.

This difficulty is only apparent, however, for we can determine the Stokes parameters of a nonpure state in exactly the same way as we must ultimately find them for a pure state: namely, by experimental measurement where the operational definitions of the Stokes parameters was discussed in connexion with Eq. (4.13).

Now, any beam of light may be considered as an incoherent superposition of a totally unpolarized beam and a completely polarized beam in some pure state. To determine the state of the polarized component, we place a general elliptical filter in the beam and vary its transmission state until we obtain maximum transmission. Since any E-filter will pass half of an unpolarized beam, the transmitted intensity is the intensity of the polarized part of the beam plus half of the unpolarized intensity. The decrease in intensity is that half of the unpolarized beam which was absorbed in the E-filter. To justify the use of the Mueller matrices or the generalized Jones calculus on nonpure beams, we need only consider them as acting on the polarized component, leaving the unpolarized part unchanged in the case of retarders, and diminishing it by half in the case of E-filters. We must remember, however, that the entire emergent beam from any E-filter will be completely polarized.

Finally, it should be remarked that Eqs. (4.18) and (4.19), which were derived in Section 4.4, remain valid for those cases where a partially polarized or unpolarized beam is incident on a series of optical devices consisting of retarders and polarizers. However, since the kets in the bracket products represent pure states, the bracket products in those expressions have no meaning when the incident light is either partially polarized or unpolarized. Consequently, in these cases we need rewrite only the left-hand side of these equations, giving

$$W = \tfrac{1}{2}(1, P_1', P_2', P_3') \begin{pmatrix} P_0 \\ P_1 \\ P_2 \\ P_3 \end{pmatrix},$$

$$W = \tfrac{1}{2}(1, P_1', P_2', P_3') \mathscr{M} \begin{pmatrix} P_0 \\ P_1 \\ P_2 \\ P_3 \end{pmatrix},$$

where the interpretation in terms of transmitted intensities or probabilities is the same as before, but these probabilities, W, may not be written as bracket products. As a simple example, let us consider again Malus' law. Assume that a P_x-filter is used to polarize incident unpolarized light of intensity I, and that a second P-filter at an orientation θ relative to the polarizer is used as an analyzer. Since unpolarized light has the Stokes parameters $(I, 0, 0, 0)$ and the Stokes parameters of the light which has traversed the analyzer is given by Eq. (4.15), we have

$$W = \tfrac{1}{2}(1, a, b, 0)\tfrac{1}{2} \begin{pmatrix} 1 & 1 & 0 & 0 \\ 1 & 1 & 0 & 0 \\ 0 & 0 & 0 & 0 \\ 0 & 0 & 0 & 0 \end{pmatrix} \begin{pmatrix} I \\ 0 \\ 0 \\ 0 \end{pmatrix},$$

where

$$a = \cos^2\theta - \sin^2\theta; \qquad b = 2\sin\theta\cos\theta$$

and the Mueller matrix (Appendix II) has been chosen such that the line of the polarizing P-filter is along the x-axis. Carrying through the matrix multiplication gives the result

$$W = \frac{I}{2}\cos^2\theta,$$

where $I/2$ is the intensity of the light transmitted by the polarizer.

PROBLEMS

4.1 Obtain the Stokes parameters for the states given in Problems 3.1 and 3.3.

Answer: $|E_1\rangle \quad P_0 = 1, P_1 = 6\sqrt{3}/13, P_2 = -\dfrac{6}{13}, P_3 = \dfrac{5}{13},$

$|E_2\rangle \quad P_0 = 1, P_1 = 0, P_2 = -\dfrac{12}{13}, P_3 = -\dfrac{5}{13}.$

4.2 Show that the operator $|E\rangle\langle E|$ is not in general idempotent if $\langle E|E\rangle \neq 1$. How can it be made idempotent? Does the proof of Eq. (4.3) depend on whether or not the operator is idempotent?

4.3 Prove for any two square matrices A and B of equal dimension that Tr $(\mathsf{AB}) = \text{Tr}\,(\mathsf{BA})$ and, hence, that

$$\langle E|\mathsf{A}|E\rangle = \text{Tr}\,(\mathsf{A}|E\rangle\langle E|) = \text{Tr}\,(|E\rangle\langle E|\mathsf{A}).$$

4.4 Expand the matrix

$$A = \begin{pmatrix} 1 & 2 \\ 3 & 4 \end{pmatrix}$$

in terms of the sigma matrices and find the Stokes parameters for this matrix. Are the Stokes parameters real?

Answer: $P_0 = 5,\ P_1 = -3,\ P_2 = 5,\ P_3 = -i.$

4.5 A matrix is anti-Hermitian if $\mathsf{A}^\dagger = -\mathsf{A}$. Prove that the Stokes parameters of an anti-Hermitian matrix are pure imaginary numbers.

4.6 For any two states E_1 and E_2 consider the two operators $|E_1\rangle\langle E_2|$ and $|E_2\rangle\langle E_1|$. If these two operators are expanded in terms of the sigma matrices, prove that the Stokes parameters of the one operator are the complex conjugates of those for the other.

4.7 Show that the matrix

$$\mathsf{M} = \cos\left(\frac{\Omega}{2}\right)\boldsymbol{\sigma}_0 + i\sin\left(\frac{\Omega}{2}\right)\boldsymbol{P}\cdot\boldsymbol{\sigma},$$

where

$$\boldsymbol{P}\cdot\boldsymbol{\sigma} = P_1\boldsymbol{\sigma}_1 + P_2\boldsymbol{\sigma}_2 + P_3\boldsymbol{\sigma}_3,$$

operating on any Jones vector leaves the length of the associated Poincaré vector unchanged if

$$1 = P_1^2 + P_2^2 + P_3^2.$$

4.8 Let the matrix of Problem 4.7 operate on any two Jones vectors and produce two new Jones vectors. Show that the angle between the Poincaré vectors associated with these states is the same for the new states as for the old.

4.9 Use the results of Problems 4.7 and 4.8 to show that, when the matrix of Problem 4.7 operates on all Jones vectors, the effect on the associated Poincaré vectors is equivalent to a rigid rotation of the Poincaré sphere about some axis. Can you specify the direction of this axis? Can you determine the angle of rotation about this axis?

4.10 Repeat Problems 3.8 and 3.9 using the Poincaré sphere to eliminate all algebraic calculations.

4.11 In Section 4.6 the individual sigma matrices were interpreted as instrument operators and the physical significance of the multiplication rule $\boldsymbol{\sigma}_i\boldsymbol{\sigma}_j = i\boldsymbol{\sigma}_k$ was discussed. Discuss this multiplication rule in terms of rotations of the Poincaré sphere.

4.12 Hamilton's quaternions are "hypercomplex" numbers with three different "complex" parts. A quaternion may be written in terms of four real numbers a, b, c, and d and three "imaginary" symbols i, j, and k, as

$$Q = a + bi + cj + dk.$$

The algebra of quaternions is defined by the multiplication table for the symbols i, j, and k given below.

	i	j	k
i	-1	k	$-j$
j	$-k$	-1	i
k	j	$-i$	-1

The row of the table determines the first symbol in the product and the column the second, e.g., $ij = k$. Why would you expect that quaternions might be useful for describing polarization states?

4.13 If the Poincaré vector of a particular polarization state is given, how would you geometrically determine the length and direction of the Poincaré vector of this beam after it has gone through a P-filter? How would you determine geometrically the Poincaré vector of this beam after it has gone through any polarization filter?

4.14 The Stokes parameters for a state $|E\rangle$ were obtained from

$$|E\rangle\langle E| = \sum_{j=0}^{3} P_j\boldsymbol{\sigma}_j,$$

where $|E\rangle$ was written as

$$|E\rangle = A|P_x\rangle + B|P_y\rangle.$$

If $|E\rangle$ had been written as

$$|E\rangle = A|R\rangle + B|L\rangle,$$

what polarization states would correspond to the Stokes vectors

$$\begin{pmatrix} 1 \\ \pm 1 \\ 0 \\ 0 \end{pmatrix} \quad \begin{pmatrix} 1 \\ 0 \\ \pm 1 \\ 0 \end{pmatrix} \quad \begin{pmatrix} 1 \\ 0 \\ 0 \\ \pm 1 \end{pmatrix}?$$

Answer: $|R\rangle$, $|L\rangle$; $|P_x\rangle \pm |P_y\rangle$; $|P_y\rangle$, $|P_x\rangle$.

4.15 If you were asked to design a series of linear and circular retarders which would alter every polarization state that passed through the series except the states

$$|E_1\rangle = \frac{1}{\sqrt{5}}(3|P_x\rangle + 4\,|P_y\rangle),\ |E_2\rangle = \frac{1}{\sqrt{5}}(4\,|P_x\rangle + 3\,|P_y\rangle),$$

why would you politely refuse?

4.16 If the Jones matrix for an optical device is Hermitian, $\mathsf{M}^+ = \mathsf{M}$, show that the corresponding Mueller matrix, as defined by Eq. (9.17), is symmetric.

4.17 A piece of optical active material is of the correct thickness to rotate a $|P_x\rangle$-state into the state $\frac{1}{\sqrt{2}}(|P_x\rangle + |P_y\rangle)$. A $|P_y\rangle$-state passes through this same piece of material and then through a linear retarder. It emerges as a $|P_y\rangle$-state once again. What fractional wave plate was it and what was the orientation of its fast axis?

Answer: Half-wave, 22.5° to x-axis.

CHAPTER 5

THE METHODS OF QUANTUM MECHANICS

Given a series of polarizers and retarders, we have seen in the previous chapters how we may predict the emergent intensity which will be measured. Likewise, we know how the Stokes parameters and, hence, the polarization state of a beam may be measured. In this chapter we are going to extend this theory to measurements of other physical quantities, such as energy and linear momentum.

We shall examine how the probable outcome of a certain type of polarization measurement may be phrased within the formalism of bras and kets. It will then be shown that it is reasonable to believe that the outcome of any measurement can be expressed in the same formalism in terms of a quantity commonly known as the "expectation value". This leads naturally to a discussion of the significance of commuting operators which in principle allow a determination of the constants of "motion" of a system. The material, which up to this point is handled mainly within the framework of finite dimensional linear vector spaces, is then extended to indicate how continuous functions may be fitted within the framework of vector spaces.

Since the purpose of this chapter is to serve as an introduction to the formulation of physics in terms of quantum mechanics, it is appropriate that the chapter concludes with one of the very important implications of this formulation, known as the uncertainty relations.

5.1 EIGENSTATES AND EIGENVALUES

The problem which we are going to examine in the following discussion is a very common one in physics, though at first glance it may appear unfamiliar. A routine example in differential equations is

$$\frac{d}{dx} f(x) = \lambda f(x), \tag{5.1}$$

where λ is a constant. The solution of this equation answers the question, "What function, when operated on by d/dx, gives back the same function multiplied by a constant?" We know the answer and in fact Eq. (5.1) may be taken as the definition of the exponential function, $f(x) = e^{\lambda x}$. The key word we have used above is "operated". There are many ways in which we may operate on a function. We may differentiate, integrate, multiply by scalars, etc. Any of these operations applied to a function $f(x)$, will in general yield a new function $g(x)$. If we let the symbol $\mathcal{O}$ stand for any admissible operation and let $\mathcal{O}f(x)$ represent the result of operating on $f(x)$ with $\mathcal{O}$, we may pose the following mathematical problem:

$$\mathcal{O}f(x) = \lambda f(x). \tag{5.2}$$

That is, do there exist functions $f(x)$ which the operation modifies by at most a multiplicative constant λ? Equations (5.1) and (5.2) are known as eigenvalue problems. If solutions, $f(x)$, exist, they are called eigenfunctions and the values of λ are called eigenvalues. Thus, $e^{\lambda x}$ is an eigenfunction of the operator d/dx with eigenvalue λ.

Eigenvalue problems are not limited to operations on functions such as $f(x)$. One may easily consider such problems where the operation is matrix multiplication. Thus, if M is a given matrix, one can consider

$$\mathsf{M}|E\rangle = \lambda|E\rangle. \tag{5.3}$$

That is, are there vectors which when multiplied by M are modified at most by scalar multiplication? If these vectors exist, they are called eigenvectors of the matrix M.

A physical interpretation of Eq. (5.3) is easily given. Since we have represented polarizing devices by matrices, this equation asks whether there exist polarization states which can pass through a given device with at most a modification in intensity or arbitrary phase. This modification is represented by the scalar multiplication. If such states exist, they will be called eigenstates of the device. As a trivial example: any state is an eigenstate of a simple absorber, since it changes only the intensity of the state and not its polarization. Eigenvalue problems are of considerable importance but often of such complexity that the eigenvalues are not easily accessible. One of the advantages of the 2×2-matrix representation of optical devices is that the eigenvalues and eigenstates for such devices can readily be obtained.

Since the optical devices we have considered can be represented by 2×2-matrices, let D be a general 2×2-matrix for which

$$\mathsf{D}|E\rangle = \lambda|E\rangle,$$

$$\begin{pmatrix} d_{11} & d_{12} \\ d_{21} & d_{22} \end{pmatrix} \begin{pmatrix} A \\ B \end{pmatrix} = \lambda \begin{pmatrix} A \\ B \end{pmatrix}, \tag{5.4}$$

where λ is the eigenvalue. Of course, this matrix equation is equivalent to the pair of simultaneous linear equations,

$$\begin{aligned} (d_{11} - \lambda)A + d_{12}B &= 0, \\ d_{21}A + (d_{22} - \lambda)B &= 0, \end{aligned} \tag{5.5}$$

which possesses a non-trivial solution only if the determinant of the coefficients of A and B is zero, i.e.,

$$\begin{vmatrix} (d_{11} - \lambda) & d_{12} \\ d_{21} & (d_{22} - \lambda) \end{vmatrix} = 0. \tag{5.6}$$

The solution of the quadratic equation in λ is

$$\lambda = \tfrac{1}{2}\{(d_{22} + d_{11}) \pm \sqrt{(d_{22} + d_{11})^2 - 4(d_{11}d_{22} - d_{12}d_{21})}\},$$

$$\lambda = \tfrac{1}{2}\{\mathrm{Tr}\,(\mathsf{D}) \pm \sqrt{(\mathrm{Tr}\,(\mathsf{D}))^2 - 4|\mathsf{D}|}\}.$$

The term $(d_{22} + d_{11})$ is just Tr D, and $(d_{11}d_{22} - d_{12}d_{21})$ is the determinant of the matrix D which we indicate by $|\mathsf{D}|$. The eigenvalues can now be written as

$$\begin{aligned} \lambda_1 &= \tfrac{1}{2}\,\mathrm{T_r}\,(\mathsf{D})\left[1 + \sqrt{1 - \frac{4|\mathsf{D}|}{(\mathrm{Tr}\ \mathsf{D})^2}}\right], \\ \lambda_2 &= \tfrac{1}{2}\,\mathrm{T_r}\,(\mathsf{D})\left[1 - \sqrt{1 - \frac{4|\mathsf{D}|}{(\mathrm{Tr}\ \mathsf{D})^2}}\right]. \end{aligned} \tag{5.7}$$

If the matrix D is diagonal, it is immediately evident from Eq. (5.6) that the eigenvalues are simply the diagonal elements. Once the eigenvalues are determined, the eigenstates can be obtained from Eq. (5.5) by substituting the eigenvalues and solving for the *ratios* A/B or B/A. To obtain A and B themselves requires a further normalization condition for the eigenvectors. As an illustration of this, consider the transmission matrix of a P-filter (Problem 3.4):

$$\mathsf{T} = \begin{pmatrix} \cos^2\theta & \cos\theta\sin\theta \\ \cos\theta\sin\theta & \sin^2\theta \end{pmatrix}.$$

Very quickly we find

$$\mathrm{Tr}\,(\mathsf{T}) = 1,$$
$$|\mathsf{T}| = 0,$$

which, from Eq. (5.7), gives this eigenvalues

$$\lambda_1 = 1,$$
$$\lambda_2 = 0.$$

Since T is the transmission matrix for a P-filter whose line is oriented at an angle θ relative to the x-axis, the eigenstate of the filter must be the P-state $|P\rangle = \cos\theta|P_x\rangle + \sin\theta|P_y\rangle$. Let us check this. According to Eq. (5.5), for $\lambda_1 = 1$,

$$(\cos^2\theta - 1)A + \cos\theta\sin\theta B = 0,$$
$$\cos\theta\sin\theta A + B(\sin^2\theta - 1) = 0,$$

which are satisfied for

$$B/A = \sin\theta/\cos\theta,$$

and the suspicion that $|P\rangle$ is the correct eigenstate is confirmed. It is easily verified that $|P'\rangle$ also corresponds to the eigenvalue $\lambda_2 = 0$, where $\langle P'|P\rangle = 0$. It is left as an exercise to show that the eigenvalues and eigenstates of instrument operators $\boldsymbol{\sigma}_i$ are given as shown in the tabulation below.

Operator	*Eigenvalue*	*Eigenstate*
$\boldsymbol{\sigma}_1$	$+1$	$\|P_x\rangle$
	-1	$\|P_y\rangle$
$\boldsymbol{\sigma}_2$	$+1$	$\|P_x\rangle + \|P_y\rangle$
	-1	$\|P_x\rangle - \|P_y\rangle$
$\boldsymbol{\sigma}_3$	$+1$	$\|R\rangle$
	-1	$\|L\rangle$

This agrees with the discussion of Section 4.6.

5.2 HERMITIAN OPERATORS

Because of their unique mathematical properties, linear Hermitian operators play an important role in physics. An operator is linear if it obeys the equation

$$\mathcal{O}[a|E_1\rangle + b|E_2\rangle] = a\mathcal{O}|E_1\rangle + b\mathcal{O}|E_2\rangle,$$

where $|E_1\rangle$ and $|E_2\rangle$ are any two kets and a and b are any scalars. It is recalled that a Hermitian matrix is one which is equal to its complex conjugate transpose. The meaning of the Hermitian property for other linear operators will be defined as they are encountered. In this section we will prove a number of the more important mathematical properties of Hermitian operators. The statements and proofs will be specifically given for Hermitian matrices but will be valid for Hermitian operators in general.

The eigenvalues of a Hermitian matrix are real

If $|E\rangle$ is an eigenvector of M with eigenvalue λ,

$$\mathsf{M}|E\rangle = \lambda|E\rangle$$

and

$$\langle E|\mathsf{M}^\dagger = \langle E|\lambda^\star.$$

Multiplying by $\langle E|$ from the left in the first equation and by $|E\rangle$ from the right in the second, we obtain

$$\langle E|\mathsf{M}|E\rangle = \lambda\langle E|E\rangle, \qquad \langle E|\mathsf{M}^\dagger|E\rangle = \lambda^\star\langle E|E\rangle.$$

Now, if $M^\dagger = M$ we must have $\lambda^\star = \lambda$.

Eigenvectors of a Hermitian matrix belonging to different eigenvalues are orthogonal

Let $|E_i\rangle$ and $|E_j\rangle$ be two eigenvectors of the Hermitian matrix M with eigenvalues λ_i and λ_j. Thus,

$$\mathsf{M}|E_i\rangle = \lambda_i|E_i\rangle, \tag{5.8}$$

$$\mathsf{M}|E_j\rangle = \lambda_j|E_j\rangle. \tag{5.9}$$

Multiplying these equations from the left by $\langle E_j|$ and $\langle E_i|$ yields

$$\langle E_j|\mathsf{M}|E_i\rangle = \lambda_i\langle E_j|E_i\rangle, \tag{5.10}$$

$$\langle E_i|\mathsf{M}|E_j\rangle = \lambda_j\langle E_i|E_j\rangle. \tag{5.11}$$

Taking the complex conjugate of the last equation and remembering that if M is Hermitian, $\mathsf{M} = \mathsf{M}^\dagger$ and $\lambda^\star = \lambda$ as well as $\langle E_i|E_j\rangle^\star = \langle E_j|E_i\rangle$, we obtain

$$\langle E_j|\mathsf{M}|E_i\rangle = \lambda_j\langle E_j|E_i\rangle. \tag{5.12}$$

Subtracting Eqs. (5.10) and (5.12), we obtain

$$(\lambda_i - \lambda_j)\langle E_i|E_j\rangle = 0. \tag{5.13}$$

Thus, if $\lambda_i \neq \lambda_j$, $\langle E_i|E_j\rangle$ must be zero and the states are orthogonal. Thus, if all of the eigenvalues of M are different, all eigenstates are mutually orthogonal. It often happens that different eigenstates of an operator have the same eigenvalue. These are known as degenerate states, and in this case our proof is inconclusive. One may show, however, that in the degenerate case a set of mutually orthogonal eigenkets can be found. Since we will not be greatly concerned with degeneracy in the remainder of this text, we will not give the general proof.

An operator M may be expressed in terms of its eigenstates and eigenvalues

The proof is best given by demonstrating the result. Let the eigenvectors of M be $|E_k\rangle$ with eigenvalues λ_k. We construct the matrix

$$\mathsf{M}' = \sum_k \lambda_k |E_k\rangle\langle E_k|. \tag{5.14}$$

Assuming that the $|E_k\rangle$ are orthonormal, we operate on $|E_i\rangle$ with M' and obtain

$$\begin{aligned}\mathsf{M}'|E_i\rangle &= \sum_k \lambda_k |E_k\rangle\langle E_k|E_i\rangle \\ &= \sum_k \lambda_k |E_k\rangle \delta_{ik} \\ &= \lambda_i |E_i\rangle.\end{aligned}$$

Thus, the eigenstates and eigenvalues of M and M' are identical. It is left to the exercises to show that this is sufficient to identify M and M'.

A necessary and sufficient condition that two operators M and N possess a common set of eigenstates is that $\mathsf{MN} = \mathsf{NM}$

Let us assume that a common set of eigenstates exist which we shall label $|\mu_i\lambda_j\rangle$, so that

$$\begin{aligned}\mathsf{M}|\mu_i, \lambda_j\rangle &= \lambda_j|\mu_i, \lambda_j\rangle, \\ \mathsf{N}|\mu_i, \lambda_j\rangle &= \mu_i|\mu_i, \lambda_j\rangle.\end{aligned}$$

Thus,

$$\mathsf{MN}|\mu_i, \lambda_j\rangle = \mu_i \mathsf{M}|\mu_i, \lambda_j\rangle = \mu_i\lambda_j|\mu_i, \lambda_j\rangle$$

and

$$\mathsf{NM}|\mu_i, \lambda_j\rangle = \lambda_j \mathsf{N}|\mu_i, \lambda_j\rangle = \lambda_j\mu_i|\mu_i, \lambda_j\rangle.$$

Since μ_i and λ_j are scalars, $\mu_i\lambda_j = \lambda_j\mu_i$, and the order in which M and N act is unimportant. Thus, $\mathsf{MN} = \mathsf{NM}$. To prove that it is necessary, we assume that $\mathsf{MN} = \mathsf{NM}$. Let $|\mu_i\rangle$ be an eigenstate of N with eigenvalue μ_i:

$$\mathsf{N}(\mathsf{M}|\mu_i\rangle) = \mathsf{MN}|\mu_i\rangle = \mu_i(\mathsf{M}|\mu_i\rangle). \tag{5.15}$$

If N is nondegenerate, $\mathsf{M}|\mu_i\rangle$ must be at most a scalar times $|\mu_i\rangle$. Thus,

$$\mathsf{M}|\mu_i\rangle = a|\mu_i\rangle,$$

and $|\mu_i\rangle$ is also an eigenvector of M. If, on the other hand, N is degenerate and there are several eigenstates with eigenvalue μ_i, then Eq. (5.15) implies at most that $\mathsf{M}|\mu_i\rangle$ is a linear combination of these degenerate states. For reasons given before we will not give the more general proof but simply state that in the degenerate case a common set of eigenstates can be found.

5.3 A CLASS OF EXPERIMENTS

We are going to examine a class of experiments which, although of relative simplicity, can serve as prototypes for any experiment designed to measure a single physical quantity. This alone is enough to recommend the investigation, but we shall see that one of these experiments is actually concerned with the measurement of angular momentum and, hence, of concrete importance.

Consider any pure polarization state with ket $|S\rangle$ and a given pair of orthogonal states, $|E\rangle$ and $|E'\rangle$. The Poincaré vectors of these states are shown in Fig. 5.1 and indicated as $\boldsymbol{E}$, $\boldsymbol{E}'$, and $\boldsymbol{S}$. For the given states $|E\rangle$ and $|E'\rangle$ a real scalar quantity associated with the state $|S\rangle$ is the projection of its Poincaré vector $\boldsymbol{S}$ along the direction of $\boldsymbol{E}$. Since we are taking all of these vectors to be of unit length, this projection is simply $\boldsymbol{S}\cdot\boldsymbol{E}$. Although this quantity is suggested by our formalism, it is not obvious that it is measurable. To show that it can be obtained from experiment, we recall that the probability that the state $|S\rangle$ will pass through an E-filter is $|\langle S|E\rangle|^2$. By Eq. (4.18) we have

$$|\langle S|E\rangle|^2 = \tfrac{1}{2}(1 + \boldsymbol{S}\cdot\boldsymbol{E}).$$

Thus,

$$\boldsymbol{S}\cdot\boldsymbol{E} = 2|\langle S|E\rangle|^2 - 1. \tag{5.16}$$

Since the right-hand side is experimentally meaningful, so is the left. In fact, Eq. (5.16) is just a redefinition of the probability scale. If $|S\rangle = |E\rangle$, then the probability of $|S\rangle$ passing through an E-filter is unity, as is $\boldsymbol{S}\cdot\boldsymbol{E}$. On the other hand, if $|S\rangle = |E'\rangle$, where $|E'\rangle$ is orthogonal to $|E\rangle$, then the

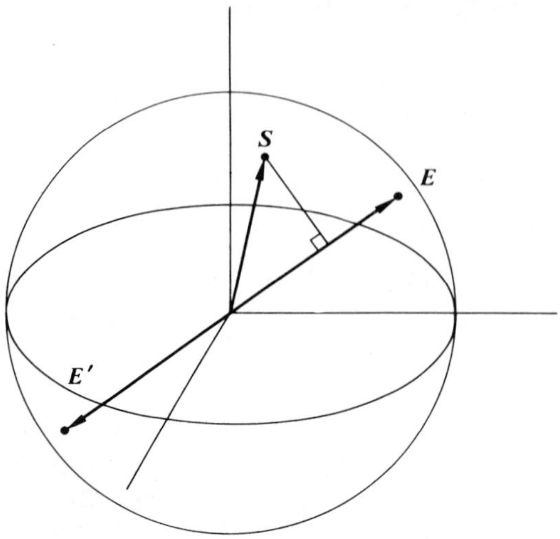

Fig. 5.1. Projection of the Poincaré vector $\boldsymbol{S}$ on the direction $E - E'$.

probability of $|S\rangle$ passing through the E-filter is zero and $\boldsymbol{S}\cdot\boldsymbol{E} = -1$. Therefore, $\boldsymbol{S}\cdot\boldsymbol{E}$ is a measure of the probability of transmission through an E-filter with $+1$ indicating complete transmission and -1 indicating complete absorption. We can now give operational instructions for the measurement of $\boldsymbol{S}\cdot\boldsymbol{E}$. The state $|S\rangle$ is sent into a resolver, such as that shown in Fig. 5.2, which splits it into the states $|E\rangle$ and $|E'\rangle$. Counters C and C' are placed in the upper and lower beams. We may arrange that these counters also perform a time averaging of their counting rates so that if

$$|S\rangle = \langle E|S\rangle|E\rangle + \langle E'|S\rangle|E'\rangle, \tag{5.17}$$

the output of counter C is $|\langle E|S\rangle|^2$ and that of counter C' is $|\langle E'|S\rangle|^2$. The outputs of these counters are then fed into devices which multiply the output of C by $+1$ and the output of C' by -1. Finally, the outputs of the multipliers are fed into an adder which takes their algebraic sum. Let us now show that this final output is simply $\boldsymbol{S}\cdot\boldsymbol{E}$. Remembering that S is a normalized state with

$$|\langle E|S\rangle|^2 + |\langle E'|S\rangle|^2 = 1,$$

we may follow the experimental procedure algebraically and obtain

$$\begin{aligned}\text{Final output} &= (+1)|\langle E|S\rangle|^2 + (-1)|\langle E'|S\rangle|^2\\ &= 2|\langle S|E\rangle|^2 - 1.\end{aligned}$$

But, by Eq. (5.16), this is just $\boldsymbol{S}\cdot\boldsymbol{E}$.

We now find another way to write $\boldsymbol{S}\cdot\boldsymbol{E}$. Consider the matrix M whose eigenvectors are just the state vectors of $|E\rangle$ and $|E'\rangle$, and whose eigenvalues are $+1$ and -1. One might doubt that such a matrix exists if an explicit form for it were not so easy to find by using Eq. (5.14). In terms of the bras and kets of E and E' we may write

$$\mathsf{M} = |E\rangle\langle E| - |E'\rangle\langle E'|. \tag{5.18}$$

Now, in terms of this matrix we prove that $\boldsymbol{S}\cdot\boldsymbol{E} = \langle S|\mathsf{M}|S\rangle$. Using the forms given in Eqs. (5.17) and (5.18), we obtain

$$\langle S|\mathsf{M}|S\rangle = |\langle E|S\rangle|^2 - |\langle E'|S\rangle|^2 = \boldsymbol{S}\cdot\boldsymbol{E}. \tag{5.19}$$

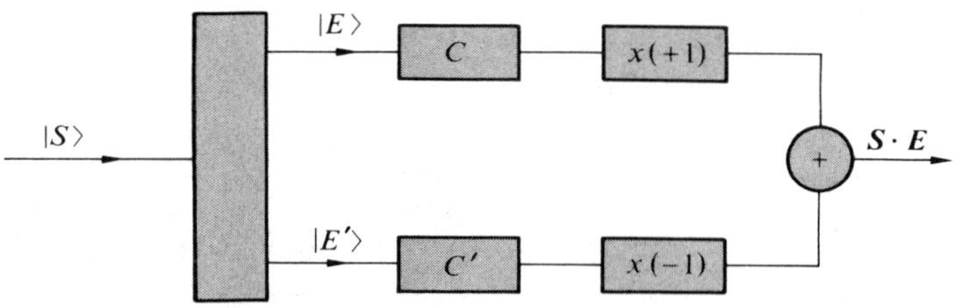

Fig. 5.2. Operational definition for the measurement of the quantity $\boldsymbol{S}\cdot\boldsymbol{E}$.

Now, although $\boldsymbol{S} \cdot \boldsymbol{E}$ is physically meaningful and may be measured, the projection of one Poincaré vector on another may be a somewhat abstract concept. In one particular case, however, the significance of Eq. (5.19) is immediate. Consider the operator

$$\mathsf{M} = \hbar\boldsymbol{\sigma}_3. \tag{5.20}$$

The eigenstates of σ_3 are $|R\rangle$ and $|L\rangle$, so we have

$$|S\rangle = \langle R|S\rangle|R\rangle + \langle L|S\rangle|L\rangle, \tag{5.21}$$

giving

$$\langle S|\mathsf{M}|S\rangle = \langle S|\hbar\boldsymbol{\sigma}_3|S\rangle = \hbar|\langle R|S\rangle|^2 - \hbar|\langle L|S\rangle|^2.$$

Since $+\hbar$ and $-\hbar$ are the spin angular momenta of R- and L-states, $\langle S|\hbar\boldsymbol{\sigma}_3|S\rangle$ is the average angular momentum along the direction of propagation which will be measured for the state $|S\rangle$, and $\hbar\boldsymbol{\sigma}_3$ is the operator corresponding to the component of angular momentum in the direction of motion. Two important points are to be found in this example. First, we have seen how in one case a measurement may be related to the projection of one vector along another. Second, we have associated with a measurement an operator M and have seen that the result of the measurement on a state $|S\rangle$ is given by $\langle S|\mathsf{M}|S\rangle$. This second point is examined in more detail below.

5.4 EXPECTATION VALUES

In the preceding section we have seen that, when $\mathsf{M} = \hbar\boldsymbol{\sigma}_3$, $\langle S|\mathsf{M}|S\rangle$ yields the measured average value of the component of angular momentum along the direction of motion of the state $|S\rangle$. For individual photons this component is either $+\hbar$ or $-\hbar$, so that for individual counts $\langle S|\mathsf{M}|S\rangle$ predicts exactly the outcome of any single count only if $|S\rangle$ is an R- or an L-state. However, in the general case there is *no better guess* of the outcome. Because of this, the value given by $\langle S|\mathsf{M}|S\rangle$ is called the expectation value for the angular momentum component.

We wish to generalize this procedure which we have given for a very particular measurement. Since we are going to generalize, let us go all the way. Assume, therefore, absolutely any physical system and some measurable quantity associated with that system—energy, momentum, polarization, etc. For simplicity, we will assume here that the measured values of this quantity can take on only a finite set of discrete values but we will later remove this restriction. Let the possible values resulting from our measurement be the set of real numbers λ_i. If the measurement results in the particular value λ_k, then the state of the system which yields this value will be called S_k and will be represented by a ket symbol $|S_k\rangle$. These are the pure states of our system in respect to the experiment which we are considering. We assume that any state can be written as a superposition of these states as

$$|S\rangle = \sum_i a_i|S_i\rangle, \tag{5.22}$$

and that the pure states are orthonormal. We now form the operator

$$\mathsf{M} = \sum_i \lambda_i|S_i\rangle\langle S_i|. \tag{5.23}$$

This operator is a matrix, of course, if we represent our bras and kets by row and column matrices, respectively. We do not rule out the possibility that other mathematical representations can be found, but these must also obey all the formal mathematical machinery we have developed. The operator M has two interesting properties. First, its eigenkets are the pure states $|S_k\rangle$ and its eigenvalues are the measured quantities λ_k,

$$\mathsf{M}|S_k\rangle = \lambda_k|S_k\rangle, \tag{5.24}$$

as we have seen in Section 5.2. The second property of M is displayed in Eq. (5.25)

$$\langle S|\mathsf{M}|S\rangle = \sum_i \lambda_i a_i^\star a_i, \tag{5.25}$$

which is easily verified with the aid of Eqs. (5.22) and (5.23). Following the usual interpretation, $a_i{}^\star a_i$ is the probability that the state $|S\rangle$ will be detected in the state $|S_i\rangle$. Furthermore, if $|S\rangle$ is detected in $|S_i\rangle$, then the measurement will result in the eigenvalue λ_i. Thus, Eq. (5.25) represents a weighted average of the possible outcomes of an experiment, and the quantity $\langle S|\mathsf{M}|S\rangle$ is called the expectation value of M because each eigenvalue λ_i of M is weighted according to the probability of measuring that value. It should be clear that, if $|S\rangle = |S_i\rangle$, the expectation value is simply the eigenvalue λ_i.

If M is looked upon as a resolver, as in Section 3.3, and if an unlimited number of physical entities is prepared such that each entity is in the same state $|S\rangle$, then the operational significance of Eq. (5.25) becomes somewhat transparent. In Fig. 5.3 M is shown as a resolver, where $|S\rangle$ is the entrant state and the D_i are counters. It is evident that the eigenvalues λ_i are dispersed among the output channels, with the result that no complete prediction

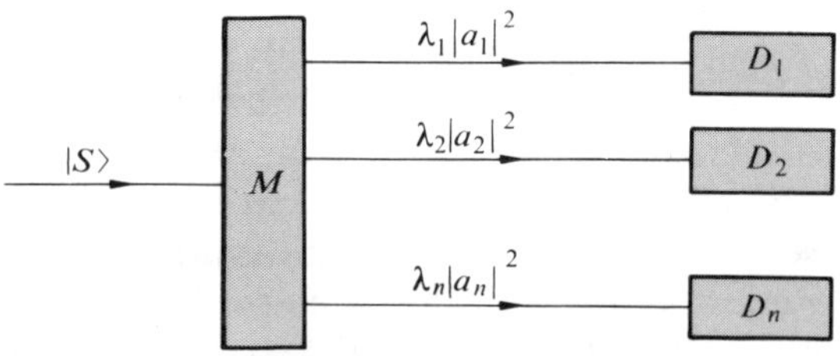

Fig. 5.3. A multichannel resolver.

can generally be made as to which exit channel any entity in the state $|S\rangle$ will use. However, since the number of counts in each detector D_i is proportional to $|a_i|^2$, we can obtain as close an approximation to the expectation value of M as we please. This means that the experimental value for one count will not generally coincide with the expectation value. This leads us to define a root mean square deviation from the expectation value.

We define the ket $|\psi\rangle$, for a Hermitian matrix M,

$$|\psi\rangle = [\mathsf{M} - \langle \mathsf{M} \rangle]|S\rangle,$$

where $\langle \mathsf{M} \rangle = \langle S|\mathsf{M}|S\rangle$, and form the product

$$\begin{aligned} \langle \psi|\psi\rangle &= \langle S|(\mathsf{M} - \langle \mathsf{M}\rangle)(\mathsf{M} - \langle \mathsf{M}\rangle)|S\rangle \\ &= \langle S|\mathsf{M}^2|S\rangle - 2\langle \mathsf{M}\rangle\langle S|\mathsf{M}|S\rangle + \langle \mathsf{M}\rangle^2 \\ &= \langle \mathsf{M}^2\rangle - \langle \mathsf{M}\rangle^2. \end{aligned}$$

We see that $\langle\psi|\psi\rangle$ is the expectation value of $(\mathsf{M} - \langle \mathsf{M}\rangle)^2$ for the state $|S\rangle$. This is just the expectation value for the square of the deviation of the individual measurements from the average of the individual measurements. The quantity $\langle\psi|\psi\rangle$ will be designated $(\Delta M)^2$. The root mean square deviation $\Delta \mathsf{M}$ is then

$$\Delta \mathsf{M} = \sqrt{\langle\psi|\psi\rangle} = \sqrt{\langle S|(\mathsf{M} - \langle \mathsf{M}\rangle^2)|S\rangle},$$

which will vanish only if

$$\mathsf{M}|S\rangle = \langle \mathsf{M}\rangle|S\rangle,$$

i.e., if $\langle \mathsf{M}\rangle$ is an eigenvalue of M corresponding to the eigenstate $|S\rangle$.

5.5 THE POSTULATES OF QUANTUM MECHANICS

Put imprecisely, quantum mechanics postulates that the formalism developed in the previous sections can be applied to any measurement. That is, for every measurement an operator M exists which yields the expectation value of the measurement through $\langle S|\mathsf{M}|S\rangle$, whose pure states can be represented by the eigenkets of M, and whose eigenvalues yield the certain result of the measurement using the pure states.

Let us see why in one way this postulate is trivially true. If we assume a complete knowledge of the possible physical states in which a system may exist, then we know the values of all measurables which will be obtained in any experiment using one of these states. To indicate how the formalism might apply, assume some system where some measurable quantity may take on values λ_i, where, just as an example, there are 25 different possible values of λ_i. Twenty-five different orthonormal kets may be constructed corresponding to these states by letting all the components of the ket $|S_i\rangle$, corresponding to the i-th state, in a column representation, be zero except the i-th one; we let this be unity. This yields 25 different 25-dimensional orthonormal ket vectors. The operator associated with the measurement is given by Eq. (5.14):

$$\mathsf{M} = \sum_{i=1}^{25} \lambda_i |S_i\rangle\langle S_i|.$$

In this sense the postulate given is trivially true and amounts to a game of playing with words. Just as it is trivially true, it is worthless, for the pursuit of physics is simply to determine the possible states of systems and to determine the values of measurables which we will obtain using these systems.

Let us look at this from another point of view. What would happen if, with no recourse to Eq. (5.14), we could by physical intuition, mathematical insight, or even a lucky guess, determine *a priori* the operator M which corresponds to a given measurable quantity? Well, first we would try to solve the eigenvalue equation

$$\mathsf{M}|S_i\rangle = \lambda_i|S_i\rangle.$$

If we could do this, we would have found the pure states $|S_i\rangle$ of the system and we would have derived the possible values the measurable could take on, the λ_i. In principle, then, the main problem of quantum mechanics is the determination of the correct operators. We shall find some of these in the following section. In practice, however, the main problem is often to solve the eigenvalue equation after we have the operator.

A necessary restriction, which we have not mentioned, is that the operators which correspond to measurables must be Hermitian. This is to ensure that its eigenvalues will be real numbers corresponding to the real numbers resulting from experiment. To summarize, then, quantum mechanics assumes

a) that to every physical system and to each measurable of that system there corresponds an Hermitian operator M;

b) that the states of the system which are pure in respect to an experiment to measure this quantity are the eigenstates $|S_i\rangle$ of M and that in the state $|S_i\rangle$ the measurement will yield the corresponding eigenvalue λ_i; and

c) that any state $|S\rangle$ may be expanded in terms of the pure states $|S_i\rangle$ and the expectation value if we use $|S\rangle$ will be $\langle S|\mathsf{M}|S\rangle$.

5.6 THE TIME OPERATOR AND COMMUTATION

Until now we have been concerned only with the polarization aspect of light, and we have ignored the possibility that the kets which describe a state must in general be functions of position and time. A very simple example is that of a P-state traversing a length of optically active material. As the line of the P-state rotates, we may think of the polarization state either as a function of its position along the material or as a function of time as it traverses the material. Furthermore, since the bracket product can be related to intensity, it must in general be a function of position and time, since some places in the universe are darker than others and the dark places and bright places are not fixed in time.

Remember that, when we write $|S\rangle$, the symbol $|\ \rangle$ indicates that we are talking about a ket and S plays the role of a label which names the state and distinguishes it from all other states which differ from it in any way, through polarization or now through space or time dependence. Let us now use a new label, α, only for polarization and write $|S\rangle = |\alpha, \boldsymbol{r}, t\rangle$ to indicate difference in polarization and space and time dependence. Use of the vector symbol $\boldsymbol{r}$ indicates only that the ket depends on the spatial position and is shorthand for $|\alpha, x, y, z, t\rangle$.

When we deal with ordinary functions of position, it sometimes happens that the different coordinates appear in such a way that a function of x, y, and z may be written as a product of functions, each depending on only one variable, $f(x, y, z) = g(x)h(y)w(z)$. In the same way, it will sometimes happen that the various dependences of a ket, or bra, may be separated from one another. If this is the case, we will write

$$|\alpha, \boldsymbol{r}, t\rangle = \psi(\boldsymbol{r})\phi(t)|\alpha\rangle.$$

This is not always possible, but it will occur often enough to make the notation useful. Note that we have included the space and time dependence by the use of scalar functions. We shall see later that this dependence is not necessarily of such a simple form.

For the moment, let us concentrate explicitly only on the time dependence and write $|S\rangle = |t\rangle$, where $|S\rangle$ may, of course, depend on other things than t. If the state varies from one time, t_1, to another time, t_2, we have $|S_1\rangle = |t_1\rangle$, $|S_2\rangle = |t_2\rangle$. Now we are going to assume that there exists a time operator which is defined by the equation

$$|S_2\rangle = \mathsf{U}(t_2, t_1)|S_1\rangle.$$

That is, there is some operator acting on a state at time t_1 which gives us the state at time t_2.

Now, although we may assume that U exists, we have not shown that it is possible to find such an operator. We must solve this problem in two stages.

First, if $|S_1\rangle$ and $|S_2\rangle$ are known, then we obviously know U, for it is given by the matrix $|S_2\rangle\langle S_1|$, as is easily shown if $\langle S_i|S_i\rangle = 1$. However, if we know only $|S_1\rangle$, we are asking whether this is sufficient to predict the state at a later time. Now, we have gone to considerable trouble to show that complete prediction is not always possible, so it would seem that U does not always exist.

Let us phrase the question a little differently. If all the information necessary to predict the statistical outcome of an experiment on a state at time t_1 is contained in $|S_1\rangle$, is it possible to determine from this the statistical outcome of the same experiment at a later time, t_2? If this much is not possible, then there is very little left of physics. We therefore assume that in principle the operator U can be found.

Let us assume that we have a state at time t_1 and a time operator which tells us how $|S_1\rangle$ develops in time. Also, let G be any operator associated with an observable quantity whose expectation value is g. Then we have the following:

$$\begin{aligned}\langle S_1|\mathsf{G}|S_1\rangle &= g_1,\\ \langle S_2|\mathsf{G}|S_2\rangle &= \langle S_1|\mathsf{U}^\dagger\mathsf{G}\mathsf{U}|S_1\rangle\\ &= g_2.\end{aligned}$$

Now, it sometimes happens that for two particular operators the order in which they are applied is of no importance. This is not always true but let us assume for the moment that it is true for G and U:

$$\begin{aligned}\mathsf{GU} &= \mathsf{UG},\\ \mathsf{G} &= \mathsf{U}^\dagger\mathsf{GU}.\end{aligned}$$

Here we have demanded that U be unitary in order that

$$\langle S_2|S_2\rangle = \langle S_1|\mathsf{U}^\dagger\mathsf{U}|S_1\rangle = \langle S_1|S_1\rangle,$$

so that the normalization of the state is time independent.

If the order in which two operators are applied is unimportant, they are said to commute. We see then that, if U and G commute, we must have

$$\langle S_2|\mathsf{G}|S_2\rangle = \langle S_1|\mathsf{G}|S_1\rangle.$$

In other words, if we measure g at time t_1, we get the same expectation value that we get at time t_2. This is just another way of saying that g is constant or that g is a "constant of the motion".

We then can say that if the operator associated with some observable commutes with the time operator U, that observable is a constant. There are many operators which do not commute with U, but in the following sections we shall show how it is possible to find some which do.

As an important example, we now find one operator which commutes with U.

In a previous chapter we often found it convenient to perform coordinate transformations on various operators and the bra and ket vectors. At that time the states and operators were considered only with respect to polarization, but the idea of coordinate transformations is of utmost importance for all observable aspects of a physical system.

Let us review briefly the formalism of such transformations. We recall that a spatial transformation can be viewed either as a physical transformation of the system or as a mathematical change of coordinates leaving the system intact. The translation of a physical system an amount x to the right is equivalent to moving the coordinate system that amount to the left. But it must be pointed out that this equivalence is true only if the entire system of interest is translated. For example, if the earth were translated within the sun's gravitational field, the physical state of the earth would change; its potential energy would change. If, on the other hand, the entire solar system were translated, then, ignoring the effect of the distant stars, we would have performed only a coordinate change. Putting this another way: if we move the entire universe, there is no way to tell that we did it; we have only changed coordinate systems.

Now, let us perform such a transformation on some total physical system. If $|S\rangle$ is the original state and $|S\rangle'$ is the state after translation, then we assume some operator T such that

$$|S\rangle' = \mathsf{T}|S\rangle,$$

where $|S\rangle'$ expresses in terms of the new coordinates the information expressed by $|S\rangle$ in terms of the old coordinates. Now, not only state vectors must transform but also any operator which acts on those states. We have seen in previous chapters that if G is any operator which can act on $|S_1\rangle$ to yield $|S_2\rangle$, then after the transformation the new operator which acts on $|S_1\rangle'$ to yield $|S_2\rangle'$ is

$$\mathsf{G}' = \mathsf{TGT}^\dagger.$$

In particular, the time operator U becomes U' after the transformation of coordinates where

$$\mathsf{U}' = \mathsf{TUT}^\dagger.$$

If in the transformation T we have moved all parts of the system of interest, the system must develop in time just as it did before. That is, the same operator that takes $|t_1\rangle$ into $|t_2\rangle$ should also take $|t_1\rangle'$ into $|t_2\rangle'$. Applying U to $|S\rangle$ should yield the same physical state as applying it to $|S\rangle'$. Within the formalism, however, the correct operator to apply to $|S\rangle'$ is U'. Thus, since it does not matter whether we use U or U', we must have $\mathsf{U} = \mathsf{U}'$. Thus,

$$\mathsf{U} = \mathsf{TUT}^\dagger,$$

$$\mathsf{TU} = \mathsf{TU}.$$

Therefore, by making the assumption that the particular location in space of a system does not affect its time development, which is the same as saying that space is everywhere the same, we find that the time operator and the translation operator commute.

5.7 AN EXAMPLE OF A TIME OPERATOR

In order to make the preceding less abstract and also to develop a little more mathematics, we shall explicitly calculate the time operator for a particular system.

All the possible pure polarization states which have been considered until now were time independent. Such states are frequently called stationary states. Consider an E-state obtained from a coherent superposition of orthogonal P-states in the interferometer shown in Fig. 2.3, where the R- and L-filters in each arm have been replaced by a P-filter in each arm. If the lines of the P-filters are at 90° relative to each other, then the interferometer forms E-states through a coherent superposition, which can be expressed as

$$|E\rangle = A_x e^{-i\phi/2}|P_x\rangle + A_y e^{+i\phi/2}|P_y\rangle. \tag{5.26}$$

There are several ways in which the state $|E\rangle$ could be forced to vary with time. One method would be to vary periodically the light absorbed by the absorbers A and A' while maintaining a constant retarder thickness. This would have the effect of making the amplitudes A_x and A_y functions of time without affecting the phase difference, ϕ. Another method would be to keep the absorbers fixed, thereby maintaining the amplitudes A_x and A_y constant in time, and to vary the retarder thickness so that the phase factor $e^{\pm i\phi/2}$ became a periodic function of time. We will restrict our attention to the latter method, to illustrate the implication and consequences for time-varying pure states.

So, then, let us assume that we have a technique for varying the retarder thickness so that the phase difference is given by $\phi = \omega t$, where the radian frequency ω is constant and t is the time. Also assume that the absorbers are not permitted to vary and that the P-filters maintain a constant orientation relative to the entire interferometer. This latter assumption implies, for example, that the P-filter in an arm of the interferometer which is pure in relation to the state $|P_x\rangle$ remains pure in relation to $|P_x\rangle$. This ensures that while $|E\rangle$ varies with time the basic kets $|P_x\rangle$ and $|P_y\rangle$ are independent of time.

Substituting $\phi = \omega t$ into Eq. (5.26), which represents the state of the light emerging from the interferometer, and differentiating the resulting equation, we have

$$\frac{\partial}{\partial t}|E(t)\rangle = |\dot{E}(t)\rangle = -\frac{i\omega}{2}[A_x e^{-i\omega t/2}|P_x\rangle - A_y e^{i\omega t/2}|P_y\rangle]. \tag{5.27}$$

Now, we notice that the term in brackets can be obtained by operating on $|E(t)\rangle$ with the sigma matrix $\boldsymbol{\sigma}_1$ of Eq. (4.9).

$$\begin{pmatrix} 1 & 0 \\ 0 & -1 \end{pmatrix} \begin{pmatrix} A_x e^{-i\omega t/2} \\ A_y e^{i\omega t/2} \end{pmatrix} = \begin{pmatrix} A_x e^{-i\omega t/2} \\ -A_y e^{i\omega t/2} \end{pmatrix}$$
$$= A_x e^{-i\omega t/2}|P_x\rangle - A_y e^{i\omega t/2}|P_y\rangle,$$

where we are using the representation

$$|P_x\rangle = \begin{pmatrix} 1 \\ 0 \end{pmatrix}, \qquad |P_y\rangle = \begin{pmatrix} 0 \\ 1 \end{pmatrix}.$$

As a result, Eq. (5.27) can be written as

$$\frac{\partial}{\partial t}|E(t)\rangle = -\frac{i\omega}{2}\boldsymbol{\sigma}_1|E(t)\rangle. \tag{5.28}$$

The equation above may ring a bell because it has the *form* of the very simple differential equation

$$\frac{dx}{dt} + ikx = 0, \tag{5.29}$$

which has the solution $x = x_0 e^{-ikt}$, where x and t are real variables and k and x_0 are real constants. Can a solution of the same form be written for Eq. (5.28)? It can, though the exponential solution, if it exists, must involve matrix operators rather than simply k or t.

We must first determine the meaning of an expression such as e^{S}, where S is a matrix. For a real variable x we know that

$$e^x = 1 + x + \frac{x^2}{2!} + \frac{x^3}{3!} + \cdots + \frac{x^n}{n!} + \cdots$$

and the significance of e^{S}, where S is a matrix, is the same as the series of e^{S}, that is,

$$e^{\mathsf{S}} = \mathsf{I} + \mathsf{S} + \frac{\mathsf{S}^2}{2!} + \cdots + \frac{\mathsf{S}^n}{n!} + \cdots,$$

where the first term in the expansion is a unit matrix. Just as e^x can be differentiated with respect to a given variable, so, too, can e^{S}. However, the derivative of a matrix such as

$$\mathsf{S} = \begin{pmatrix} s_{11} & s_{12} \\ s_{21} & s_{22} \end{pmatrix}$$

involves the derivative of each and every matrix element, i.e.,

$$\dot{\mathsf{S}} = \begin{pmatrix} \dot{s}_{11} & \dot{s}_{12} \\ \dot{s}_{21} & \dot{s}_{22} \end{pmatrix},$$

where the dot over the matrix elements indicates differentiation with respect to a given variable, say d/dt. The series expansion for e^x can be differentiated term by term, but we must know how to differentiate the product of two matrices. Suppose that we have two matrices A and B whose product is C, i.e., $\mathsf{AB} = \mathsf{C}$. From Appendix I we know that

$$c_{ik} = \sum_j a_{ij} b_{jk},$$

and by differentiating both sides of the above equation we have

$$\dot{c}_{ik} = \sum (\dot{a}_{ij} b_{jk} + a_{ij} \dot{b}_{jk}),$$

which implies that

$$\frac{d}{dt} \mathsf{C} = \frac{d\mathsf{A}}{dt} \mathsf{B} + \mathsf{A} \frac{d\mathsf{B}}{dt} \cdot$$

If $\mathsf{A} = \mathsf{B} = \mathsf{S}$, then

$$\frac{d\mathsf{S}^2}{dt} = \mathsf{S} \frac{d\mathsf{S}}{dt} + \frac{d\mathsf{S}}{dt} \mathsf{S}.$$

In general, a matrix S does not commute with its derivative and, hence,

$$\frac{d\mathsf{S}^2}{dt} \neq 2\mathsf{S} \frac{d\mathsf{S}}{dt} \cdot$$

However, the matrices we shall be using are of the form

$$\mathsf{S} = t^n \mathsf{C},$$

where C is a matrix of constants. In this case we can write

$$\frac{d\mathsf{S}^2}{dt} = 2\mathsf{S} \frac{d\mathsf{S}}{dt},$$

and from this it can be shown by induction that

$$\frac{d\mathsf{S}^n}{dt} = n\mathsf{S}^{n-1} \frac{d\mathsf{S}}{dt} \cdot$$

It is left as an exercise to show from the series expansion of e^{S} that

$$\frac{d}{dt} e^{\mathsf{S}} = e^{\mathsf{S}} \frac{d\mathsf{S}}{dt} \cdot$$

Since Eqs. (5.28) and (5.29) have the same form, we are strongly tempted to write the solution of Eq. (5.28) as

$$|E(t)\rangle = e^{-i\omega t \boldsymbol{\sigma}_1/2} |E(0)\rangle,$$

and, indeed, it is the solution, as we can verify. This says that the time development of $|E(t)\rangle$ is related to $|E(0)\rangle$ through the operator $e^{-i\omega t\sigma_1/2}$. For this example

$$\mathsf{U}(t, 0) = e^{-i\omega t\sigma_1/2}.$$

A good idea of what the solution

$$|E(t)\rangle = e^{-i\omega t\sigma_1/2}|E(0)\rangle$$

means physically in this case can be obtained through the Poincaré sphere representation. Recall that the Stokes parameters are given by

$$\begin{aligned} P_0 &= A_x^2 + A_y^2, \\ P_1 &= A_x^2 - A_y^2, \\ P_2 &= 2A_xA_y \cos \omega t, \\ P_3 &= 2A_xA_y \sin \omega t, \end{aligned}$$

where $\phi = \omega t$, of course, is the difference in phase between the components. Both A_x and A_y are constant and only P_2 and P_3 are functions of time. In terms of the Poincaré sphere, the physical situation is shown in Fig. 5.4. The endpoint of the vector P is restricted to move in a circle on the sphere. Though $|E(t)\rangle$ varies in time, it is a pure state at each and every instant of time, since it satisfies the requirement

$$P_0^2 = P_1^2 + P_2^2 + P_3^2.$$

In addition, since P_2 and P_3 are periodic functions of time, the vector $\boldsymbol{P}$ precesses about the P_1-axis at the rate of ω rad/sec. It is evident from this

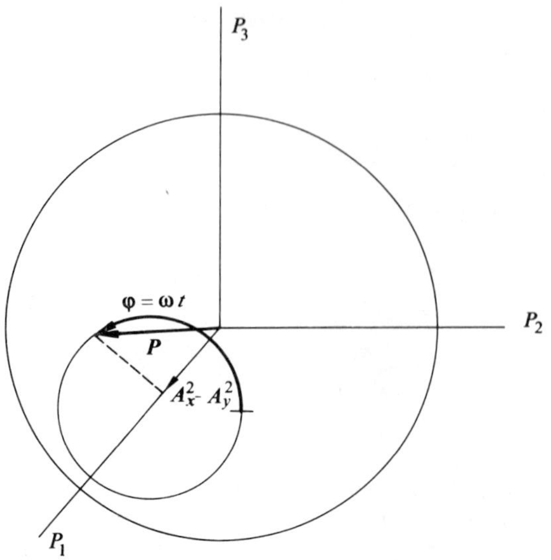

Fig. 5.4. The effect of the time operator $e^{i\omega t\sigma_1/2}$.

Poincaré sphere description that the Stokes parameter P_1 is a "constant of the motion" in the sense described in Section 5.5. If it is a constant of the motion, then the operator $\boldsymbol{\sigma}_1$, which corresponds to the observable P_1, should commute with the time operator:

$$e^{i\omega t\boldsymbol{\sigma}_1/2}.$$

By using the series expansion for the exponential and remembering that $\boldsymbol{\sigma}_1^N = \boldsymbol{\sigma}_0$ if N is even and $\boldsymbol{\sigma}_1^N = \boldsymbol{\sigma}_1$ if N is odd, one can obtain

$$\mathsf{U} = \cos\left(\frac{\omega t}{2}\right)\boldsymbol{\sigma}_0 + i\sin\left(\frac{\omega t}{2}\right)\boldsymbol{\sigma}_1.$$

It is obvious that $\boldsymbol{\sigma}_1$ commutes with itself and with $\boldsymbol{\sigma}_0$. As a result, $\boldsymbol{\sigma}_1$ commutes with U and, hence, the observable, P_1, associated with $\boldsymbol{\sigma}_1$ should be a constant of the motion, as we have shown.

5.8 THE LINEAR MOMENTUM OPERATOR

We have seen that for every operator which commutes with the time operator there is associated a constant of the motion, and we have shown that the translation operator commutes with U. We must now find the constant associated with T. Rather than consider a general translation, let us examine an infinitesimal translation of our system to the right along the x-axis. What we mean by a small translation is that the translation operator can be written as the identity operator plus another very small term. Thus, we write

$$\mathsf{T} = \mathsf{I} - \frac{i\Delta x}{\hbar}\mathsf{P}_x.$$

The presence of the Δx is to ensure that we can make the second part as small as we wish. The $i/\hbar$ is included so that we will not have to alter the units of the operator P_x later.

If we translate a function an amount Δx to the right, then from Fig. 5.5 we see that the translated function at x must have the same value as the old

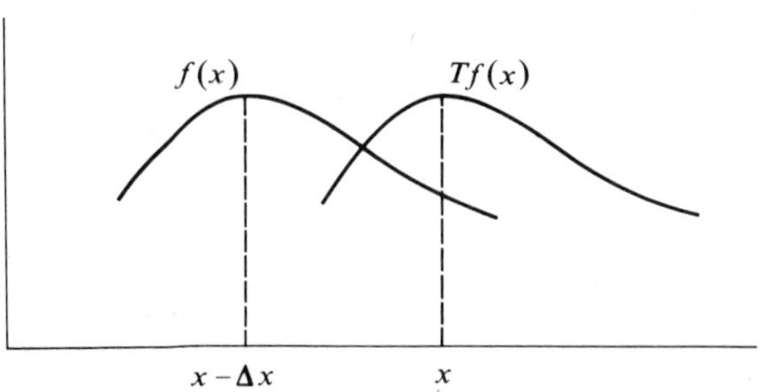

Fig. 5.5. Spatial translation of the function $f(x)$.

function did at $x - \Delta x$. If Δx is very small, we can write by the methods of elementary calculus

$$f(x - \Delta x) = f(x) - \Delta x \frac{\partial}{\partial x} f(x),$$

and we see that we must have

$$\mathsf{P}_x = -i\hbar \frac{\partial}{\partial x}.$$

If T commutes with U, it is easy to show that P_x must also, and we are going to find the constant associated with P_x rather than T. Whatever it may be, we know that it will be an eigenvalue of P_x and that the corresponding eigenstate of P_x will be that pure state which possesses that eigenvalue. To find these, we solve

$$-i\hbar \frac{\partial}{\partial x} \psi(x) = C\psi(x),$$

which has a solution

$$\psi(x) = Ae^{iCx/\hbar}.$$

We have seen in Chapter 2 that light traveling along the x-axis has a linear momentum of p_x per photon and has a periodicity along the x-axis of $\lambda = h/p_x$. But $e^{iCx/\hbar}$ has a periodicity of h/C. Thus, we identify the eigenvalues of P_x with the x-component of linear momentum, p_x, and P_x as a linear momentum operator. We now have

$$\psi(x) = Ae^{ip_x x/\hbar}.$$

In a similar fashion the operator for the other components of linear momentum are

$$\mathsf{P}_y = -i\hbar \frac{\partial}{\partial y}, \qquad \mathsf{P}_z = -i\hbar \frac{\partial}{\partial z}.$$

These may be combined to yield the vector linear momentum operator

$$\mathsf{P} = -i\hbar\, \boldsymbol{\nabla} = -i\hbar \left(\boldsymbol{e}_x \frac{\partial}{\partial x} + \boldsymbol{e}_y \frac{\partial}{\partial y} + \boldsymbol{e}_z \frac{\partial}{\partial z} \right),$$

where $\boldsymbol{\nabla}$ is the gradient operator and $\boldsymbol{e}_x$, $\boldsymbol{e}_y$, and $\boldsymbol{e}_z$ are unit vectors along the three axes. The eigenstate for the operator P may be seen to be

$$\psi(x, y, z) = A \exp \left[\frac{i}{\hbar} (p_x x + p_y y + p_z z) \right]$$

and represents a photon whose momentum is not along any particular axis but is given by

$$\boldsymbol{p} = p_x \boldsymbol{e}_x + p_y \boldsymbol{e}_y + p_z \boldsymbol{e}_z.$$

5.9 THE TIME DEPENDENCE

It is obvious that any operator commutes with itself and in particular that U commutes with U. We can therefore ask what constant of the motion is associated with the time operator? We could proceed by analogy with the previous section, but let us take a somewhat different approach. If we consider the development of a system over an infinitesimal time interval, we may write

$$|t + \Delta t\rangle = |t\rangle + \frac{\partial}{\partial t}|t\rangle \Delta t.$$

In terms of the time operator,

$$|t + \Delta t\rangle = \mathsf{U}(t + \Delta t, t)|t\rangle.$$

Writing

$$\mathsf{U}(t + \Delta t, t) = \mathsf{I} - \frac{i\Delta t}{\hbar}\mathsf{H},$$

we find

$$\mathsf{H} = i\hbar \frac{\partial}{\partial t}.$$

We then set up the eigenvalue equation for the time dependence of $|S\rangle$ and obtain

$$i\hbar \frac{\partial}{\partial t}\phi(t) = C\phi(t),$$

$$\phi(t) = Ae^{-iCt/\hbar}.$$

Now, any phenomenon which is periodic in space at a given instant of time must also be periodic in time at a given point if it travels past that point with some constant velocity, c. In elementary courses it is shown that the relation between the space periodicity, λ, and the time periodicity, τ, is given by $\lambda = \tau c$. Since we must have $\lambda = h/|\boldsymbol{p}|$,

$$\tau = h/|\boldsymbol{p}|c.$$

But $|\boldsymbol{p}|c = E$, the energy of the photon. We thus have

$$\phi(t) = Ae^{-iEt/\hbar}.$$

The eigenvalue of H is thus identified as the energy of the photon and H as the energy operator. The total space and time dependence of the photon state having linear momentum $\boldsymbol{p}$ and energy E can then be written

$$f(\boldsymbol{r}, t) = \psi(\boldsymbol{r})\phi(t) = A \exp\left(\frac{i}{\hbar}\right)(p_x x + p_y y + p_z z - Et).$$

It is often convenient to define the wave vector $\boldsymbol{k} = (1/\hbar)\boldsymbol{p}$ with magnitude $|\boldsymbol{k}| = 2\pi/\lambda$ and the symbol ω by $E = \hbar\omega$. Using these symbols, we may write

$$f(\boldsymbol{r}, t) = Ae^{i(\boldsymbol{k}\cdot\boldsymbol{r} - \omega t)}.$$

5.10 FUNCTIONS AS VECTORS

The functions which we have obtained as eigenstates of the momentum and energy operators differ from those states which we have previously considered in that the eigenvalues form a continuous range rather than take on only discrete values. Also, the operators which we have found are differential operators rather than matrices. We now want to see how all of this can be fitted into the formalism developed for the manipulation of bras and kets. What we are going to show is that in many respects the concept of a function and that of a vector are similar. We are going to find ways in which a function may be interpreted as a vector. The most obvious way to do this is by examining the concept of an integral of two functions (which may be complex) $f^\star(x)$ and $g(x)$.

Now, it is well known that the definite integral of the product of these two functions may be defined in terms of a sum in the following manner:

$$\int_a^b dx\, f^\star(x)g(x) = \lim_{\substack{\Delta x_i \to 0 \\ N\to\infty}} \sum_{i=1}^{N} f^\star(x_i)g(x_i)\Delta x_i.$$

Consider the scalar product (bracket product) of the vectors

$$\langle S_1|S_2\rangle = \sum_i A_i{}^\star B_i\langle e_i|e_i\rangle,$$

where

$$|S_1\rangle = \Sigma\, A_i|e_i\rangle, \qquad |S_2\rangle = \Sigma\, B_j|e_j\rangle.$$

By comparing this with the expression for the definite integral, we see the correspondence between the two expressions. We can interpret the integral of a product as a "scalar product" of two vectors. We can interpret a function as a vector which has an infinite number of components. Each "component" of the function is its value for a certain value of x, and instead of labeling the components by the names of coordinates we label them by the value of x at which the function takes on this value. This analogy between functions and vectors is carried over into the terminology of functions. For example, if

$$\int_a^b dx\, f^\star(x)g(x) = 0,$$

the two functions are said to be orthogonal in the interval (a, b).

We now see how to interpret a bracket product if the bra and ket are functions. If $|S_1\rangle = f(x)$ and $|S_2\rangle = g(x)$, then

$$\langle S_1|S_2\rangle = \int_a^b dx\, f^\star(x)g(x),$$

where the range of x from a to b is determined by the range over which the physical meaning of $|S_1\rangle$ and $|S_2\rangle$ is defined. We also see that when a ket vector is a function, the bra vector which is the adjoint of the ket is given by the complex conjugate of the ket function.

It is fairly obvious how to interpret something like $\langle S_1|\mathsf{P}_x|S_2\rangle$, where $\mathsf{P}_x = -i\hbar(\partial/\partial x)$. We have

$$\langle S_1|\mathsf{P}_x|S_2\rangle = \frac{\hbar}{i}\int_a^b dx\, f^\star(x)\frac{\partial}{\partial x}g(x).$$

There is only one difficulty left, and that is the meaning of an adjoint operator when the operator operates on a function. Let us recall what we mean by the adjoint $\mathsf{M}^\dagger$ of a matrix M operating on a ket $|S\rangle$. By definition $[|S\rangle]^\dagger = \langle S|$, and $\mathsf{M}^\dagger$ is defined by

$$\langle S|\mathsf{M}^\dagger = [\mathsf{M}|S\rangle]^\dagger.$$

It is not clear, however, what the meaning would be for a differential operator placed to the right of the function on which it is to operate. For vectors and matrices we can write

$$\langle S|\mathsf{M}^\dagger|S\rangle = [\mathsf{M}|S\rangle]^\dagger|S\rangle$$

and thus define the adjoint operator $P^\dagger$ for functions by

$$\int_a^b dx f^\star(x)\mathsf{P}^\dagger g(x) = \int_a^b dx[\mathsf{P}f(x)]^\star g(x).$$

If for a matrix operator $\mathsf{M}^\dagger = \mathsf{M}$, M is called a Hermitian operator. If it happens that

$$\int_a^b dx[\mathsf{P}f]^\star g = \int_a^b dx\, f^\star \mathsf{P}g,$$

then P is said to be Hermitian.

It is left to the exercises to show that the linear momentum and the time operators which we have found are Hermitian in this sense.

5.11 CONTINUOUS SUPERPOSITION OF STATES

We have seen that the function $Ae^{i(\boldsymbol{k}\cdot\boldsymbol{r}-\omega t)}$ is simultaneously an eigenket (or now an eigenfunction) of both the linear momentum operator and the energy operator, and is a pure state with respect to both energy and linear momentum measurements. The total ket of the state, of course, should also include polarization and be written

$$|S\rangle = Ae^{i(\boldsymbol{k}\cdot\boldsymbol{r}-\omega t)}|\alpha\rangle,$$

but we shall continue to ignore polarization for a while.

Any state may be written as a superposition of pure states, but until now we have dealt only with a finite discrete set of different pure states. Now, the set of pure states is infinite, depending on a continuous variable, $\boldsymbol{k}$. The extension should be apparent. We must go from the discrete sum

$$|S\rangle = \Sigma\, A_i|S_i\rangle$$

to an integration

$$|S\rangle = \int d\boldsymbol{k}\; A(\boldsymbol{k})|S(\boldsymbol{k})\rangle. \tag{5.30}$$

The symbol $\int d\boldsymbol{k}$ in Eq. (5.30) has been used as shorthand to indicate a triple integration over the three independent components of $\boldsymbol{k}$—k_x, k_y, and k_z:

$$\int d\boldsymbol{k} \rightarrow \int_{-\infty}^{+\infty}\int_{-\infty}^{+\infty}\int_{-\infty}^{+\infty} dk_x\, dk_y\, dk_z.$$

Let us examine this superposition in more detail in the case of one dimension, where $\boldsymbol{k} = k\boldsymbol{e}_x$. We have

$$|x, t\rangle = \int_{-\infty}^{+\infty} dk\; a(k)\, e^{i(kx-\omega t)}.$$

But $\omega = ck$, so

$$|x, t\rangle = \frac{1}{\sqrt{2\pi}}\int_{-\infty}^{\infty} A(k, t)\, e^{ikx}\, dk,$$

where

$$A(k, t) = \sqrt{2\pi}\, a(k)\, e^{-ikct}.$$

We should like to examine $|x, t\rangle$ mainly in respect to the variables x and k, so we could either neglect t or set it equal to some particular value. We could then write

$$|x\rangle = f(x) = \frac{1}{\sqrt{2\pi}}\int_{-\infty}^{\infty} A(k)\, e^{ikx}\, dk. \tag{5.31}$$

Whatever the function $A(k)$ may be, so long as the integral exists and is not infinite, we are able to make a new function of the variable x out of it.

We can pose the following mathematical problem. Is it possible to find two different functions $A(k)$ which yield the same function $f(x)$? The rather surprising answer is that we cannot, and the pairing of functions through this method is one-to-one. The truly amazing thing, however, is that the whole process can be turned inside out; and if we know $f(x)$, the function $A(k)$ can be found by writing

$$A(k) = \frac{1}{\sqrt{2\pi}} \int_{-\infty}^{\infty} f(x)\, e^{-ikx}\, dx. \tag{5.32}$$

A proof of this is given in Appendix IV.

Of course, for this to be true there must be certain restrictions on the functions $f(x)$ and $A(k)$, but these will almost always be met by the "nice" functions with which we will be dealing. The pair of functions $f(x)$ and $A(k)$ are said to be a Fourier transform pair. To write a function $f(x)$ as a Fourier transform is to decompose it in a sum (integral) of simple exponential functions of spatial period $\lambda = 2\pi/k$. The function $A(k)$ tells us the extent to which $f(x)$ is composed of the function e^{ikx} with k lying between k and $k + dk$.

As an example, we are going to work out one particular state which is not one of the pure states we have found but a Fourier superposition. Instead of depending only on the coordinate x, let us try to find a state which depends only on the distance from the origin r. We may imagine the point r to be located on the z-axis with no loss of generality. The simplest choice which we can make for the function $A(k)$ is that it be a constant. In addition, let us impose the restriction that for all states which we will add together the different k's differ only in direction and not in magnitude. Thus, the tips of all k-vectors will lie on the surface of a sphere of radius k about the origin. The proper region of integration in k-space is therefore the surface of this sphere. Thus,

$$|S\rangle = \int A e^{ikr\cos\theta}\, ds.$$

But, from Fig. 5.6,

$$ds = 2\pi k^2 \sin\theta\, d\theta\, d\phi.$$

The integration may be carried out to yield

$$\frac{2\pi Ak}{-ir}\, e^{ikr\cos\theta}\Bigg|_{\theta=0}^{\theta=\pi}.$$

If we reinsert the time dependence and then evaluate the limits, we obtain

$$|S\rangle = \frac{(2\pi)^{3/2} Ak}{ir} \cos(kr)\, e^{-i\omega t}.$$

Actually, this state function will be of very little interest to us. However,

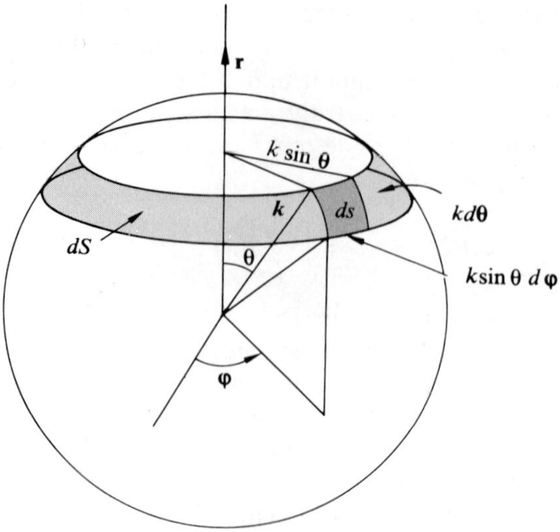

Fig. 5.6. The spherical area element $ds = k^2 \sin\theta\, d\theta\, d\phi$. The integration over ϕ leaves the element $dS = 2\pi k^2 \sin\theta\, d\theta$.

the cosine can be written as the difference of two exponentials, and part of the above function,

$$|S'\rangle = \frac{(2\pi)^{3/2} Ak}{ir} e^{i(kr-\omega t)}, \tag{5.33}$$

will be of great importance after we show in what way it can be considered a state function. First let us state that there is no *simple* way to obtain this function from a Fourier superposition, as in Eq. (5.30). However, we may identify Eq. (5.33) as an acceptable state function in another way. If one computes the expectation value for the component of linear momentum along any given direction using the state function of Eq. (5.33), one obtains zero. This is to be expected, since the state was obtained by adding with equal amplitudes states with all possible directions of $\boldsymbol{k}$. Yet all of these $\boldsymbol{k}$-vectors had the same magnitude. One suspects that this function might be an eigenstate, not of the linear momentum operator but of the operator for the magnitude of linear momentum, whose square would be

$$\begin{aligned} |\mathsf{P}|^2 &= \mathsf{P} \cdot \mathsf{P} = (-i\hbar\boldsymbol{\nabla}) \cdot (-i\hbar\boldsymbol{\nabla}) \\ &= -\hbar^2 \nabla^2. \end{aligned}$$

If one writes out the Laplacian

$$\nabla^2 = \frac{\partial^2}{\partial x^2} + \frac{\partial^2}{\partial y^2} + \frac{\partial^2}{\partial z^2},$$

it is simple to show that the state function of Eq. (5.33) satisfies

$$-\hbar^2\nabla^2|S'\rangle = \hbar^2 k^2|S'\rangle.$$

Thus, this is an acceptable state function which would be associated with a source radiating photons of fixed linear momentum $|\hbar\boldsymbol{k}|$ uniformly in all directions, i.e., a monochromatic point source.

5.12 SPATIAL LOCALIZATION

With the results of the preceding sections we have reached a point where we may explicitly write the state function as a function of space and time, and we can interpret the bracket product as an integral. We now examine once again the physical interpretation of the bracket product.

When we were concerned only with polarization states, we interpreted $\langle S|S\rangle = 1$ to mean that the probability that $|S\rangle$ would pass through any filter plus the probability that $|S\rangle$ would be absorbed was unity. Thus, the probability that $|S\rangle$ will be detected in one state or another, or simply detected, is unity. It would be natural, when speaking of position, to interpret

$$\langle \boldsymbol{r}, \alpha|\boldsymbol{r}, \alpha\rangle = \langle\alpha|\alpha\rangle \int d\boldsymbol{r}\, f^*(\boldsymbol{r}) f(\boldsymbol{r})$$

as the probability that a photon would be found somewhere and in some polarization state. Much as was done in Eq. (5.30), we use the symbol $d\boldsymbol{r}$ to indicate a volume element often written $dV = dx\,dy\,dz$. The integral represents

$$\int d\boldsymbol{r} \to \int_{-\infty}^{+\infty}\int_{-\infty}^{+\infty}\int_{-\infty}^{+\infty} dx\,dy\,dz.$$

With the above interpretation we would set

$$\langle \boldsymbol{r}, \alpha|\boldsymbol{r}, \alpha\rangle = 1.$$

This is not in general possible. For example, if the state in question is a pure state of energy and momentum, we have

$$\langle S|S\rangle = \langle\alpha|\alpha\rangle \int d\boldsymbol{r}\, e^{-i(\boldsymbol{k}\cdot\boldsymbol{r}-\omega t)}\, e^{+i(\boldsymbol{k}\cdot\boldsymbol{r}-\omega t)} = \langle\alpha|\alpha\rangle \int d\boldsymbol{r}.$$

If the state is defined over all space, then the integral is to be taken over all space and is infinite. It is thus impossible to interpret these particular state functions as functions of a single photon. They can at best be interpreted as functions for a beam of photons which fills all space uniformly and, hence, carries an infinite energy.

If $|S\rangle$ cannot be normalized to unity but represents a beam, $f^*f\,d\boldsymbol{r}$ will be related to the intensity of the beam in the volume element $d\boldsymbol{r}$. If the state can be normalized, $f^*f\,d\boldsymbol{r}$ will be taken as the probability of finding the photon within the element dr. Although we have indicated the reasonableness of this interpretation, there are difficulties which must be pointed out. First of all, finding a photon at some point in space means that the photon must interact with matter to make its presence known. However, the interaction between photons and matter is an electromagnetic one. It is determined, therefore, by the distribution of charges and currents throughout all of space. These distributions have so far been completely neglected. However, it is still reasonable that the probability that an interaction will take place should be large in those regions where $f^\star f\, dr$ is large.

We now examine the implications of making $f^\star f\, dr$ large in a given region of space and very small elsewhere, that is, localizing the photon. The one-dimensional case will illustrate the situation. We ignore polarization, and as we are still interested in the state function at some particular instant of time, we continue to write

$$f(x) = \frac{1}{\sqrt{2\pi}} \int_{-\infty}^{\infty} dk\, A(k) e^{ikx}.$$

As an example of a localized state function we consider the case illustrated in Fig. 5.7. The probability density $f^\star f$ has the constant value $1/w$ in the region $-w/2 \leqslant x \leqslant w/2$ and is zero elsewhere. It is thus normalized to $\langle S|S\rangle = 1$. By Eq. (5.32) we may solve for $A(k)$:

$$\begin{aligned} A(k) &= \frac{1}{\sqrt{2\pi}} \int_{-\infty}^{\infty} f(x) e^{-ikx}\, dx \\ &= \frac{1}{\sqrt{2\pi}} \int_{-w/2}^{w/2} \frac{1}{\sqrt{w}} e^{-ikx}\, dx \\ &= \frac{\sqrt{w}}{\sqrt{2\pi}} \frac{\sin wk/2}{wk/2}. \end{aligned}$$

Let Δk be a measure of the region over which the function $A(k)$ is large. A plot of this function is shown in Fig. 5.8. A reasonable measure of Δk would be the distance between the first zeros of the function or $\Delta k = 2(2\pi/w)$. Since the region over which the state function $f(x)$ is large is $\Delta x = w$, we have $\Delta x\, \Delta k = 4\pi$. The implication is that if we try to localize the state function by making Δx small, we must make Δk large. What is the physical significance of this?

In the Fourier superposition of states each pure state e^{ikx} is a pure state of linear momentum $p_x = k\hbar$. By superimposing these states we obtain a nonpure state whose linear momentum is not well defined but has a dispersion $\Delta p_x = \hbar\, \Delta k$. Thus, in the example given an attempt to localize the function $f(x)$, and, hence, the photon, to a region Δx leads to a dispersion in the linear momentum. Likewise, attempting to limit the linear momentum dispersion

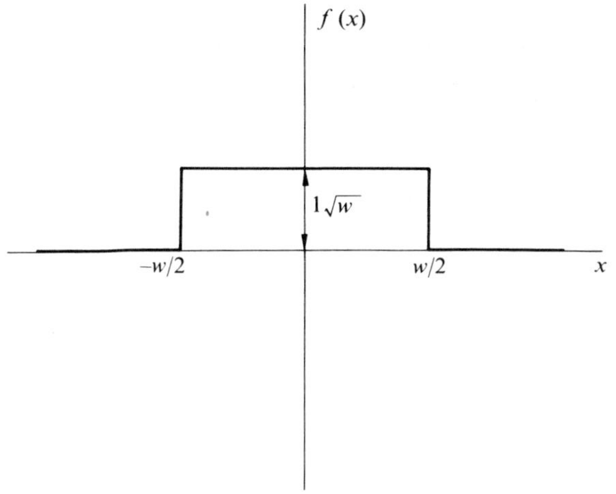

Fig. 5.7. A simple case of a localized state function.

must result in a large uncertainty Δx for the location of the photon. The pure state of linear momentum, e^{ikx}, is completely unlocalized and has an infinite dispersion Δx.

The result we have obtained is not due to the particular state function we have chosen to examine. We proceed to examine the relation between the dispersions in position and momentum measurements for any state. Let the x-dependence of the state function be written as any function $|S\rangle = f(x)$. The probability that this state will be detected between x and

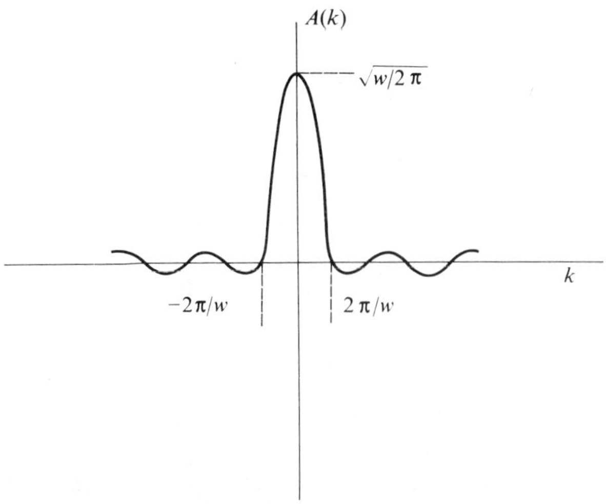

Fig. 5.8. The function $A(k)$ which is the Fourier transform of the function $f(x)$ shown in Fig. 5.7.

$x + dx$ is $f^\star(x)f(x)\,dx$. The expectation value for the x-location of the photon is obtained by weighting each value of x by the probability of measuring that value and is given by

$$\langle x\rangle = \int dx\, f^\star(x)\, x\, f(x).$$

We thus identify the variable x itself as the operator associated with the measurement of the x-coordinate at which the photon will be detected. In bracket notation the expectation value of x becomes

$$\langle x\rangle = \langle S|x|S\rangle,$$

where the bracket product must be interpreted as an integral over the coordinate x. If many similar states are prepared, the dispersion in the values of $\langle x\rangle$ which will be measured is as given in Section 5.4:

$$(\Delta x)^2 = \langle S|x^2 - \langle x\rangle^2|S\rangle. \tag{5.34}$$

Measurement of the linear momentum of the same state $|S\rangle$ will result in an expectation value for this momentum given by

$$\langle S|p_x|S\rangle = \langle p_x\rangle,$$

where $p_x = -i\hbar(\partial/\partial x)$. If many measurements are made, the dispersion in the values obtained will be

$$(\Delta p_x)^2 = \langle S|p_x^2 - \langle p_x\rangle^2|S\rangle. \tag{5.35}$$

In Section 5.14 it will be shown that these definitions of the dispersions Δx and Δp_x lead to Heisenberg's uncertainty relation:

$$\Delta x\, \Delta p_x \geqslant \tfrac{1}{2}\hbar. \tag{5.36}$$

The other coordinates and linear momenta satisfy similar relationships:

$$\Delta y\, \Delta p_y \geqslant \tfrac{1}{2}\hbar,$$
$$\Delta z\, \Delta p_z \geqslant \tfrac{1}{2}\hbar.$$

An example commonly used to illustrate these results is the diffraction of a light beam by a slit. As shown in Fig. 5.9, a wide beam is incident from the left on a slit of width d. The plane containing the slit is taken along the x-axis and perpendicular to the beam direction. The uncertainty in the x-coordinate of an incident photon may be taken as the width of the beam. If this is very large, we may make the uncertainty, Δp_x, of the x-component of the incident momentum very small without violating Eq. (5.36). We may therefore say that the x-component of momentum is essentially zero and that the beam is traveling parallel to the z-axis. Placing the slit in the beam localizes the beam to a region in x of the slit width. If this is made small, then by Eq. (5.36) the uncertainty in the x-component of linear momentum

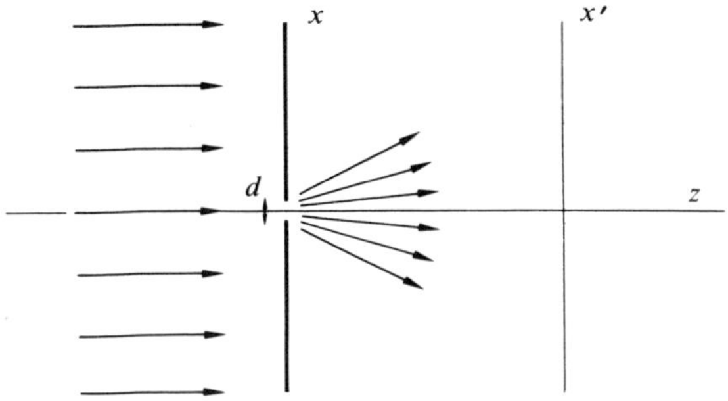

Fig. 5.9. Diffraction by a slit.

must now become large. On the right of the slit the beam should therefore spread out. This is indeed what happens. The calculation of this spreading is given in the following chapter and it is left to the problems of that chapter to show that the spreading is in agreement with Eq. (5.36).

Equation (5.36) tells us that we may not prepare states which simultaneously have arbitrary small dispersions in position and in linear momenta. It is sometimes stated that simultaneous measurements of position and momentum cannot result in arbitrarily small uncertainties or dispersions. The second form of the statement follows from the first. If the states cannot be prepared, the measurements cannot be made. However, the first form is often more closely related to experiment. In the slit experiment it is assumed that after leaving the slit the x-component of linear momentum is not further modified. It can therefore be measured by the position of absorption of the photon in a photographic emulsion in the x'-plane. This measurement is not simultaneous with the position determination which was made by the slit at an earlier time. The slit did take an incident state with small dispersion in x-momentum and attempt to prepare from it a state which simultaneously had small dispersion in the x-position. This cannot be done.

5.13 TEMPORAL LOCALIZATION

Let us again consider the full one-dimensional superposition

$$|S\rangle = \frac{1}{\sqrt{2\pi}} \int_{-\infty}^{\infty} dk\, A(k) e^{i(kx-\omega t)},$$

but examine this function at a particular point in space as a function of time. Without loss of generality we take this point to be the origin $x = 0$ and obtain

$$f(t) = \frac{1}{\sqrt{2\pi}} \int_{-\infty}^{\infty} d\omega\, \mathscr{A}(\omega) e^{-i\omega t}, \tag{5.37}$$

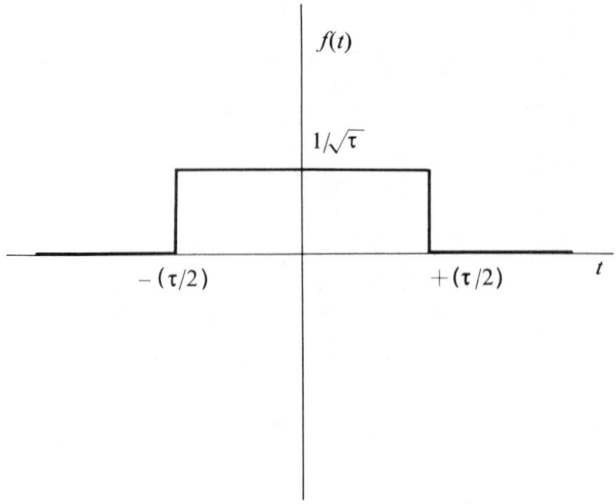

Fig. 5.10. State function for a photon localized within the time interval $\Delta t = \tau$.

where we have set

$$\mathscr{A}(\omega) = \frac{1}{c} A\left(\frac{\omega}{c}\right).$$

Now, the probability that the photon will be detected between time t and time $t + dt$ is given by $f^{\star}(t)f(t)\,dt$. We may attempt to localize a photon in time by using a state function, as shown in Fig. 5.10. The function takes on the value $1/\sqrt{\tau}$ for $-(\tau/2) \leqslant t \leqslant \tau/2$ and is zero at other times. The probability that the photon will be detected at the origin within the time interval $\Delta t = \tau$ around $t = 0$ is unity. As usual, we may use the inverse transform and obtain $\mathscr{A}(\omega)$:

$$\begin{aligned} \mathscr{A}(\omega) &= \frac{1}{\sqrt{2\pi}} \int_{-\infty}^{\infty} f(t)\, e^{i\omega t}\, dt \\ &= \frac{1}{\sqrt{2\pi}} \int_{-\tau/2}^{\tau/2} \frac{1}{\sqrt{\tau}}\, e^{i\omega t}\, dt \\ &= \sqrt{\frac{\tau}{2\pi}}\, \frac{\sin \omega\tau/2}{\omega\tau/2}. \end{aligned}$$

The function $\mathscr{A}(\omega)$ is shown in Fig. 5.11.

In order to obtain a state function localized in time, we have been forced to superimpose many pure states, $e^{-i\omega t}$. The resulting dispersion $\Delta\omega$ in ω may be taken as the interval between the first zeros of $\mathscr{A}(\omega)$ and is given by $\Delta\omega = 2(2\pi/\tau)$. We obtain for the product of the dispersions $\Delta t\, \Delta\omega = 4\pi$. Just as an uncertainty in k corresponds to an uncertainty in linear momentum,

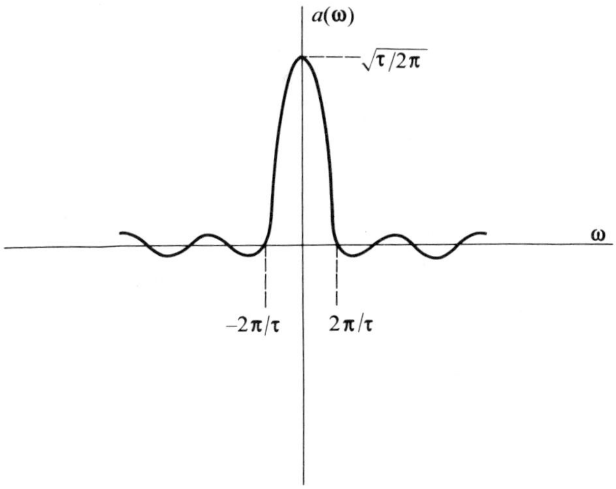

Fig. 5.11. The function $\mathscr{A}(\omega)$, the Fourier transform of $f(t)$, as shown in Fig. 5.10.

an uncertainty in ω corresponds to an uncertainty in energy $\Delta E = \hbar\,\Delta\omega$, so that $\Delta t\,\Delta E = 4\pi\hbar$. In the following section we shall show that, regardless of what time dependence is chosen for the state function, the dispersions must obey

$$\Delta t\,\Delta E \geqslant \tfrac{1}{2}\hbar. \tag{5.38}$$

In the general case one defines the dispersion Δt by

$$(\Delta t)^2 = \langle S|t^2 - \langle t\rangle^2|S\rangle,$$

where

$$\langle t\rangle = \langle S|t|S\rangle.$$

The dispersion ΔE is defined, using the energy operator $E = i\hbar(\partial/\partial t)$, by

$$(\Delta E)^2 = \langle S|-\hbar^2\frac{\partial^2}{\partial t^2} - \langle E\rangle^2|S\rangle,$$

with

$$\langle E\rangle = \langle S|i\hbar\frac{\partial}{\partial t}\,|\,S\rangle.$$

The interpretations of Eqs. (5.36) and (5.38) differ in some respects. Equation (5.36) applies to the preparation of states at a given moment in time. Equation (5.38) of its nature is concerned with states considered over an explicitly nonzero time interval. In fact, Eq. (5.38) directly relates the length of time over which an energy measurement is made to the dispersion of energies which will be measured. One can attempt to confine a light beam to a pulse by means of a shutter which is open for the time interval Δt. Following the shutter a detector may be placed which measures the energy of photons striking it. The dispersion ΔE which is measured and the open shutter time Δt obey Eq. (5.38). Note that Δt is *not* the time during which the detector registers the absorption of a single photon. This time can be very much smaller than the Δt allowed by Eq. (5.38).

Equation (5.38) then relates the temporal duration of a state to its energetic purity. In principle, a perfectly monoenergetic beam would have to exist for an infinite length of time. Because of the smallness of $\hbar$, however, a fairly monoenergetic beam is not an impossibility. For photon energies in the visible region a pulse duration of 1 microsecond (10^{-6} sec) is compatible with a relative energy dispersion, $\Delta E/E$, of one part in a billion (10^{-9}).

An instructive application of Eq. (5.38) is obtained by allowing the light source itself to act as the shutter which determines Δt. By this is meant that isolated sources are self-extinguishing and cease to emit when their internal energy is consumed. Consider a large group of individual radiators (atoms, molecules, etc.) which radiate photons of energy $\hbar\omega$, in decaying from an excited energy state, $E_{\text{ex.}}$, to the ground state, E_0. If these sources are independent radiators, the intensity of radiation $I(t)$ follows an exponential decay as shown in Fig. 5.12 and obeys

$$I(t) = I_0 e^{-t/\tau}.$$

The intensity decreases as the population of the higher energy state is depleted by radiative transmissions. The decay time, τ, at which the intensity reaches $1/e$ of its initial values may be taken both as a measure of the lifetime of the higher state and the length of time in which the radiation intensity is appreciable. Thus, the decay itself is equivalent to a shutter which is open for a time

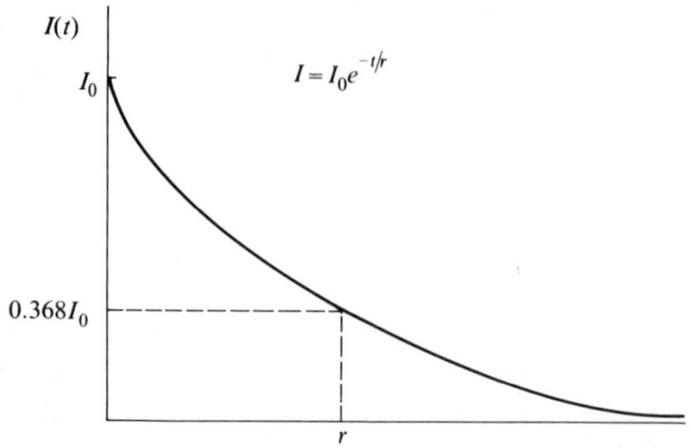

Fig. 5.12. The exponential decay of intensity from an isolated collection of independent sources.

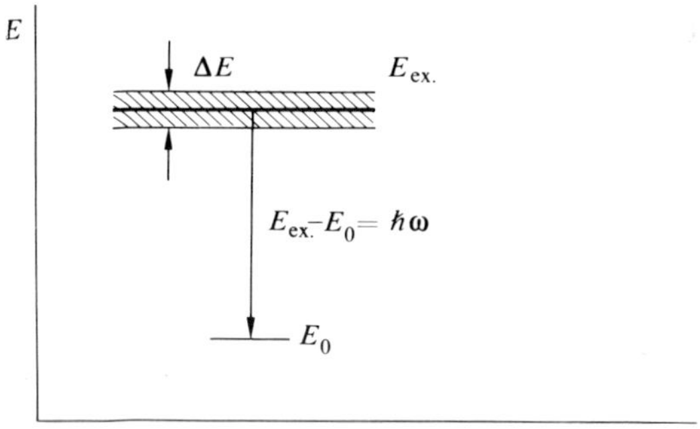

Fig. 5.13. A two-level energy system which decays by emission of a photon of energy $\hbar\omega$. If the lifetime of the excited state is τ, the energy of this state is uncertain by $\Delta E > \hbar/2\tau$.

interval of approximately τ. The dispersion in photon energies, ΔE, should obey Eq. (5.38) in the form $\tau\Delta E \geqslant \frac{1}{2}\hbar$. This relation is in agreement with much experimental evidence.

If there is a dispersion ΔE among the emitted photons, then there must be a dispersion of the energies $E_{ex.} - E_0 = \hbar\omega$ as shown in Fig. 5.13. Equation (5.38) can be applied to the radiation sources themselves as well as to the photons they emit. In fact (along with Eq. 5.34) it appears to apply to all physical systems. If one attempts to prepare many similar physical systems with the same energy, E, with lifetime Δt, the dispersion in energies, ΔE, and the lifetime Δt must obey Eq. (5.38).

5.14 COMMUTATION AND UNCERTAINTY

In this section we derive a general relation of which Eqs. (5.36) and (5.38) are special cases. We consider therefore any two Hermitian operators, M and N. These operators are associated with certain measurable attributes of a state. If many measurements are performed on similarly prepared states, $|S\rangle$, the average values obtained will be the expectation values

$$\langle \mathsf{M}\rangle = \langle S|\mathsf{M}|S\rangle,$$
$$\langle \mathsf{N}\rangle = \langle S|\mathsf{N}|S\rangle.$$

A measure of the dispersions of the individual values about the average values will be taken as being given by

$$(\Delta\mathsf{M})^2 = \langle S|\mathsf{M}^2 - \langle\mathsf{M}\rangle^2|S\rangle,$$
$$(\Delta\mathsf{N})^2 = \langle S|\mathsf{N}^2 - \langle\mathsf{N}\rangle^2|S.$$

We investigate the extent to which the product of the dispersions $\Delta\mathsf{M}\,\Delta\mathsf{N}$ may be minimized by the choice of the state $|S\rangle$ upon which the measurements are made.

Proceeding formally, we define the two kets

$$|\psi_1\rangle = (\mathsf{M} - \langle\mathsf{M}\rangle)|S\rangle,$$
$$|\psi_2\rangle = (\mathsf{N} - \langle\mathsf{N}\rangle)|S\rangle.$$

The product of the dispersions may then be written

$$\Delta\mathsf{M}\,\Delta\mathsf{N} = \sqrt{\langle\psi_1|\psi_1\rangle\langle\psi_2|\psi_2\rangle}.$$

Now, define the ket $|\psi\rangle$ by

$$|\psi\rangle = \frac{1}{\sqrt{\langle\psi_1|\psi_1\rangle}}\,|\psi_1\rangle \pm \frac{i}{\sqrt{\langle\psi_2|\psi_2\rangle}}\,|\psi_2\rangle,$$

which leads to

$$\langle\psi|\psi\rangle = 2 \pm \frac{i}{\Delta\mathsf{M}\,\Delta\mathsf{N}}\,[\langle\psi_1|\psi_2\rangle - \langle\psi_2|\psi_1\rangle].$$

The bracket product $\langle\psi|\psi\rangle$ is real and positive, so we must have

$$2 \geqslant \frac{1}{\Delta\mathsf{M}\,\Delta\mathsf{N}}\left|\langle\psi_1|\psi_2\rangle - \langle\psi_2|\psi_1\rangle\right|,$$

$$\Delta\mathsf{M}\,\Delta\mathsf{N} \geqslant \frac{1}{2}\left|\langle\psi_1|\psi_2\rangle - \langle\psi_2|\psi_1\rangle\right|.$$

From the definitions of $|\psi_1\rangle$ and $|\psi_2\rangle$ this becomes

$$\Delta\mathsf{M}\,\Delta\mathsf{N} \geqslant \tfrac{1}{2}|\langle S|\mathsf{MN} - \mathsf{NM}|S\rangle|, \tag{5.39}$$

where the right-hand member is identically zero if M and N commute. It is important to remember that Eq. (5.39) is an inequality and not an equality. For certain states $|S\rangle$ the product of the dispersions may be very large compared to the right-hand side of Eq. (5.39). However, it is impossible to prepare states for which the product of the dispersions will be smaller than the value given above. The application of Eq. (5.39) to the cases of Sections 5.13 and 5.14 is easy. If M represents the position operator x and N represents the linear momentum operator $p_x = -i\hbar(\partial/\partial x)$, we have

$$\begin{aligned}
&\langle S|x\left(-i\hbar\frac{\partial}{\partial x}\right) - \left(-i\hbar\frac{\partial}{\partial x}\right)x|S\rangle \\
&= \langle S|-i\hbar x\frac{\partial}{\partial x} + i\hbar + i\hbar x\frac{\partial}{\partial x}|S\rangle \\
&= \langle S|i\hbar|S\rangle \\
&= i\hbar.
\end{aligned}$$

Thus, Eq. (5.39) yields Eq. (5.36):

$$\Delta x \, \Delta p_x \geqslant \hbar/2.$$

The remaining uncertainty relations follow as easily and their derivation from Eq. (5.39) is left to the problems.

Though Eq. (5.39) may be applied to any two Hermitian operators, it should be apparent from the previous sections that the physical interpretation of the result must be very carefully examined in each case.

PROBLEMS

5.1 Prove that the eigenvalues of an anti-Hermitian operator are imaginary.

5.2 Prove that the linear momentum operator $p_x = -i\hbar(\partial/\partial x)$ is Hermitian. Assume that all state functions vanish at infinity.

5.3 Prove that a necessary and sufficient condition that two matrices M and N be identical is that $\langle S|\mathsf{M}|S'\rangle = \langle S|\mathsf{N}|S'\rangle$ for all $|S\rangle$ and $|S'\rangle$.

5.4 Show that two matrices are identical if they have a common set of eigenfunctions and eigenvalues. [*Hint:* Use the result of Problem 5.2 and expand $|S\rangle$ and $|S'\rangle$ in the common eigenfunctions.]

5.5 Show that the eigenvalues of any projection operator are 0, 1.

5.6 If two projection operators commute, prove that their product is a projection operator. Find a simple example to prove that the converse of this statement is not true.

5.7 Give a counter-example to show that a matrix with real eigenvalues need not be Hermitian.

5.8 Find a simple matrix S whose elements are functions of time for which $\mathsf{S}(d\mathsf{S}/dt) \neq (d\mathsf{S}/dt)\mathsf{S}$.

5.9 Consider the product of two exponential operators $e^{a\sigma_i}$, $e^{b\sigma_j}$. What special precaution must be taken if this product is to be written $e^{a\sigma_i + b\sigma_j}$?

5.10 Show that the expectation value of the x-component of linear momentum associated with the state $|S'\rangle$ given by Eq. (5.33) is zero.

5.11 Consider the matrix

$$\boldsymbol{\sigma}_{\theta\phi} = \begin{pmatrix} \cos\theta & e^{-i\phi}\sin\theta \\ e^{i\phi}\sin\theta & -\cos\theta \end{pmatrix},$$

where θ and ϕ are the usual angles defined for spherical coordinates (see Fig. 6.2). Is this matrix Hermitian? Is it unitary? Prove that the eigenvectors of this matrix are

$$|S_1\rangle = \begin{pmatrix} \cos\dfrac{\theta}{2} \\ e^{i\phi}\sin\dfrac{\theta}{2} \end{pmatrix}, \qquad |S_2\rangle = \begin{pmatrix} -\sin\dfrac{\theta}{2}\,e^{-i\phi} \\ \cos\dfrac{\theta}{2} \end{pmatrix},$$

with eigenvalues $+1$ and -1.

5.12 Compare the matrix $\boldsymbol{\sigma}_{\theta\phi}$ with the form of the matrix operator given in Problem 4.7. What are the Stokes parameters associated with $\boldsymbol{\sigma}_{\theta\phi}$? What are the components of a vector whose direction in real space are given by the angles θ and ϕ in spherical coordinates?

Answer: $0,\ 2\cos\theta,\ 2\cos\phi\cos\theta,\ 2\sin\phi\cos\theta$.

5.13 The matrix $\mathsf{S}_{\theta\phi}$ defined by $\mathsf{S}_{\theta\phi} = (\hbar/2)\boldsymbol{\sigma}_{\theta\phi}$ may be taken as being the operator associated with the component of spin angular momentum along the $\theta\phi$-direction of a nonrelativistic electron. If the electron is in state $|S_1\rangle$ of Problem 5.11, the expectation value of the angular momentum in the $\theta\phi$-direction is $+\hbar/2$ with probability $+1$. If the electron is in state $|S_2\rangle$, this projection is $-\hbar/2$.

a) Assume that the system (the electron) is in the state described by

$$|S\rangle = \begin{pmatrix}1\\0\end{pmatrix}.$$

What is the expectation value of the operator $\mathsf{S}^{\theta\phi}$? If $\mathsf{S}^{\theta\phi}$ is considered as a resolver, what is the probability of measuring eigenvalue $+\hbar/2$?

b) For a system in the state $|S\rangle$ of (a) for what particular values of θ and ϕ is the angular momentum certainly $+\hbar/2$? Why is it reasonable to associate the state $|S\rangle$ with an electron whose spin angular momentum is along the z-direction?

Answer: (*a*) $\hbar/2\cos\theta,\ \cos^2\theta/2$. (*b*) $\theta = 0$.

5.14 Redraw Fig. 5.2, adding new boxes as necessary so that the final output is still $\boldsymbol{S}\cdot\boldsymbol{E}$ when the output of the counters C and C' is the total number of counts rather than a time average, as supposed in Section 5.3.

5.15 A beam of light encounters two identical half-wave plates in succession. The first plate is fixed and the second plate rotates about the beam axis in such a way that the angle ϕ between the fast axes of the two plates is given by $\phi = \omega t$. If $|0\rangle$ is the state of the emergent beam at $t = 0$ and $|t\rangle$ the state at time t, find the time operator $\mathsf{U}(t, 0)$ such that $|t\rangle = \mathsf{U}(t, 0)\,|0\rangle$. [*Hint:* $|0\rangle$ is identical with the original state before the plates. Which of the sigma matrices commutes with $\mathsf{U}(t, 0)$? What is the constant of the motion associated with this sigma matrix. Interpret $\mathsf{U}(t, 0)$ in terms of the Poincaré sphere.]

Answer: $\begin{pmatrix}\cos 2\omega t & -\sin 2\omega t\\ \sin 2\omega t & \cos 2\omega t\end{pmatrix},\ \boldsymbol{\sigma}_3.$

5.16 An experiment is performed to measure the probability that a $\boldsymbol{P}$-state whose line is at angle θ to the x-axis will pass through a P_x-filter. If P_x is the matrix for the filter, the expectation value of the experiment is given by Malus' law, $\langle P|\mathsf{P}_x|P\rangle = \cos^2\theta$. Show that the dispersion $\Delta\mathsf{P}_x$ is given by $\Delta\mathsf{P}_x = \cos\theta\sin\theta$.

5.17 It is impossible to prepare a beam which is simultaneously in two different P-states, whose lines are not parallel. Taking the matrices M and N in Eq. (5.39) to be the matrices for two $\boldsymbol{P}$-filters, P and P', compute $\Delta\mathsf{P}\Delta\mathsf{P}'$, using Eq. (5.39). Show that if the correct states $|S\rangle$ are chosen, $\Delta\mathsf{P}\Delta\mathsf{P}'$ becomes zero. Compute the product of the dispersions, using the results of Problem 5.16, and obtain the same result. Why does this result not imply that a beam may simultaneously be in two different P-states? Consider these results in the light of the closing remarks of this chapter.

5.18 Use Eq. (5.39) to obtain Eq. (5.38).

CHAPTER 6

DIFFRACTION AND IMAGE FORMATION

Realizing that the state function will generally be a function of position, it is apparent that this function will also depend on the location and geometry of various objects which exist within the field. In this chapter we shall investigate the effect of such objects on the state function. We assume an incident state given within some region, described by a state function $|S\rangle_{\text{inc.}}$. We now modify the incident state by adding one or more objects to the region (or removing them from it). Of course, the state function will change. Call the new state the modified state, $|S\rangle_{\text{mod.}}$. The basic diffraction problem is the determination of $|S\rangle_{\text{mod.}}$ when $|S\rangle_{\text{inc.}}$ is known and when the nature of the added objects is specified. In general, this is an exceedingly difficult problem. In fact, the mathematical specification in any rigorous way of these added objects is often difficult if not impossible. We shall, therefore, examine cases of relative simplicity for which $|S\rangle_{\text{inc.}}$ and $|S\rangle_{\text{mod.}}$ are assumed to be scalar functions of position and for which polarization effects may be neglected.

The results obtained will be applied to the extremely practical problem of image formation by optical systems. We shall see that the formation of an image may be viewed as a double diffraction process. This viewpoint yields a powerful control on the quality and type of image formed for a given object.

6.1 GREEN'S THEOREM METHOD

In the abstract theory of vector functions, where no physical interpretation is implied, it is possible to prove certain theorems relating volume integrals to integrals over surfaces. One of these theorems, known as Green's theorem, is given below:

$$\int d\mathbf{r}\,(\psi\nabla^2\phi - \phi\nabla^2\psi) = \int(\psi\boldsymbol{\nabla}\phi - \phi\boldsymbol{\nabla}\psi)\cdot\mathbf{n}\,dA. \tag{6.1}$$

In this equation ϕ and ψ are arbitrary scalar functions of position, except that they must possess first and second partial derivatives with respect to all spatial coordinates so that the Laplacian and gradient operators are meaningful. They must also take on only such values as will leave the integrals finite. In short, the physicist would say that they must be "nice" functions.

Since this theorem holds for any two scalar functions, it certainly holds for any two scalar functions which happen to be photon state functions.

We now assume that the functions ϕ and ψ are state functions which are known to obey

$$\nabla^2 f(\boldsymbol{r}) = -k^2 f(\boldsymbol{r}), \tag{6.2}$$

or are monochromatic Fourier superpositions of the form

$$f(\boldsymbol{r}) = \frac{1}{(2\pi)^{3/2}} \int d\boldsymbol{k}\, A(\boldsymbol{k})\, e^{i\boldsymbol{k}\cdot\boldsymbol{r}}, \tag{6.3}$$

where the magnitude of $\boldsymbol{k}$ is constant independent of the direction of $\boldsymbol{k}$. It is left as an exercise to show that the function $f(\boldsymbol{r})$ as defined by Eq. (6.3) satisfies Eq. (6.2) when the magnitude of $\boldsymbol{k}$ is constant.

Making these assumptions, we easily see that the left-hand side of Eq. (6.1) vanishes and we obtain

$$\int(\psi\nabla\phi - \phi\nabla\psi)\cdot\boldsymbol{n}\,dA = 0. \tag{6.4}$$

Before proceeding, it might be well to indicate where we are going. By restricting the state functions we shall consider, to those which obey Eq. (6.2), we have greatly narrowed this class of function. To see how much we have limited this class we examine the following question. Assume that a function is defined within a finite volume and that we are given the value of this function and its gradient over the surface which bounds the volume. What do we know about the function inside the boundary? In general, nothing! But, if we also demand that the function obey Eq. (6.2), we shall see that we know everything about it inside the boundary. The importance of this result for the diffraction problem is that we may replace the search for $|S\rangle_{\text{mod.}}$ over a volume of space by a search for this function over a surface bounding that volume.

Continuing, we consider a point in space surrounded by some surface. We divide this surface into two parts. The first is a small sphere surrounding the point at which we wish to compute the state function. If this sphere is small enough, the average of the function over this surface will equal the value of the function at the center of the sphere. The geometry is shown in Fig. 6.1. Here A and A' make up together a surface bounding the volume. A' is the spherical surface surrounding the point of interest located at $\boldsymbol{r}_0$. An arbitrary point within the volume or on either A or A' is located by $\boldsymbol{r}_1$

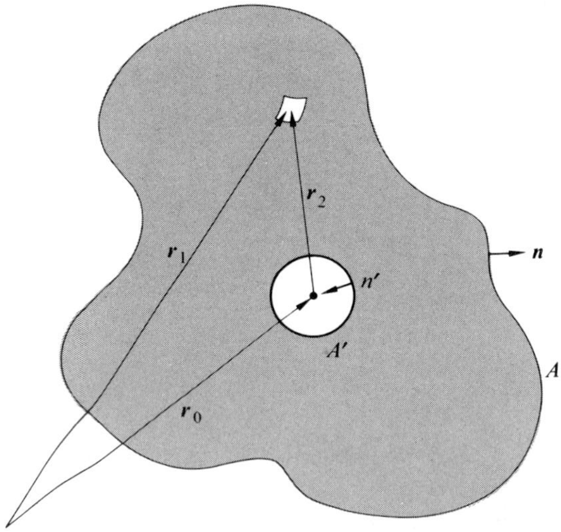

Fig. 6.1. The geometry and notation used in the derivation of Eq. (6.5).

with $\boldsymbol{r}_1 = \boldsymbol{r}_0 + \boldsymbol{r}_2$. Choose the function ψ in Green's theorem to be the state function which exists in this region and choose the function ϕ to be

$$\phi = \frac{1}{r}\, e^{ikr},$$

where $r = |\boldsymbol{r}_2|$. Splitting the integral over the total surface into two integrals, one over A and one over A', we have

$$\int_A \left[\psi\boldsymbol{\nabla}\left(\frac{1}{r}\, e^{ikr}\right) - \frac{1}{r}\, e^{ikr}\, \boldsymbol{\nabla}\psi\right] \cdot \boldsymbol{n}\, dA + \int_{A'} \left[\psi\, \boldsymbol{\nabla}\left(\frac{1}{r}\, e^{ikr}\right) - \frac{1}{r}\, e^{ikr}\, \boldsymbol{\nabla}\psi\right] \cdot \boldsymbol{n}'\, dA' = 0.$$

Since the second integral is over a small sphere, we write the gradient operator in spherical coordinates:

$$\boldsymbol{\nabla} = \boldsymbol{e}_r \frac{\partial}{\partial r} + \boldsymbol{e}_\theta \frac{1}{r}\frac{\partial}{\partial \theta} + \boldsymbol{e}_\phi \frac{1}{r \sin\theta}\frac{\partial}{\partial \phi}.$$

The coordinates and the unit vectors are defined in standard fashion, as shown in Fig. 6.2. Taking the dot product with the inward-directed normal $\boldsymbol{n}'$, we replace both operations by $-\partial/\partial r$. Also, assuming that the sphere A' can be made vanishingly small and that ψ does not vary appreciably over its surface, we replace ψ by its value at the center, ψ_0. We also replace

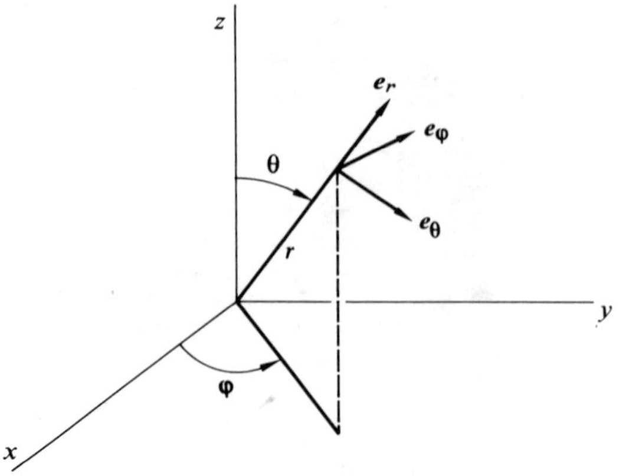

Fig. 6.2. Spherical coordinates.

$\partial\psi/\partial r$ by its maximum value over the surface, certainly without decreasing its contribution to the integral. The second integral then becomes

$$\int \left[\psi_0/r^2 - ik\psi_0/r + \frac{1}{r}\left(\frac{\partial\psi}{\delta r}\right)_{\text{max.}}\right] e^{ikr}\, dA'.$$

In the limit as kr approaches zero, e^{ikr} may be replaced by unity. Since r is constant over the sphere, the entire integrand is constant. The area of the sphere being $4\pi r^2$, we obtain for the integral

$$4\pi\psi_0 - ik4\pi r\psi_0 + \left(\frac{\partial\psi}{\partial r}\right)_{\text{max.}} 4\pi r.$$

As r approaches zero, only the first term remains. Finally,

$$\psi_0 = -\frac{1}{4\pi}\int_A \left[\psi\nabla\left(\frac{1}{r}\, e^{ikr}\right) - \frac{1}{r}\, e^{ikr}\, \nabla\psi\right] \cdot \boldsymbol{n}\, dA. \tag{6.5}$$

Thus, if ψ and $\nabla\psi$ are known over the surface which bounds the volume V, we may compute ψ_0 at any point in V.

6.2 APERTURES IN PLANE SCREENS

In many cases of practical interest the propagation of light is limited by the presence of a plane screen whose transmission and absorption properties are functions of position on the screen. In this section Eq. (6.5) will be adapted to this case, beginning with the situation in which various portions of the screen either totally transmit or totally absorb. We are thus concerned with an opaque screen with one or more apertures cut into it.

In order to make the problem more specific, assume that the incident state is the spherically symmetric state $|S\rangle_{\text{inc.}} = A/r_1\, e^{ikr}$ arising from a

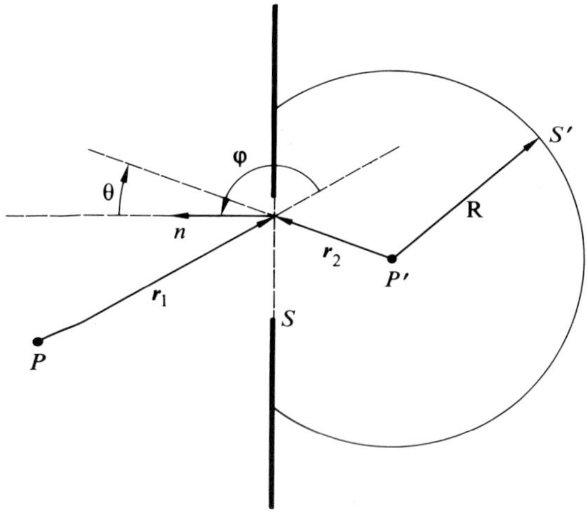

Fig. 6.3. Geometry and notation used for the study of scalar diffraction by apertures in plane screens.

point source, P, located to the left of the screen. In Fig. 6.3 the surface of integration, A, is broken into two parts, S and S'. S' is that portion of a spherical surface of radius R, centered on the observation point P', which lies to the right of the screen. The portion S is that portion of the screen intersected by S', plus the aperture. In order to use Eq. (6.5), we now require a knowledge of the state function over this surface. Since we do not know the effect of the screen on the incident function, we do not possess this knowledge. We will assume that the total function over the aperture is simply the incident one and that any differences near the edge of the aperture or elsewhere can be neglected. We thus expect that the results derived may not be valid near the aperture, and this is so. In fact, near the aperture the equations to be derived do not yield the assumed function upon which they are based. However, those values obtained for points far from the aperture prove to be in very good agreement with experiment.

In addition to knowing ψ over S, we must also know it over S'. There are two methods of handling this part of the problem. Both allow the radius of the sphere, R, to become infinite. The first technique assumes that the disturbance at P' due to the field on a section of S' decreases as $1/R^2$. However, since the surface area of S' increases as R^2, one must take some care in showing that these two competing influences do lead to a vanishingly small contribution from S'. For a proof of this the student is referred to the bibliography.*

---

* Stone, John M., *Radiation and Optics*, McGraw-Hill (1963).

The second method is more physical in flavor. In any real situation the light has not been incident for an infinite length of time. Hence, it has been able to propagate to the right of the screen only a finite distance. We may simply place the surface S' past this distance and the disturbance will be identically zero over S'. This method has difficulties also. Any wave which has not existed for all time cannot be written as a simple monochromatic Fourier superposition. It must be composed of a set of simple monochromatic functions with a finite range of k-values. However, if the incident wave does start at some instant in time, we may assume that after a sufficiently long time any transient behavior will disappear and the function at P' will approach the monochromatic case. Since we can place S' as far away as we wish, we can make this approximation as good as we wish. Thus, over the aperture we assume $\psi = A/r_1 e^{ikr_1}$ and over the remainder of both surfaces we assume $\psi = 0$. Equation (6.5) thus becomes

$$\psi_0 = -\frac{1}{4\pi}\int_{\text{aperture}} \left[A\,\frac{e^{ikr_1}}{r_1}\,\nabla\left(\frac{e^{ikr_2}}{r_2}\right) - \frac{1}{r_2}\,e^{ikr_2}\,\nabla\,\frac{A}{r_1}e^{ikr_1}\right]\cdot \boldsymbol{n}\,dS.$$

Since the exponential functions are spherically symmetric about P and P', we may write

$$\nabla\,\frac{e^{ikr_1}}{r_1} = \boldsymbol{e}_{r_1}\,\frac{\partial}{\partial_{r_1}}\left(\frac{e^{ikr_1}}{r_1}\right) = \boldsymbol{e}_{r_1}\left(ik - \frac{1}{r_1}\right)\frac{e^{ikr_1}}{r_1},$$

$$\nabla\,\frac{e^{ikr_2}}{r_2} = \boldsymbol{e}_{r_2}\,\frac{\partial}{\partial r_2}\left(\frac{e^{ikr_2}}{r_2}\right) = \boldsymbol{e}_{r_2}\left(ik - \frac{1}{r_2}\right)\frac{e^{ikr_2}}{r_2}.$$

From Fig. 6.3 it is seen that

$$\boldsymbol{e}_{r_2}\cdot\boldsymbol{n} = \cos\theta, \qquad \boldsymbol{e}_{r_1}\cdot\boldsymbol{n} = \cos\phi.$$

We may then write

$$\psi_0 = -\frac{1}{4\pi}\int dS\,\frac{Ae^{ik(r_1+r_2)}}{r_1 r_2}\left\{\left(ik - \frac{1}{r_2}\right)\cos\theta - \left(ik - \frac{1}{r_1}\right)\cos\phi\right\}.$$

If we are concerned with visible light, $k \simeq 10^5\ \text{cm}^{-1}$, and $1/r$ will usually not be larger than $10^{-1}\ \text{cm}^{-1}$. Thus, we ignore the $1/r_1$- and $1/r_2$-terms and obtain the Kirchoff-Fresnel formula for the diffraction of a monochromatic scalar wave:

$$\psi_0 = -\frac{ik}{4\pi}\int dS\,\frac{Ae^{ik(r_1+r_2)}}{r_1 r_2}\,[\cos\theta - \cos\phi].$$

The term involving the cosines of θ and ϕ is known as the obliquity factor.

Even though we have made a number of approximations in obtaining this result, the integrals which arise are usually of great difficulty. Although it will by no means make the integrals trivial, there is one final approximation which we make. Usually, P and P' are far enough from the screen and close enough to the axis, passing through the aperture, normal to the screen, to enable us to approximate:

$$\cos\phi \simeq -1, \qquad \cos\theta \simeq 1.$$

Finally, we obtain

$$\psi_0 = -\frac{ik}{2\pi}\int dS\,\frac{Ae^{ik(r_1+r_2)}}{r_1 r_2}. \tag{6.6}$$

6.3 FRESNEL DIFFRACTION

In the following sections we will consider the case of Fraunhofer diffraction in which both P and P' will essentially be placed an infinite distance from the screen. In this section we will apply Eq. (6.6), as it stands, with both the source point and the observation point at some finite distances from the screen. However, we are assuming that these distances are large enough to make the obliquity factor constant and equal to 2. In this approximation the possible variation in r_1 and r_2 is so small that their product in the denominator may be considered constant and taken through the integral sign. Since k may be extremely large ($10^5\ \text{cm}^{-1}$), even a small variation in r_1 or r_2 can produce a significant change in the exponent $k(r_1 + r_2)$. Thus, the exponential cannot be considered constant.

In the spirit of a rather long example, we will consider now a rectangular aperture. We will take this aperture to lie in the $\xi\eta$-plane and the source point, P, to be on the z-axis. The observation point will have coordinates x and y. The geometry is as shown in Fig. 6.4. Writing out r_1 and r_2 we have

$$\begin{aligned} r_1 &= [r_{10}^2 + \xi^2 + \eta^2]^{1/2}, \\ r_2 &= [r_{20}^2 + (\xi - x)^2 + (\eta - y)^2]^{1/2}. \end{aligned}$$

Assuming that the aperture is small compared to r_{10} and r_{20}, these expressions may be approximated, using $\sqrt{1+\alpha} \simeq 1 + \alpha/2$, which is valid for $\alpha \ll 1$:

$$\begin{aligned} r_1 &\simeq r_{10} + (\xi^2 + \eta^2)/2r_{10}, \\ r_2 &\simeq r_{20} + [(\xi - x)^2 + (\eta - y)^2]/2r_{20}. \end{aligned}$$

Thus, the sum in the exponent becomes

$$\begin{aligned} r_1 + r_2 = r_{10} + r_{20} &+ \frac{1}{2}\frac{(x^2 + y^2)}{r_{20}} + \xi^2\frac{1}{2}\left(\frac{1}{r_{10}} + \frac{1}{r_{20}}\right) \\ &+ \eta^2\frac{1}{2}\left(\frac{1}{r_{10}} + \frac{1}{r_{20}}\right) - \frac{\xi x}{r_{20}} - \frac{\eta y}{r_{20}}. \end{aligned} \tag{6.7}$$

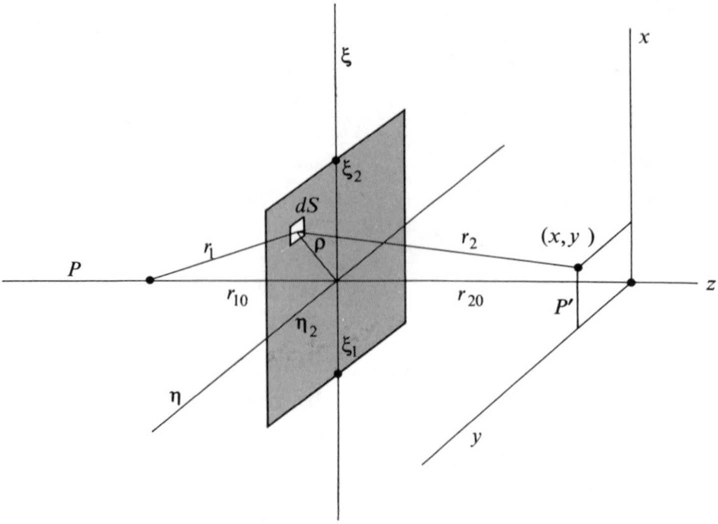

Fig. 6.4. The special case of a rectangular aperture in a plane screen.

If we complete the square for both quadratic terms in ξ and η, we obtain

$$r_1 + r_2 = r_{10} + r_{20} + \frac{(x^2 + y^2)}{2(r_{20} + r_{10})} + \frac{r_{10} + r_{20}}{2r_{10}r_{20}} \left[\left(\xi - \frac{r_{10}x}{r_{10} + r_{20}} \right)^2 + \left(\eta - \frac{r_{10}y}{r_{10} + r_{20}} \right)^2 \right]. \tag{6.8}$$

Equation (6.6) may now be written

$$\psi_0 = -\frac{ik}{2\pi} \frac{A'}{r_{10}r_{20}} \int_{\xi_1}^{\xi_2} d\xi \exp \left[\frac{ik(r_{10} + r_{20})}{2r_{10}r_{20}} \left(\xi - \frac{r_{10}x}{r_{10} + r_{20}} \right)^2 \right] \times \int_{\eta_1}^{\eta_2} d\eta \exp \left[\frac{ik(r_{10} + r_{20})}{2r_{10}r_{20}} \left(\eta - \frac{r_{10}y}{r_{10} + r_{20}} \right)^2 \right],$$

with

$$A' = A \exp \{ik[r_{10} + r_{20} + (x^2 + y^2)/2(r_{20} + r_{10})]\}.$$

If we change the variables of integration to

$$\xi' = \xi - \frac{r_{10}x}{r_{10} + r_{20}}, \qquad \eta' = \eta - \frac{r_{10}y}{r_{10} + r_{20}},$$

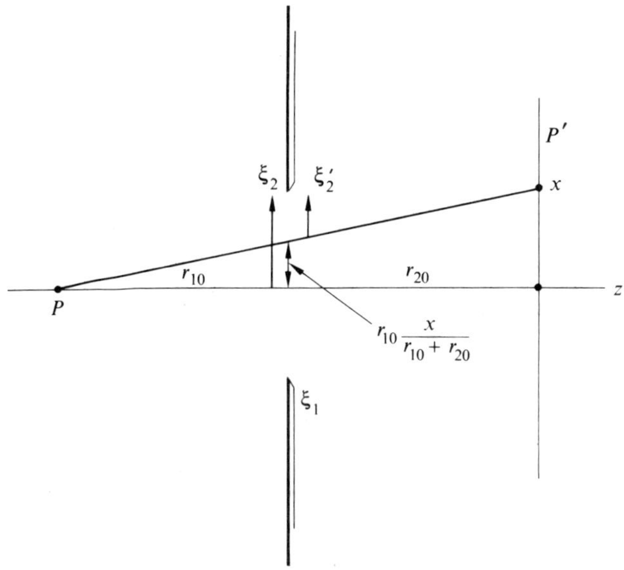

Fig. 6.5. The transformation from the variable ξ to the variable ξ' is equivalent to a transformation which sets the x-coordinate of point P' equal to zero.

then the integrals have the same form as would be obtained by setting x and y equal to zero but with the limits of integration changed to

$$\xi'_{1,2} = \xi_{1,2} - \frac{r_{10}x}{r_{10} + r_{20}},$$

$$\eta'_{1,2} = \eta_{1,2} - \frac{r_{10}y}{r_{10} + r_{20}}.$$

From Fig. 6.5 it is evident that this transformation is equivalent to rotating the z-axis to join P and P', ignoring the fact that the aperture plane is then not perfectly normal to the z-axis. An alternative point of view is that the diffraction pattern as a whole shifts in the manner that one would expect from simple geometric optics, when the source or aperture is shifted. Since this is the case, we may confine our attention to the case where x and y are zero. Finally, to get the integrals into standard form we make the substitutions

$$\alpha = \xi\sqrt{2(r_{10} + r_{20})/\lambda r_{10}r_{20}},$$
$$\beta = \eta\sqrt{2(r_{10} + r_{20})/\lambda r_{10}r_{20}}, \tag{6.9}$$

and obtain

$$\psi_0 = -\frac{iA'}{2(r_{10} + r_{20})}\int_{\alpha_1}^{\alpha_2} e^{i\pi\frac{\alpha^2}{2}}\,d\alpha \int_{\beta_1}^{\beta_2} e^{i\pi\frac{\beta^2}{2}}\,d\beta. \tag{6.10}$$

6.4 THE CORNU SPIRAL

The integrals in Eq. (6.10) cannot be evaluated in terms of elementary functions but must be computed by numerical methods. Since the imaginary exponential can be written in terms of the sine and cosine, this evaluation is usually done in terms of the Fresnel integrals,

$$
\begin{aligned}
C(u) &= \int_0^u \cos\frac{\pi t^2}{2}\,dt, \\
S(u) &= \int_0^u \sin\frac{\pi t^2}{2}\,dt.
\end{aligned}
\tag{6.11}
$$

In terms of these integrals Eq. (6.10) becomes

$$
\begin{aligned}
\psi_0 = &-\frac{i}{2}\,\psi_{00}\,[C(\alpha_2) - C(\alpha_1) + iS(\alpha_2) - iS(\alpha_1)] \\
&\times [C(\beta_2) - C(\beta_1) + iS(\beta_1) - iS(\beta_1)],
\end{aligned}
\tag{6.12}
$$

where

$$\psi_{00} = \frac{A'}{r_{10} + r_{20}}$$

is the value the state function would have at P' if the screen were completely removed, ignoring a phase difference between A and A'. For a numerical tabulation of the Fresnel integrals the reader is referred to the literature.* It is possible to present the contents of such a tabulation in graphical form, as is done in Fig. 6.6, which represents a complex space in which the real axis takes on the values of $C(u)$ and the imaginary axis those of $S(u)$. The resulting curve, which is known as Cornu's spiral, is obtained by considering u as a parameter. For any value of u one can immediately read off the values of $C(u)$ and $S(u)$. We now wish to show how one may use the Cornu spiral to obtain a diffraction pattern by graphical means. If we think of

$$\int_0^\alpha e^{i\pi t^2/2}\,dt$$

as a summation of infinitesimal complex vectors $e^{i\pi t^2/2}\,dt$, then the complex number which is the value of this definite integral is simply the vector drawn from the origin to the point on the spiral labeled by α. Likewise, since

$$\int_{\alpha_1}^{\alpha_2} = \int_0^{\alpha_2} - \int_0^{\alpha_1},$$

the integral from α_1 to α_2 is the complex vector drawn from the point labeled by α_1 to that labeled by α_2. This is illustrated in Fig. 6.7. In addition, the

---

* Jahnke, E., and J. Emde, *Tables of Functions*, Dover Publications, New York, (1945).

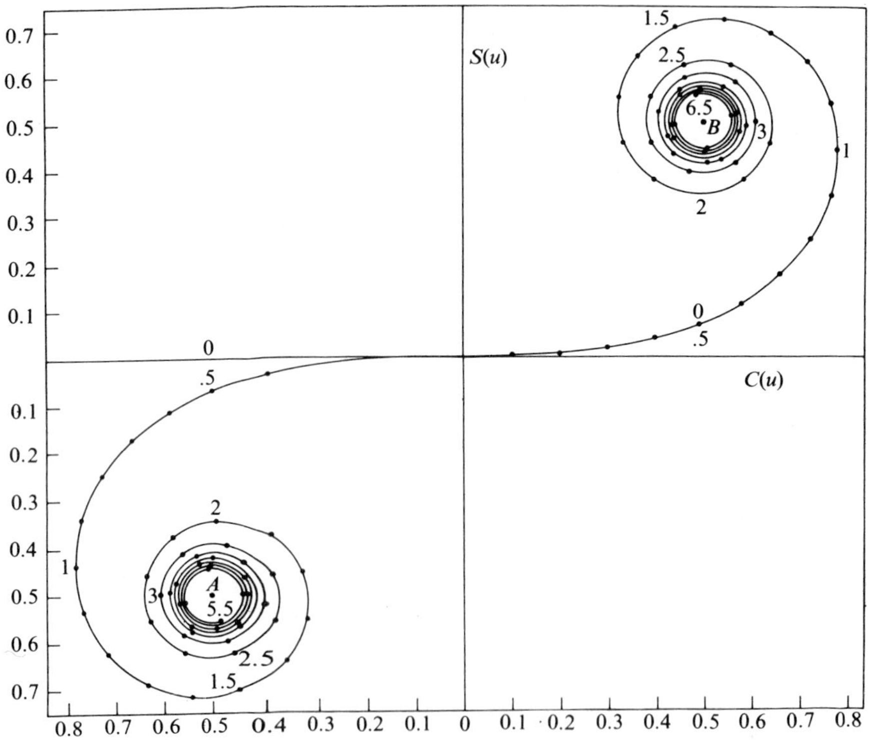

Fig. 6.6. The Cornu spiral.

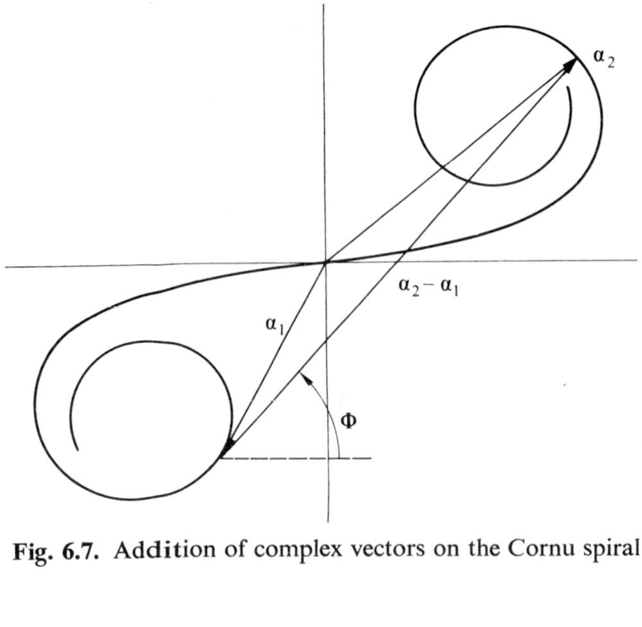

Fig. 6.7. Addition of complex vectors on the Cornu spiral.

length of the complex vector $e^{i\pi t^2/2}\,dt$ is simply dt. Thus, the path length along the spiral from the origin to the point labeled by α is just α. Since these vectors are a representation for complex numbers, it is important to remember that the length of the vector gives the magnitude of the complex number and that its phase angle is given by the angle between the vector and the real axis, which is indicated by ϕ in Fig. 6.7. All of this holds equally well for the evaluation of the integral in β. After the appropriate values of α and β have been obtained from Eq. (6.9), it is only necessary to obtain the related two complex vectors and multiply them together. The product of two such vectors is, of course, a vector whose length is the product of the lengths and whose phase angle is the sum of the phase angles.

In the following sections we are going to examine several special cases which will appear to violate the assumptions under which Eq. (6.10) was derived. Before proceeding we wish to discuss the extent to which Eq. (6.10) is really limited by these assumptions. In purely geometric optics there is no diffraction. Under the assumption of rectilinear propagation any aperture will cast a shadow whose boundary marks a discontinuous jump in intensity. Scalar diffraction theory replaces this perfectly defined shadow with a continuous transition in intensity and an imperfectly defined shadow. Yet from everyday experience one knows that the geometric theory cannot be all bad. Shadows and boundaries do exist. We are led to expect, therefore, that the intensity in a diffraction pattern is determined mainly by some small area in the diffracting screen lying near a line joining the source point and the observation point, and is little influenced by those parts of the screen which are far from this line. In the Cornu spiral it is seen that as α and β become large the spiral closes in to the value $\frac{1}{2}$ for both $C(u)$ and $S(u)$, at point B, or to the value $-\frac{1}{2}$ at point A. Once α and β are large, making them larger has little effect. The meaning of this is that if large values of α and β can be obtained from Eq. (6.9) without violating the assumptions which have been made, making them larger will violate the assumptions and introduce errors, but errors which can be small enough to be ignored. For example, assume a small square aperture, with P and P' both 1 meter away from the screen, such that the aperture subtends 1° from P'. We find for visible light, $\lambda = 5 \times 10^{-5}$ cm, that α and β are approximately 50. Thus, as we now enlarge the aperture to a size which would violate the assumption of a constant obliquity factor, we will not introduce a large error. In fact, in certain cases we introduce no error at all. If we remove the screen, then the edges of the aperture are at infinity and $\alpha_2 = \beta_2 = \infty$, $\alpha_1 = \beta_1 = -\infty$. Thus, in Eq. (6.12) the two complex vectors are both $(1 + i)$, giving

$$\psi_0 = \psi_{00},$$

which is, of course, the exact result.

6.5 THE STRAIGHT EDGE AND SLIT

The two remaining cases which we will consider are one-dimensional. That is, we will allow the α dimensions of the aperture to become infinite. In this case the complex vector associated with α is $(1 + i)$ and Eq. (6.12) becomes

$$\psi_0 = \frac{1 - i}{2} \psi_{00}[C(\beta_2) - C(\beta_1) + iS(\beta_2) - iS(\beta_1)]. \tag{6.13}$$

We will return to this expression which yields the diffraction pattern of an infinite slit shortly. Now, however, we will also remove one edge of the slit to infinity, leaving a single straight edge. If we remove the edge at η_1, we obtain

$$\psi_0 = \frac{1 - i}{2} \psi_{00}[\tfrac{1}{2}(1 + i) + C(\beta_2) + iS(\beta_2)].$$

If the observation point is very far to the right of the remaining edge, then, as in the preceding example, we have a vector from A to B and we obtain the full incident intensity. As we move the observation point to the left, or bring the edge at η_2 closer to the axis, the tip of the vector moves around $\boldsymbol{B}$, oscillating in length until it finally passes through the origin of the spiral. At this point β_2 is zero and we are at the location of the geometric shadow. Continuing into the shadow, the oscillation ceases and the magnitude of the vector decreases monotonically. The measured intensity, which is $|\psi_0|^2$, is given in Fig. 6.8, with $\psi_{00} = 1$.

Returning now to Eq. (6.13), we may consider the pattern due to two straight edges forming a single slit. As we move the observation point across the pattern, we are changing η_1 and η_2, but the difference $\eta_2 - \eta_1$ must, of course, be constant. Thus, on the Cornu spiral the length of the vector will vary but the distance along the spiral will remain constant. To obtain the slit pattern, we first compute $\beta = \beta_2 - \beta_1$ and slide an arc of

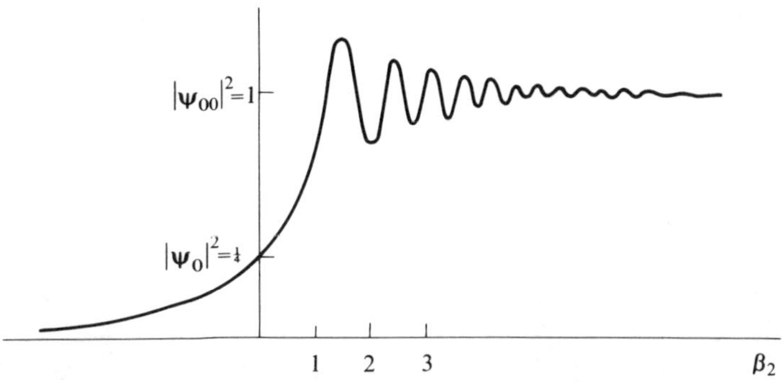

Fig. 6.8. The diffraction pattern of a straight edge. (After Towne, *Wave Phenomena*, Addison-Wesley, Reading, Mass., 1967.)

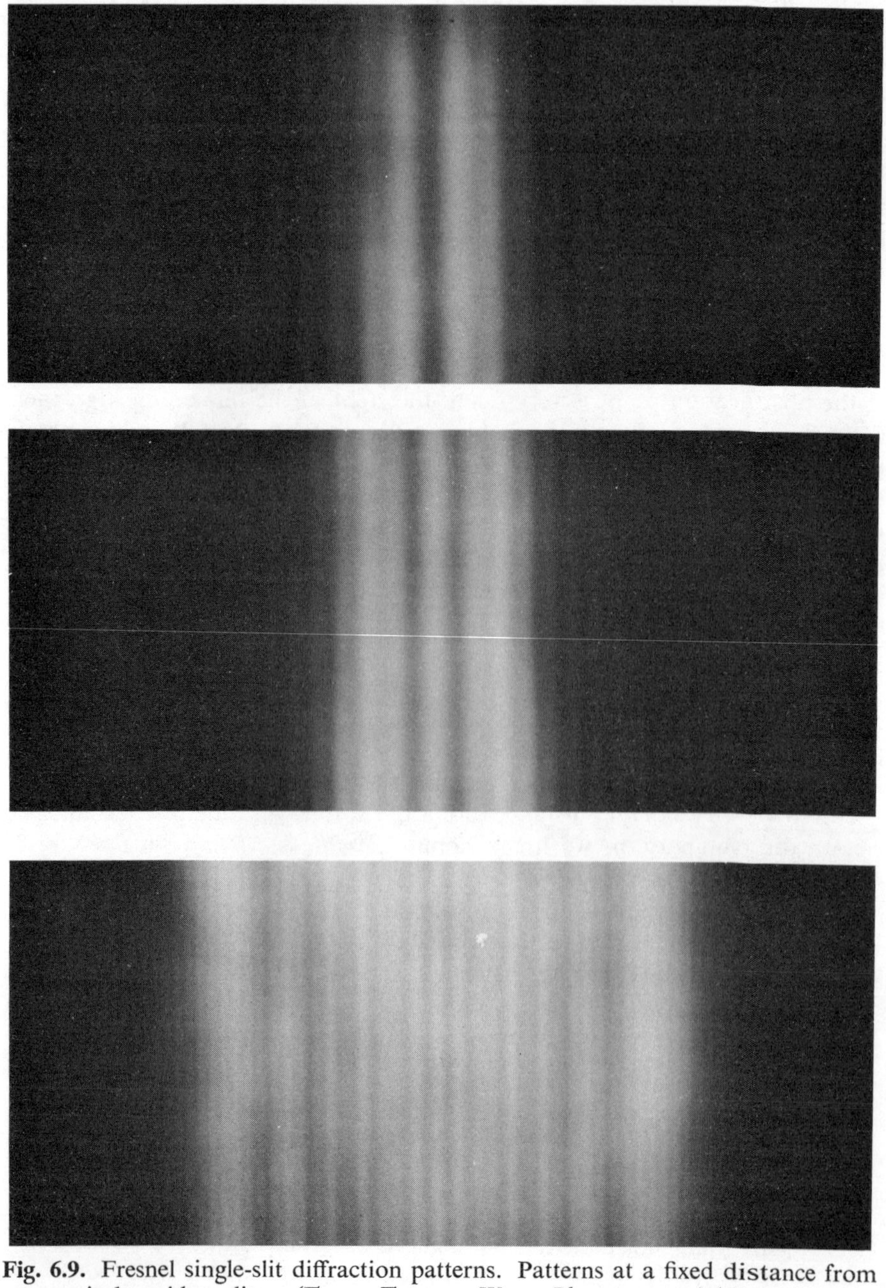

Fig. 6.9. Fresnel single-slit diffraction patterns. Patterns at a fixed distance from progressively wider slits. (From Towne, *Wave Phenomena*, Addison-Wesley, Reading, Mass., 1967.)

this length along the spiral, taking the distance between the ends of the arc as the magnitude of the state function. Figure 6.9 gives a series of typical single slit diffraction patterns differing only in the width of the slit. The slit width increases from top to bottom. Notice that the center of the pattern is not necessarily a maximum or a minimum of the pattern but can be either, depending on the slit width.

6.6 FRAUNHOFER DIFFRACTION

The integrals obtained in the case of Fresnel diffraction posed particularly difficult mathematical problems because of the quadratic dependence of the exponent on the variables ξ and η. We now show that this dependence can be reduced to a linear one by the use of an appropriate lens system. This case is known as Fraunhofer diffraction. It can be shown that this is equivalent to allowing the distances r_{10} and r_{20} to become infinite.

We regroup the terms in Eq. (6.6):

$$\psi_0 = -\frac{ik}{2\pi}\int dS\left(A\,\frac{e^{ikr_1}}{r_1}\right)\left(\frac{e^{ikr_2}}{r_2}\right). \tag{6.14}$$

The first term in brackets is the value of the state function at the element dS in the aperture. In the preceding section we considered the extreme case, where the aperture was totally open and the screen totally absorbing. Let us now consider the more general case, where the transmission through the aperture can take on any value. This can be accomplished by placing a screen over the aperture, whose transmission properties are a function of position. It would then be reasonable to include in the integral a function which adjusts the amplitude and phase to its new value at each point within the aperture. Doing this, we rewrite Eq. (6.14) as

$$\psi_0 = -\frac{ikA}{2\pi}\int dS\,u(\xi,\eta)\,\frac{e^{ik(r_1+r_2)}}{r_1r_2}, \tag{6.15}$$

where the amplitude of the complex function $u(\xi, \eta)$ is a measure of the absorption properties of the screen at the point ξ, η and the phase of this function determines any retardation of the wave at that point.

In going from Eq. (6.14) to Eq. (6.15) it must be remembered that Eq. (6.14) demands that the screen be illuminated by a single point source. This ensures that the phase difference between different portions of the screen is fixed or, in the sense of Chapter 4, that the illumination is coherent. This will be important for much of what follows. We also point out that Eq. (6.15) implies that a knowledge of the state function over the aperture is sufficient to determine the diffraction pattern. The Green's theorem method, however, assumed a knowledge of both the function and its gradient. It

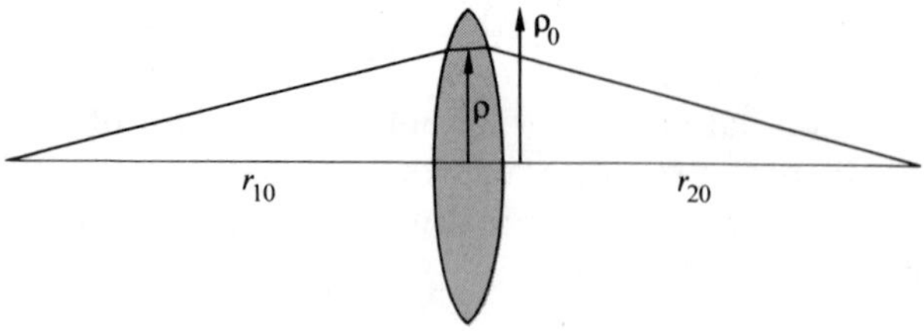

Fig. 6.10. The parameter ρ for a ray passing through a thin lens. The quantity ρ_0 is the radius of the lens.

would seem now that this much knowledge is not really necessary. This is indeed the case, and the details of obtaining Eq. (6.14) without presupposing a knowledge of the gradient of the incident state are left to the problems.

Within the diffracting aperture we now place a single converging thin lens of focal length f, chosen such that the points P and P' are conjugate, i.e.,

$$\frac{1}{f} = \frac{1}{r_{10}} + \frac{1}{r_{20}} \cdot \tag{6.16}$$

We consider this lens as a plate of varying thickness which produces various phase shifts in the light it transmits. We must therefore calculate the thickness of the lens as a function of the distance from the axis. In Fig. 6.10 the lens is shown with its thickness and its radius, ρ_0, greatly exaggerated. In an actual case ρ_0 is small enough compared to r_{10} and r_{20} to enable all rays passing through the lens to be taken as parallel to the axis. We now calculate the thickness of the lens in terms of the radii of curvature of its surfaces. This is most easily accomplished by splitting the lens along its median plane and calculating the thickness of each half separately. The left-hand half is shown in Fig. 6.11, again exaggerated. The thickness, t, a distance ρ from the axis is written $t = T - \delta$. If R is the radius of curvature of the spherical surface, this may be written

$$t = (R - \sqrt{R^2 - \rho_0^2}) - (R - \sqrt{R^2 - \rho^2}),$$

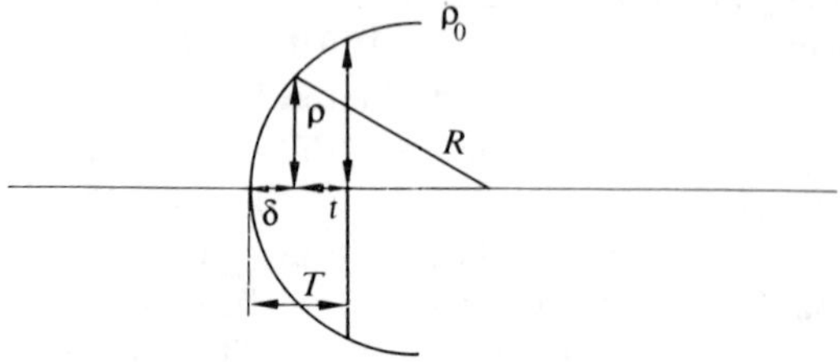

Fig. 6.11. A plano-convex lens with exaggerated thickness.

using the Pythagorean theorem. Factoring R from the radicals, we obtain

$$t = R\sqrt{1 - \left(\frac{\rho}{R}\right)^2} - R\sqrt{1 - \left(\frac{\rho_0}{R}\right)^2},$$

and if both ρ and ρ_0 are small compared to R, we may use the approximation $\sqrt{1 \pm \alpha} \simeq 1 \pm (\alpha/2)$, which is valid for $\alpha \ll 1$ and obtain

$$t = \frac{1}{2R}(\rho_0^2 - \rho^2).$$

For the original lens whose first and second surfaces have radii R_1 and R_2 we thus obtain

$$t = t_1 + t_2 = \frac{\rho_0^2 - \rho^2}{2}\left(\frac{1}{R_1} - \frac{1}{R_2}\right).$$

The minus sign arises from the sign conventions introduced in Chapter 1 for the radii. If the lens is of refractive index n, the difference in the optical path length through the lens a distance ρ from the axis and the same path in air is

$$\Delta L = (n - 1)\frac{\rho_0^2 - \rho^2}{2}\left(\frac{1}{R_1} - \frac{1}{R_2}\right),$$

or, using Eq. (1.30),

$$\Delta L = \frac{\rho_0^2 - \rho^2}{2}\frac{1}{f}. \tag{6.17}$$

The phase shift introduced by this thickness is $k\Delta L$. After insertion of the lens in or near the diffracting aperture, Eq. (6.15) becomes

$$\psi_0 = -\frac{ikA}{2\pi}\iint d\xi\, d\eta\, \frac{e^{ik(r_1 + r_2 + \Delta L)}}{r_1 r_2}\, u(\xi, \eta). \tag{6.18}$$

We now approximate $r_1 + r_2$ as in Eq. (6.7), but instead of completing the square as in Eq. (6.8) we regroup the terms to obtain

$$\begin{aligned} r_1 + r_2 = r_{10} &+ \left[r_{20} + \frac{1}{2r_{20}}(x^2 + y^2)\right] \\ &+ \left[\frac{1}{2}\left(\frac{1}{r_{10}} + \frac{1}{r_{20}}\right)(\xi^2 + \eta^2)\right] - \frac{1}{r_{20}}(\xi x + \eta y). \end{aligned} \tag{6.19}$$

If r is taken to be the distance from the center of the lens to the point x, y, in the diffraction pattern, its magnitude is given by

$$r = \sqrt{r_{20}^2 + x^2 + y^2},$$

which in the usual fashion may be approximated by

$$r = r_{20} + \frac{1}{2r_{20}}(x^2 + y^2).$$

This is recognized as the first term in brackets in Eq. (6.19). Noting that $\rho^2 = \xi^2 + \eta^2$ and using Eqs. (6.16) and (6.17), we are able to write

$$r_1 + r_2 + \Delta L = \left(r_{10} + \frac{\rho_0^2}{2f}\right) + r - \frac{1}{r_{20}}(\xi x + \eta y).$$

After we drop the constant terms $r_{10} + \rho_0^2/2f$ in the exponent, Eq. (6.18) becomes

$$\psi(x, y) = -\frac{ikA}{2\pi}\frac{e^{ikr}}{r_{10}r_{20}}\iint d\xi\, d\eta\, u(\xi, \eta)\, e^{-\frac{ik}{r_{20}}(\xi x + \eta y)}, \tag{6.20}$$

where the integration is to be taken over the aperture. If we make the change of variables

$$\alpha = \frac{k\xi}{r_{20}}, \qquad \beta = \frac{k\eta}{r_{20}},$$

and define $u(\alpha, \beta)$ to be equal to $u(\xi, \eta)$ over the aperture and zero elsewhere, we obtain

$$\psi(x, y) = -\frac{ikA}{2\pi}\frac{e^{ikr}}{r_{10}}\frac{r_{20}}{k^2}\int_{-\infty}^{+\infty}\int_{-\infty}^{+\infty} d\alpha\, d\beta\, u(\alpha, \beta)\, e^{-i(\alpha x + \beta y)}. \tag{6.21}$$

Thus, the diffraction pattern found in the image plane of the lens, except for a multiplicative factor, is the Fourier transform of the distribution over the aperture. The significance of this remark, for the theory of image formation, is profound, as we shall see.

6.7 THE RECTANGULAR APERTURE

We assume a rectangular aperture symmetric about the z-axis with edges located so that α and β at the edges take on the values $\pm\alpha_0$ and $\pm\beta_0$. For an open aperture $u(\alpha, \beta)$ is set equal to unity over the aperture and zero elsewhere. If all the constants in Eq. (6.21) are grouped into a constant C, the integrals become

$$\begin{aligned}\psi(x, y) &= C\int_{-\alpha_0}^{\alpha_0}\int_{-\beta_0}^{\beta_0} d\alpha\, d\beta\, e^{-i\alpha x}\, e^{-i\beta y}\\ &= C'\,\frac{\sin \alpha_0 x}{\alpha_0 x}\,\frac{\sin \beta_0 y}{\beta_0 y},\end{aligned}$$

where $C' = 4\alpha_0\beta_0 C$. If the intensity of the incident state is adjusted so that

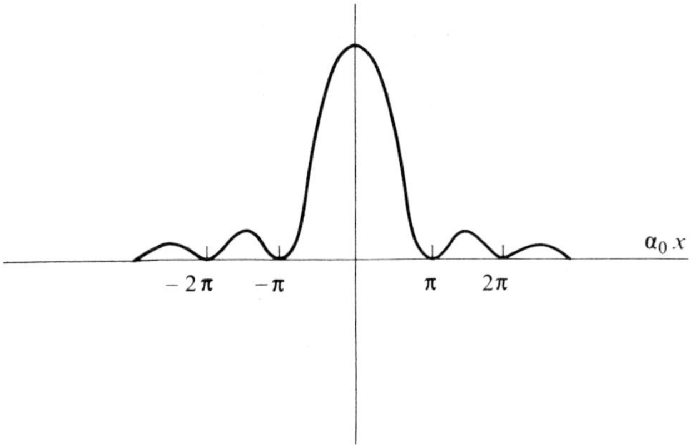

Fig. 6.12. Fraunhofer diffraction pattern of a single slit. (After Rossi, *Optics*, Addison-Wesley, Reading, Mass., 1957.)

the intensity in the diffraction pattern at $x = 0$, $y = 0$ is unity, the intensity in the diffraction pattern is then given by

$$I(x, y) = \left(\frac{\sin \alpha_0 x}{\alpha_0 x}\right)^2 \left(\frac{\sin \beta_0 y}{\beta_0 y}\right)^2. \tag{6.22}$$

A plot of the function $(\sin \alpha x/\alpha x)^2$ is given in Fig. 6.12. The zeros of this function obviously occur for $\alpha x = n\pi$, $n \neq 0$. It is left to the problems to show that the locations of the secondary maximums are given by the solutions of the transcendental equation

$$\tan \alpha x = \alpha x. \tag{6.23}$$

The first few solutions of this equation are given in Table 6.1.

Table 6.1

Maximum and minimum values of the function $\sin \alpha x_n/\alpha x_n$ and the associated values of $(\sin \alpha x_n/\alpha x_n)^2$

n	αx_n	$\sin \alpha x_n/\alpha x_n$	$(\sin \alpha x_n/\alpha x_n)^2$
1	0.0	1.0	1.0
2	4.4934	−0.2171	0.0472
3	7.7253	+0.1284	0.0166
4	10.9041	−0.0913	0.00834
5	14.0662	+0.0709	0.00503
10	29.8116	−0.0335	0.00112
15	45.5311	+0.0220	0.000484

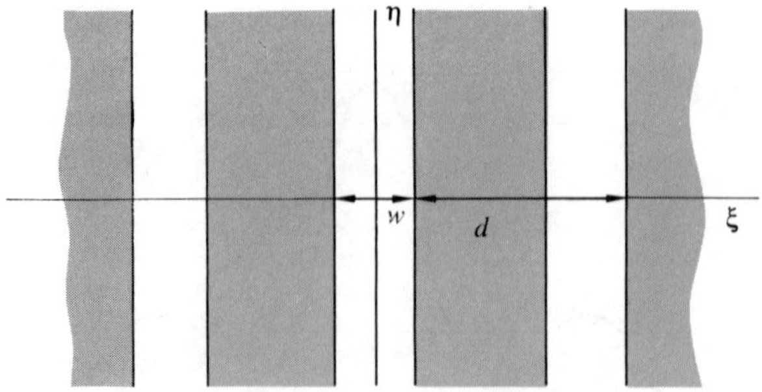

Fig. 6.13. A model for a diffraction grating. Slits of width w are separated by opaque regions of width $w - d$.

6.8 THE DIFFRACTION GRATING

A simple model of a transmission diffraction grating consists of alternating opaque and transparent strips, as shown in Fig. 6.13. The width of each slit is w and the distance between slits is d. Since the problem is basically one-dimensional, we ignore the integral over β. If the constant terms are dropped, Eq. (6.21) becomes

$$\psi(x) = \int d\alpha\, u(\alpha)\, e^{-i\alpha x}$$

$$= \sum_{n=-N}^{n=+N} \int_{nb-\frac{a}{2}}^{nb+\frac{a}{2}} d\alpha\, e^{-i\alpha x},$$

where each integral in the sum is over one slit with $a = kw/r_{20}$, $b = kd/r_{20}$. If there are N slits to the right and to the left of the center slit, the total number of slits, $\mathcal{N}$, is given by $\mathcal{N} = 2N + 1$. Performing the integration, we obtain

$$\psi(x) = a \frac{\sin xa/2}{xa/2} \sum_{n=-N}^{N} e^{-ixnb}.$$

The sum is recognized as a geometric series with ratio e^{-ixb}. Making use of the rule for such a sum given below,

$$\sum_{n=-N}^{N} z^n = z^{-N} \frac{1 - z^{2N+1}}{1 - z},$$

we obtain

$$\psi(x) = a \frac{\sin xa/2}{xa/2} \frac{\sin \mathcal{N} xb/2}{\sin xb/2}, \tag{6.24}$$

Which may be normalized such that

$$I(x) = \left(\frac{\sin xa/2}{xa/2}\right)^2 \left(\frac{\sin \mathcal{N} xb/2}{\sin xb/2}\right)^2. \tag{6.25}$$

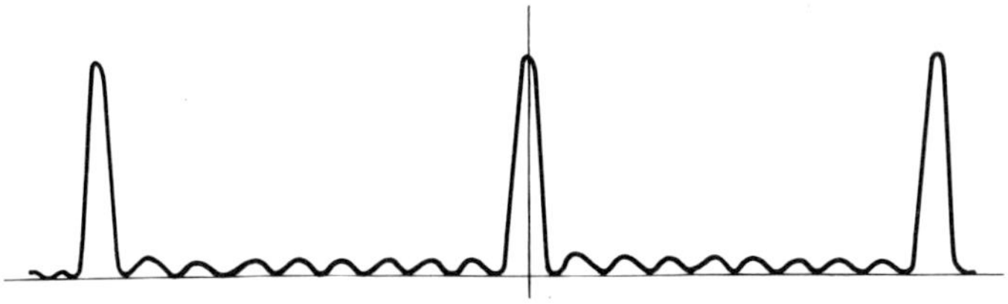

Fig. 6.14. Fraunhofer diffraction pattern of a grating with $\mathcal{N} = 10$. (From Rossi, *Optics*, Addison-Wesley, Reading, Mass., 1957.)

If $\mathcal{N}$ is set equal to one, the first portion of this pattern is recognized as the Fraunhofer diffraction pattern of a single slit. It is to be compared with the x-dependence of the rectangular aperture pattern and is seen to be the function already tabulated in Table 6.1. If, on the other hand, $\mathcal{N}$ is large, then the first portion of the function is slowly varying compared to the second. For the moment we shall consider that it may be treated as a constant. A plot of the second portion of the function is given in Fig. 6.14. The large principal maxima occur when the sine functions in both numerator and denominator are zero. This occurs when $xb/2 = n\pi$. At these points the value of the function is $\mathcal{N}^2$. It is left as an exercise to show that the secondary maxima are located by the solutions of the transcendental equation

$$\tan \frac{\mathcal{N}bx}{2} = \mathcal{N} \tan \frac{bx}{2} \cdot \tag{6.26}$$

There are $\mathcal{N} - 1$ zeros located between each pair of principal maxima at

$$x = 2n\pi/\mathcal{N}b \quad (n = 1, 2, 3, \ldots, n - 1).$$

If $\mathcal{N}$ is very large, the value of $xb/2$ for the first few secondary maxima near the central principal maximum is small and one may approximate, with increasing accuracy as $\mathcal{N}$ increases,

$$\tan \frac{bx}{2} = \sin \frac{bx}{2} = \frac{bx}{2} \cdot$$

If we make this approximation in the denominator of Eqs. (6.24) and (6.25), the second portion of the function becomes that tabulated in Table 6.1 multiplied by the constant $\mathcal{N}$.

For large $\mathcal{N}$ the secondary maxima are either so close to the principal maxima or of such small amplitude as to be indiscernible and the pattern is one of sharp peaks. If the slowly varying portion is included, the amplitude of these peaks would decrease going away from the central maximum, as shown in Fig. 6.15.

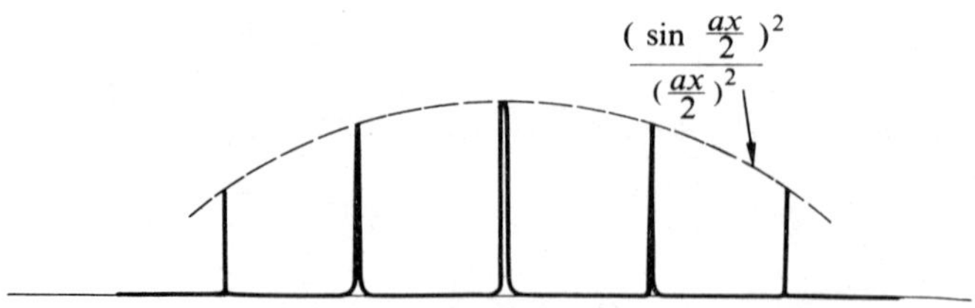

Fig. 6.15. A grating diffraction pattern for large $\mathcal{N}$ including the single-slit pattern.

6.9 THE CIRCULAR APERTURE

We now consider a circular aperture of radius R. In terms of the variables α and β, the integration will be over a circle of radius kR/r_{20}, as is indicated in Fig. 6.16. By the symmetry of the problem, if the pattern along the x-axis is known, it is known along any other radius in the xy-plane. We therefore set $y = 0$ in Eq. (6.21). The integral may be reduced to an integration over a single variable α if the element of integration,

$$dA = 2\beta\, d\alpha = 2\sqrt{\left(\frac{kR}{r_{20}}\right)^2 - \alpha^2}\, d\alpha,$$

shown in Fig. 6.16 is used. Once again we ignore the constant in Eq. (6.21) and obtain

$$\psi(x, 0) = 2\int_{-kR/r_{20}}^{kR/r_{20}} d\alpha\, e^{-i\alpha x}\sqrt{\left(\frac{kR}{r_{20}}\right)^2 - \alpha^2}.$$

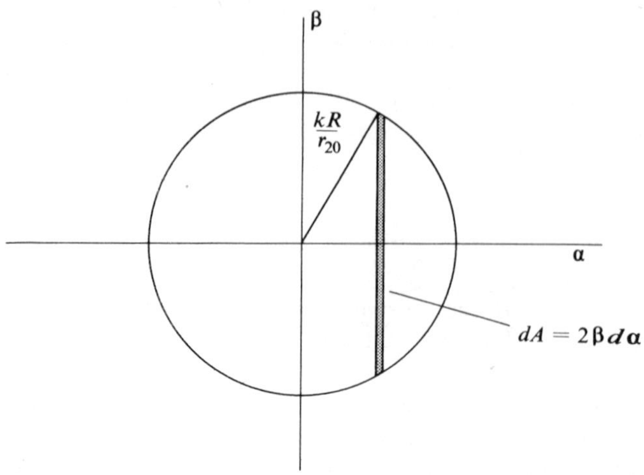

Fig. 6.16. Geometry used in deriving the Fraunhofer diffraction pattern of a circular aperture.

Making the change of variable

$$\gamma = kRx/r_{20}, \qquad u = r_{20}/kR,$$

we obtain

$$\psi(x, 0) = 2\left(\frac{kR}{r_{20}}\right)^2 \int_{-1}^{+1} e^{-i\gamma u} \sqrt{1 - u^2}\, du,$$

which may be written

$$\psi(x, 0) = 4\left(\frac{kR}{r_{20}}\right)^2 \int_{0}^{1} \sqrt{1 - u^2} \cos(\gamma u)\, du.$$

The last integral cannot be evaluated in terms of elementary functions. It can be given in terms of the transcendental function $J_1(\gamma)$, which is known as Bessel's function of the first kind of order one. In fact, it is possible to define this function by

$$J_1(\gamma) = \frac{2\gamma}{\pi} \int_0^1 \sqrt{1 - u^2} \cos(\gamma u)\, du.$$

Again dropping multiplicative constants, we may write the intensity in the diffraction pattern as

$$I(x, 0) = \left(\frac{2J_1(\gamma)}{\gamma}\right)^2, \qquad \gamma = kRx/r_{20}. \tag{6.27}$$

As it happens, the intensity given by Eq. (6.27) is normalized to an intensity of unity at the center of the pattern, $x = 0$. A plot of Eq. (6.27) is given in Fig. 6.17. The diffraction pattern of a circular aperture is thus a bright

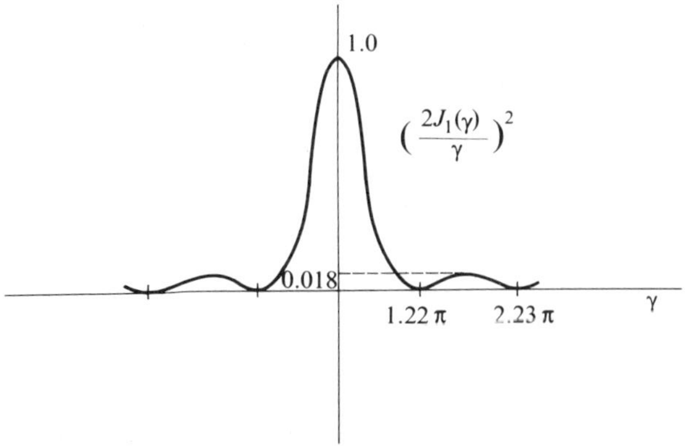

Fig. 6.17. The function $\left(\dfrac{2J_1(\gamma)}{\gamma}\right)^2$.

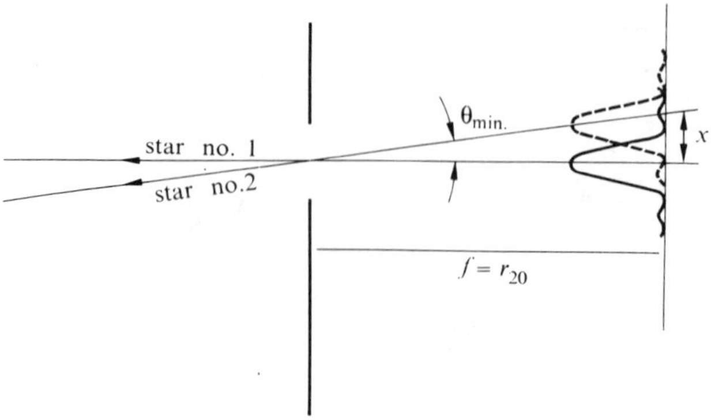

Fig. 6.18. Overlap of the diffraction pattern of two stars.

central spot, known as the Airy disk, surrounded by alternate dark and light rings. The first dark ring occurs for $\gamma = 1.22\pi$.

If one considered the point source which illuminates the circular aperture to be a star, and the lens in the aperture to be the objective lens of a telescope, then the image of the star in the focal plane of the objective would be the diffraction pattern given in Eq. (6.27). A reasonable criterion, due to Rayleigh, for distinguishing two very close stars is that the center of one pattern can be no closer to the center of the other than the first dark ring. Thus, the angular separation of the stars must be the angular separation of the dark ring from the center of the pattern as viewed from the objective. This is shown in Fig. 6.18. Since $\theta_{\text{min.}}$ is small, we may set $x/r_{20} = \theta_{\text{min.}}$. We thus have

$$1.22\pi = kRx/r_{20}$$

or

$$\theta_{\text{min.}} = 1.22\lambda/D,$$

where $D = 2R$ is the diameter of the telescope objective. Photographs of two diffraction patterns satisfying Rayleigh's criterion are given in Fig. 6.19.

6.10 IMAGE FORMATION IN COHERENT LIGHT

Consider the image-forming system shown in Fig. 6.20. S is a point source which coherently illuminates the aperture of lens L_1. The plane P' containing the lens L_2 and the plane containing the source S are taken to be conjugate. Planes P and P'' are likewise taken to be conjugate for lens L_2. A transparency or slide, the object, which may vary in thickness and in absorption, is placed next to L_1 in the object plane, P. According to simple geometric optics, an inverted image of this object will be formed in the plane P'' by lens L_2. The

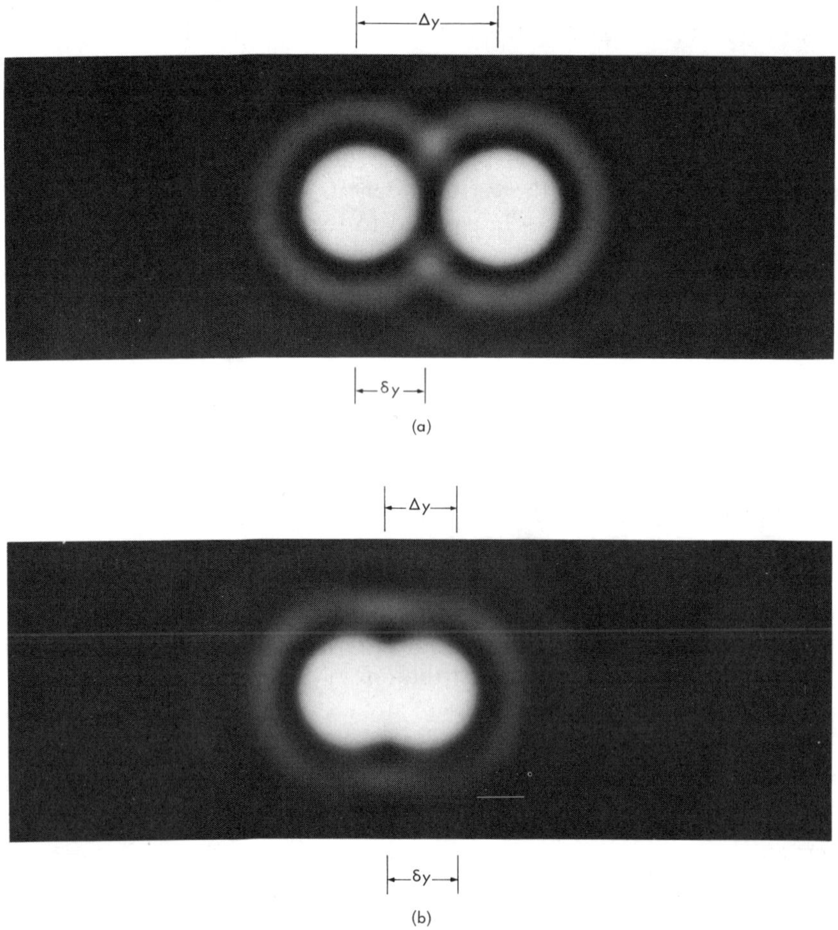

Fig. 6.19. (a) Images of two point sources with first dark rings tangent; (b) images of two point sources which are just resolved according to the Rayleigh criterion. (From Towne, *Wave Phenomena*, Addison-Wesley, Reading, Mass., 1967.)

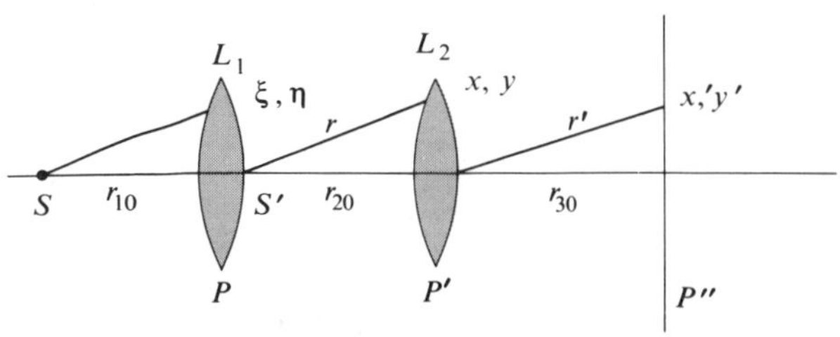

Fig. 6.20. A particular image-forming system. An object placed in the plane of P is illuminated by a point source at S and an image is formed in the plane P'' by lens L_2.

magnification will be r_{30}/r_{20}. We wish to interpret this image formation in terms of Eq. (6.20). Taking the coordinates of a point in plane P to be ξ and η, the function $u(\xi, \eta)$ describes the object. $\psi(x, y)$ is the value of the state function at a point in the aperture of L_2 whose coordinates are x and y. The distance from the vertex of L_1 to the point x, y is r. Finally, the function $\phi(x', y')$ is the state function at point x', y' in the image plane P''; r' is the distance to this point from the vertex of lens L_2.

Equation (6.20) may now be applied to the diffraction process between plane P and P', and is written

$$\psi(x, y) = \frac{Ae^{ikr}}{r_{20}} F(x, y), \tag{6.28}$$

where

$$F(x, y) = -\frac{ik}{2\pi}\frac{1}{r_{10}} \iint d\xi\, d\eta\, u(\xi, \eta)\, e^{-ik/r_{20}(\xi x + \eta y)}. \tag{6.29}$$

Now, to the order of approximation which was used in deriving Eq. (6.20), $r_{20} = r$ and the term $Ae^{ikr}/r_{20} = Ae^{ikr}/r$ and represents the state function which would be obtained over the aperture of L_2 if a point source were placed at the vertex of L_1. Thus, placing an object described by $u(\xi, \eta)$ in the plane P with a point source at S yields exactly the same distribution in P' as if we placed a point source at S' and an object described by $F(x, y)$ in the plane P'. Thought of in this way, the distribution in plane P'' may immediately be written down, since the planes P and P'' are also conjugate. The distribution in P'' is

$$\phi(x', y') = \frac{Ae^{ikr'}}{r_{30}} F'(x', y'), \tag{6.30}$$

where

$$F'(x', y') = -\frac{ik}{2\pi}\frac{1}{r_{20}} \iint dx\, dy\, F(x, y)\, e^{-ik/r_{30}(xx' + yy')}. \tag{6.31}$$

The distribution in the image plane P'', $F'(x', y')$, may now be written in terms of the distribution in the object plane, $u(\xi, \eta)$, by placing Eqs. (6.28) and (6.29) in Eqs. (6.30) and (6.31). This yields

$$\phi(x', y') = -\left(\frac{k}{2\pi}\right)^2 \frac{A \exp(ikr')}{r_{10}r_{20}r_{30}}$$

$$\iint_P d\xi\, d\eta \iint_{\mathscr{A}} dx\, dy\, u(\xi, \eta) \exp\left[-i\left(\frac{k\xi}{r_{20}} + \frac{kx'}{r_{30}}\right)x\right] \exp\left[-i\left(\frac{k\xi}{r_{20}} + \frac{ky'}{r_{30}}\right)y\right], \tag{6.32}$$

where P and $\mathscr{A}$ beneath the integrals indicate that the integral over ξ and η is to be taken over the entire plane P, whereas that in x and y is to be taken only over the aperture of lens L_2. This distinction will be examined later but at this point we assume that the aperture is large enough to contain most of the initial diffraction pattern, $F(x, y)$, of the object $u(\xi, \eta)$. We may then extend the integral to be over the entire xy-plane. Referring to Appendix IV, we may write

$$\iint_{-\infty}^{+\infty} dx\, dy \exp\left[-i\left(\frac{k\xi}{r_{20}} + \frac{kx'}{r_{30}}\right)x\right] \exp\left[-i\left(\frac{k\eta}{r_{20}} + \frac{ky'}{r_{30}}\right)y\right] = (2\pi)^2\, \delta\left(\frac{k\xi}{r_{20}} + \frac{kx'}{r_{30}}\right) \delta\left(\frac{k\eta}{r_{20}} + \frac{ky'}{r_{30}}\right).$$

Placing this in Eq. (6.32) and performing the final integration over ξ and η, we obtain

$$\phi(x', y') = -A\,\frac{r_{20}}{r_{10}r_{30}}\, e^{ikr'}\, u\left(-\frac{r_{20}}{r_{30}}x', -\frac{r_{20}}{r_{30}}y'\right). \tag{6.33}$$

Thus, the value of the state function in the image at the point x', y', except for multiplicative constants, is the value in the object function at point

$$\xi = -r_{20}/r_{30}x', \qquad \eta = -r_{20}/r_{30}y'.$$

The image faithfully reproduces the object, is inverted, and is magnified in the ratio r_{30}/r_{20}. This is all in agreement with simple geometric optics.

As has been pointed out, Eq. (6.20) implies that a diffraction pattern in the Fraunhofer case is basically the Fourier transform of the state function over the diffracting aperture. We have seen in this section that image formation in coherent illumination may be viewed as a double diffraction process. A diffraction pattern of the object was formed in the aperture of lens L_2 and the diffraction pattern of this pattern gave the image in plane P''. This is the essence of Abbe's theory of image formation. The detailed analysis of other systems of lens will, of course, vary, but in all cases some intermediate plane will be found over which the state function is the Fourier transform of the object and the final image is formed by a second transformation.

The practical importance of this viewpoint is that by insertion of various filters in the plane of the first transform one may control the manner in which the different portions of the first transform recombine to form the final image. Some examples of this will be examined in the following section.

6.11 SPATIAL FILTERING IN COHERENT ILLUMINATION

As an example of how a final image may be modified by the use of filters which alter the Fourier transform of the object, the following simple case is considered. As the object we take the diffraction grating discussed in Section 6.8. This is represented in Fig. 6.21. For simplicity the lenses L_1 and L_2 have not been shown and we have set $r_{20} = r_{30}$. According to Section 6.8,

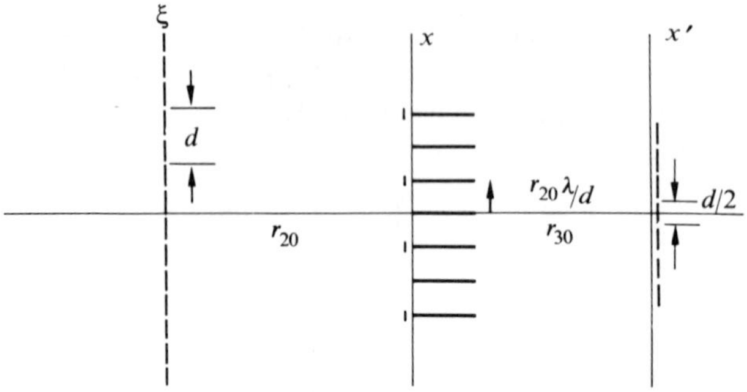

Fig. 6.21. Image formation with a diffraction grating as the object. In the figure the distances r_{20} and r_{30} have been set equal.

the diffraction pattern in plane P' will be a series of bright lines spaced a distance $r_{20}\lambda/d$ apart. We now place an opaque mask in the xy-plane blocking off every other line. Thus, only the central line and alternate lines on each side of it are allowed to recombine to form the final image. Since these peaks are spaced twice as far apart as the original pattern, they are characteristic of an initial grating whose slit spacing is $d/2$ rather than d. The image formed in the final plane will thus not be that of the original grating, but of a grating whose slits are twice as close together. Hence, the rather surprising result: by removing something at the intermediate plane we have added something to the final image. In particular, we have doubled the number of slits.

Now consider the effect of the finite diameter of the aperture of lens L_2. We have seen that the spatial period of the final image of the grating depends on the spacing of the peaks in the intermediate pattern. Thus, it is reasonable to say that if the final image is to have the same period as the initial object, the aperture of lens L_2 must at least be large enough to pass the central maximum and the first maximum on each side. If the number of slits, $\mathcal{N}$, in the grating is large, we may approximate these three peaks in the diffraction pattern by delta functions, and write the distribution

$$F(x) = A\left[\delta\left(x + \frac{2\pi r_{20}}{kd}\right) + \delta(x) + \delta\left(x - \frac{2\pi r_{20}}{kd}\right)\right].$$

The final transform gives the distribution in x' as

$$\begin{aligned} F'(x') &\propto \int dx\, F(x)\, e^{-ik/r_{30}xx'} \\ &= A\left[e^{i2\pi\frac{r_{20}}{r_{30}}\frac{x'}{d}} + 1 + e^{-i2\pi\frac{r_{20}}{r_{30}}\frac{x'}{d}}\right] \\ &= A\left[1 + 2\cos\left(2\pi\frac{r_{20}}{r_{30}}\frac{x'}{d}\right)\right], \end{aligned}$$

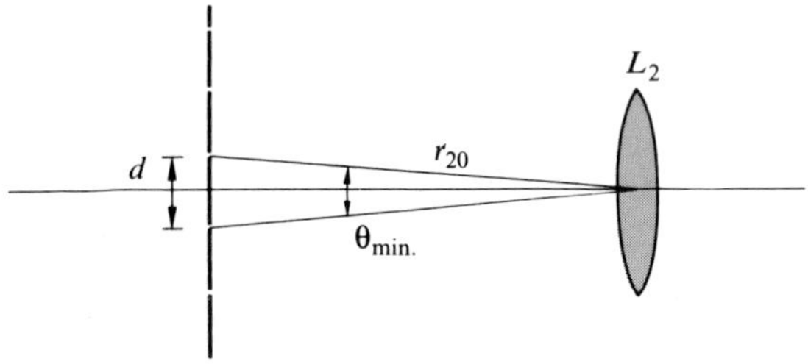

Fig. 6.22. The angle subtended by two slits as viewed from lens L_2.

which has a spatial period equal to that of the original grating with magnification r_{30}/r_{20}. If the diameter of L_2 is smaller than that considered above, only the central maximum will be transmitted and the pattern in the x-plane will be one of uniform illumination. Thus, if portions of the original object —in this case, the slits of a grating—are to be resolved in the final image, the diameter of the image forming lens must satisfy

$$D \geqslant 2\frac{2\pi r_{20}}{kd} = 2\frac{\lambda r_{20}}{d}.$$

From Fig. 6.22, we see that the angle subtended from the lens by two slits is given by $\theta = d/r_{20}$. Combining these two equations, we find that the condition to be satisfied in order that the two slits be resolved by the lens is

$$\theta \geqslant 2\lambda/D,$$

which may be compared to Rayleigh's criterion:

$$\theta \geqslant 1.22\lambda/D.$$

In the preceding discussion we have examined the image formation process for an object with a basic single period. However, any object may be thought of as a combination of simple spatial variations by considering its Fourier transform. It is this transform which is spread out for us in the plane of lens L_2. The value of the state function at the point x, y tells us the extent to which the object is composed of the simple exponential variations $e^{ik\xi x/r_{20}}$ and $e^{ik\eta y/r_{20}}$ of spatial periods $r_{20}\lambda/x$ and $r_{20}\lambda/y$. In particular, the value at $x = 0$, $y = 0$ gives the contribution of infinite spatial period, which is simply a uniform illumination. Now, an image of low contrast may be thought of as being an image of high contrast plus an excessive uniform background illumination One should be able to improve the contrast of an image, therefore, by removing a portion of this background This may be accomplished by placing a small opaque disk over the center of lens L_2. This is illustrated in Fig. 6.23. Figure 6.23(a) is a photograph of an image

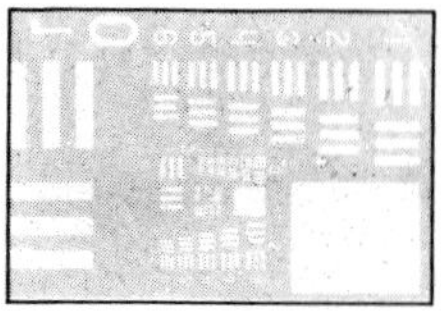

Object

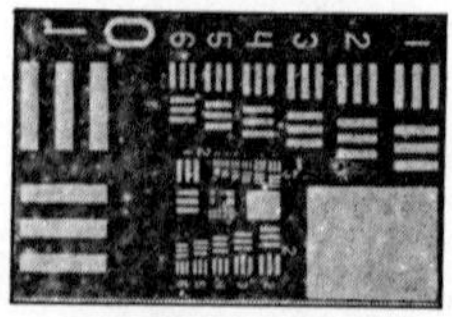

Image

Fig. 6.23. (a) An image of low contrast; (b) Higher-contrast image formed by the same system with the central portion of the lens obscured. (From O'Neill, *Introduction to Statistical Optics*, Addison-Wesley, Reading, Mass., 1963.)

formed by a lens system as shown in Fig. 6.20. Figure 6.23(*b*) is a second photograph taken after a small black dot had been placed in the center of lens L_2.

The objects considered until now have been amplitude objects, i.e. $u(\xi, \eta)$ is a real function. The case of phase objects, which are transparent but which may vary in thickness or in index of refraction, is of great importance in microscopy. For a pure phase object $|u(\xi, \eta)|^2 = 1$ and Eq. (6.33) tells us that the intensity distribution in the image will be uniform. Writing

$$u(\xi, \eta) = e^{in(\xi, \eta)},$$

we further assume that the retardations, n, introduced by the phase object are small, $n(\xi, \eta) \ll 1$, so that we may approximate the exponential by

$$u(\xi, \eta) = 1 + in(\xi, \eta) - \frac{n^2(\xi, \eta)}{2}. \tag{6.34}$$

If Eq. (6.34) is placed in Eq. (6.29), the first of the three integrals yields the diffraction pattern of the aperture in the $\xi\eta$-plane. This will be concentrated in a small region around the origin in the xy-plane. The distribution in the remainder of the plane will be determined almost wholly by the remaining two integrals. Furthermore, under the assumption of small $n(\xi, \eta)$, the contribution of these integrals to the pattern at $x = 0$, $y = 0$ may be neglected. The center of the aperture in the xy-plane is now covered with a small plate which introduces a retardance of $\lambda/4$ and a fractional absorption, f. This is seen as equivalent to replacing Eq. (6.34) by

$$u(\xi, \eta) = if + in(\xi, \eta) - \frac{n^2(\xi, \eta)}{2}. \tag{6.35}$$

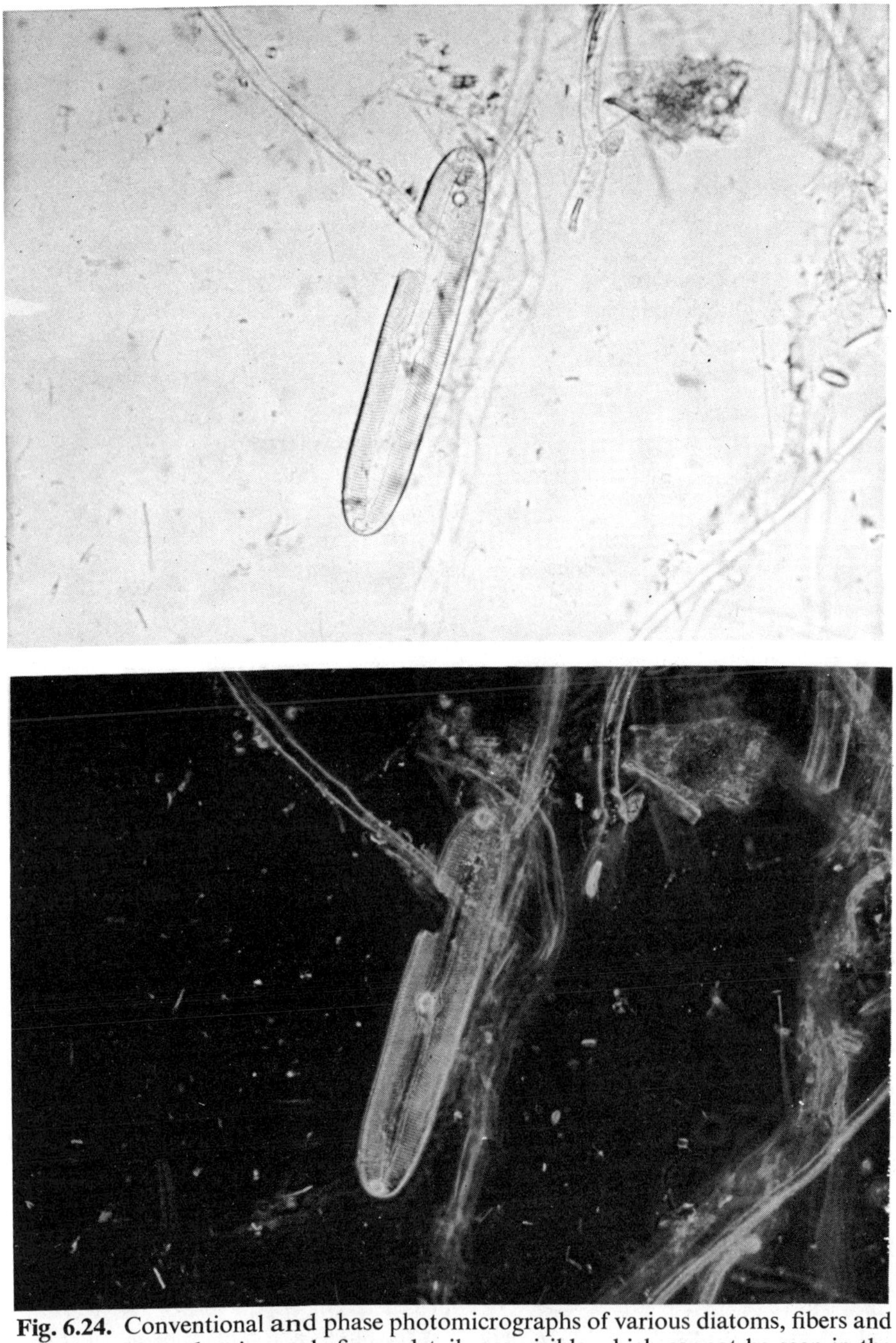

Fig. 6.24. Conventional and phase photomicrographs of various diatoms, fibers and bacteria. Note that in each figure details are visible which cannot be seen in the other. (Courtesy of T. J. Lowery and R. Hawley.)

By Eqs. (6.35) and (6.33), the phase object $u(\xi, \eta)$ without the retarder plate would produce an image $\phi(x', y')$, given by

$$\phi(x', y') = -\frac{Ar_{20}}{r_{10}r_{30}} e^{ikr'} \left[1 + in\left(-\frac{r_{20}}{r_{30}} x', -\frac{r_{20}}{r_{30}} y'\right) - \frac{1}{2} n^2 \left(-\frac{r_{20}}{r_{30}} x', -\frac{r_{20}}{r_{30}} y'\right)\right],$$

with $|\phi(x', y')|^2 = A^2(r_{20}/r_{10}r_{30})^2$, neglecting terms in n^2 and higher. With the retarder plate the image will be given by

$$\phi'(x', y') = -\frac{Ar_{20}}{r_{10}r_{30}} e^{ikr'}[if + in - \tfrac{1}{2}n^2],$$

with $|\phi'(x', y')|^2 = (f^2 + 2fn)|\phi(x', y')|^2$ to the same order. If one defines the contrast in the final image as

$$C = (I - I_{\text{min.}})/I_{\text{min.}},$$

where $I_{\text{min.}}$ is the lowest intensity in the image with $n = 0$, one obtains

$$C = 2n/f.$$

It is seen that even for very small retardations in the original object, e.g., $n = 10^{-3}$, an image of appreciable contrast can be obtained. If f is taken as 10^{-2}, the contrast is 20%.

Although the optical construction of a commercial phase contrast microscope differs from the optical system of Fig. 6.20, the optical principle involved is as described here. In Fig. 6.24 photomicrographs are shown of the object with and without the insertion of the retarder plate.

PROBLEMS

6.1 Verify Eq. (6.1) by applying the divergence theorem, vector identity 14, to the function $\Psi\Delta\phi - \phi\Delta\Psi$.

6.2 Using the definite integrals of Appendix VII, verify that the Fresnel integrals of Eq. (6.11) satisfy

$$C(\infty) = S(\infty) = 1/2, \; C(-\infty) = S(-\infty) = -1/2.$$

6.3 If an aperture in a plane screen is symmetric about some axis, show directly from Eq. (6.21) that the diffraction pattern is symmetric.

6.4 A square aperture is cut in a screen whose side is 0.3 mm. A monochromatic point source of wavelength $\lambda = 5000$ Å is placed 20 cm to the left of the aperture and a screen is placed 30 cm to the right. Let the observation point be taken at the corner of the geometric shadow of the aperture. Show by using the Cornu spiral that the intensity at the observation point is 0.022 of the intensity at the same point when the diffracting screen is removed.

6.5 The center of a circular aperture is blocked out until only a very thin annular region is allowed to transmit. Without a detailed calculation show that in the Fraunhofer case minima of the annular diffraction pattern occur at the location of the maxima of the original circular aperture pattern.

6.6 Consider the Fraunhofer pattern of a single slit. Show that the width of the central maximum is in qualitative agreement with the uncertainty relation, Eq. (5.36), for the position and momentum of a photon.

6.7 Show that, by Eq. (6.21), a shift in the location of the diffracting aperture does *not* result in a shift in the diffraction pattern. Interpret this result physically.

6.8 Consider the case of Fresnel diffraction with both source and observation point on the z-axis. Let the observation point be moved along the axis until it is at a point of maximum intensity on the axis. If the source distance, r_{10}, and the observation point distance, r_{20}, are now varied, the observation point being kept at an intensity maximum, show that, approximately,

$$\frac{1}{r_{10}} + \frac{1}{r_{20}} = \text{constant}.$$

From Eq. (6.10) what variation must be ignored to obtain this result?

6.9 Show that the secondary maxima in the Fraunhofer diffraction pattern of a rectangular aperture are located by the solution of $\tan \alpha x = \alpha x$, $\tan \beta y = \beta y$.

6.10 If $|\boldsymbol{k}| = k$ is constant, show that $f(\boldsymbol{r})$ of Eq. (6.3) satisfies Eq. (6.2).

6.11 Babinet's principle for scalar diffraction states that complementary screens produce complementary diffraction patterns. We take complementary screens to mean that one screen is opaque where the other is transparent and vice versa. If Ψ_1 and Ψ_2 are the state functions produced by complementary screens, $\Psi_1 + \Psi_2 = 1$. Use Babinet's principle to show how the Cornu spiral may be used to compute the diffraction pattern of a rectangular obstacle.

6.12 Show from Eq. (6.4) that a knowledge of the gradient of the incident state function over a diffracting aperture is unnecessary if a function ϕ is picked which identically vanishes over the aperture. Such a function is

$$\phi = \frac{e^{ikr}}{r} - \frac{e^{ikr'}}{r'}.$$

The distance r is defined as r_2, as in Section 6.1, and r' is as shown in Fig. P6.1. The point Q is located symmetrically across the screen from the observation point P'. Using the function ϕ defined above, apply Eq. (6.4) to apertures in plane screens and obtain

$$\Psi_0 = -\frac{ik}{2\pi}\iint dS\Psi\frac{e^{ikr}}{r},$$

where the integration is over the aperture. The same approximations should be used as were made in Section 6.1.

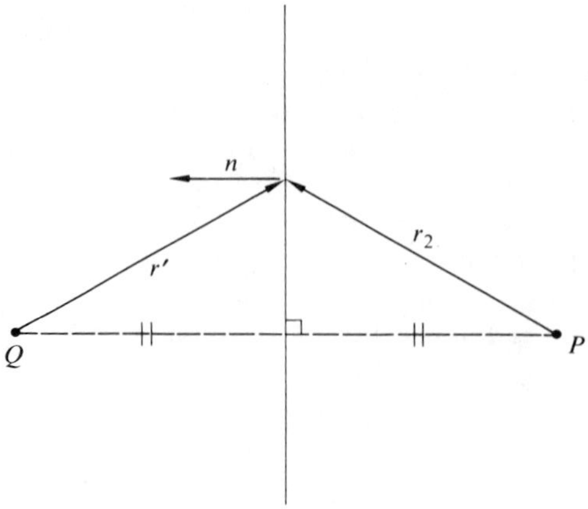

Figure P6.1

6.13 Another model for a diffraction grating may be constructed as follows. A slit of width $2Nd$ is covered with a mask such that when uniformly illuminated from the left the intensity transmitted varies across the slit according to

$$\cos\frac{2\pi\xi}{d} = \cos\frac{2\pi r_{20}}{kd}\alpha.$$

Show that the intensity in the Fraunhofer pattern is given by

$$\frac{1}{x}\frac{\sin\left(\dfrac{kd\mathcal{N}}{2r_{20}}x\right)}{1-\left(\dfrac{kdx}{2\pi r_{20}}\right)^2},$$

where $\mathcal{N} = 2N + 1$. Note that the denominator vanishes at $x = \lambda r_{20}/d$. Interpret this result.

6.14 Show that the same techniques which converted the phase object of Eq. (6.34) into an amplitude image will convert the amplitude object

$$\mu(\xi, \eta) = 1 + n(\xi, \eta)$$

into a phase image. Assume that the amplitude variation of the object is small, $n(\xi, \eta) \ll 1$.

CHAPTER 7

THE ELECTROMAGNETIC FIELD AND COHERENCE

In 1873, Maxwell published his *Treatise on Electricity and Magnetism.* It has since been impossible to deny that light is somehow a manifestation of the electromagnetic field. In fact, by the turn of the century, light was thought to be nothing more than a self-propagating class of electromagnetic field. What forced this identification? Allow Maxwell to answer:

> ". . . The properties of the electromagnetic medium are, insofar as we have gone, similar to those of the luminferous medium, but the best way to compare them is to determine the velocity with which an electromagnetic disturbance would propagate through the medium. If this should be equal to the velocity of light, we would have strong reason to believe that the two media, occupying as they do the same space, are really identical."*

Today we would prefer to delete the reference to the media, but the message remains. Theory does predict these velocities to be equal, and the study of optics is forever married to the study of the electromagnetic field.

We now know that there exist physical entities, e.g., neutrinos, which also travel with this velocity, but which are not identified with light. The final witness to our marriage is the interaction of light with matter. Light interacts with the electric and magnetic properties of matter, i.e., with charge and current distributions. Neutrinos do not interact in this way. The honeymoon was quickly over. In October, 1900, Planck published his theory of black body radiation. Five years later Einstein raised Planck's mathematical trick of quantization to the status of accepted physical theory, using it to explain the photoelectric effect.

Maxwell's theory is of its nature a continuous one. There is no way of obtaining from it bundles of energy which are discretely absorbed or emitted.

* "Ether", *Encyclopedia Britannica,* 9th Edn, Vol. VIII. Published posthumously in 1878.

Our problem is to superimpose on Maxwell's theory a discrete interpretation of the nature of radiation. In the following chapters we will examine some of the aspects of this problem.

7.1 THE CLASSICAL FIELD

Maxwell's equations, which completely describe the classical electric and magnetic fields are given below in MKS units. The vectors $\boldsymbol{E}$ and $\boldsymbol{H}$ are the electric and magnetic intensities. The electric displacement and the magnetic flux density are represented by the symbols $\boldsymbol{D}$ and $\boldsymbol{B}$. The quantities ρ and $\boldsymbol{J}$ are the scalar charge density and the vector current density.

$$\nabla \times \boldsymbol{E} = -\frac{\partial \boldsymbol{B}}{\partial t}, \tag{7.1}$$

$$\nabla \cdot \boldsymbol{D} = \rho, \tag{7.2}$$

$$\nabla \times \boldsymbol{H} = \frac{\partial \boldsymbol{D}}{\partial t} + \boldsymbol{J}, \tag{7.3}$$

$$\nabla \cdot \boldsymbol{B} = 0. \tag{7.4}$$

In the absence of dielectric and magnetic materials

$$\boldsymbol{D} = \varepsilon_0 \boldsymbol{E}, \tag{7.5}$$

$$\boldsymbol{B} = \mu_0 \boldsymbol{H}, \tag{7.6}$$

where ε_0 and μ_0 are the permittivity and permeability of free space.

The classical force on a charged particle moving with velocity $\boldsymbol{v}$, is given by the Lorentz force

$$\boldsymbol{F} = q[E + (\boldsymbol{v} \times \boldsymbol{B})], \tag{7.7}$$

which can be written as a force density in terms of ρ and $\boldsymbol{J}$

$$f = \rho E + \boldsymbol{J} \times \boldsymbol{B}. \tag{7.8}$$

It is assumed that these equations are familiar and that their experimental origins need not be discussed.

7.2 THE WAVE EQUATION AND THE SPEED OF LIGHT

In free space ρ and $\boldsymbol{J}$ are set equal to zero and we may eliminate $\boldsymbol{B}$ and $\boldsymbol{D}$ from Maxwell's equations, using Eqs. (7.5) and (7.6). We then obtain

$$\nabla \times \boldsymbol{E} = -\mu_0 \frac{\partial \boldsymbol{H}}{\partial t}, \tag{7.9}$$

$$\nabla \cdot \boldsymbol{E} = 0, \tag{7.10}$$

$$\nabla \times \boldsymbol{H} = \varepsilon_0 \frac{\partial \boldsymbol{E}}{\partial t}, \tag{7.11}$$

$$\nabla \cdot \boldsymbol{H} = 0. \tag{7.12}$$

Taking the curl of Eqs. (7.9) and (7.11) and using vector identity 11 from Appendix V, we obtain

$$-\nabla^2 \boldsymbol{E} = -\mu_0 \frac{\partial}{\partial t}(\nabla \times \boldsymbol{H}), \tag{7.13}$$

$$-\nabla^2 \boldsymbol{H} = \varepsilon_0 \frac{\partial}{\partial t}(\nabla \times \boldsymbol{E}), \tag{7.14}$$

where the divergence terms have been eliminated by using Eqs. (7.10) and (7.12). Eliminating the curl from the right-hand side by using Eqs. (7.9) and (7.11), we obtain the well-known wave equations for the vectors $\boldsymbol{E}$ and $\boldsymbol{H}$:

$$\nabla^2 \boldsymbol{E} = \frac{1}{c^2}\frac{\partial^2 \boldsymbol{E}}{\partial t^2}, \tag{7.15}$$

$$\nabla^2 \boldsymbol{H} = \frac{1}{c^2}\frac{\partial^2 \boldsymbol{H}}{\partial t^2}. \tag{7.16}$$

These represent six equations, one for each component of $\boldsymbol{E}$ and $\boldsymbol{H}$. These equations may not be solved independently, however, for the various components are still interrelated by Maxwell's equations. We have, of course, set

$$c = (\varepsilon_0 \mu_0)^{-1/2}.$$

Let us seek solutions of the wave equations which have a simple exponential time dependence. That is, we seek solutions for the components which can be written* as

$$\begin{aligned} \boldsymbol{E}(\boldsymbol{r}, t) &= \boldsymbol{E}_0(\boldsymbol{r})\, e^{-i\omega t}, \\ \boldsymbol{H}(\boldsymbol{r}, t) &= \boldsymbol{H}_0(\boldsymbol{r})\, e^{-i\omega t}. \end{aligned} \tag{7.17}$$

Placing this form into the wave equation, we find after cancelling the exponential

$$\nabla^2 F_i + k^2 F_i = 0, \tag{7.18}$$

where F_i stands for any of the six components and $|\boldsymbol{k}| = \omega/c$. Particularly simple solutions of Eq. (7.18) are $e^{i\boldsymbol{k}\cdot\boldsymbol{r}}$, giving the space and time dependence as

$$F_i = A_i\, e^{i(\boldsymbol{k}\cdot\boldsymbol{r} - \omega t)}.$$

---

* An equally acceptable choice for the time dependence would be $e^{+i\omega t}$. However, the choice of sign in the exponent largely determines all sign conventions for polarization states. To be consistent with the conventions developed in previous chapters we must choose the minus sign here.

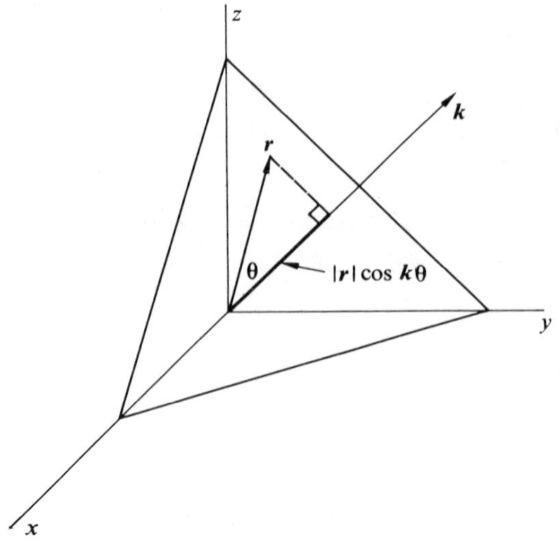

Fig. 7.1. A plane of constant phase for the function $e^{i\boldsymbol{k}\cdot\boldsymbol{r}}$.

Frequently one chooses either the real or imaginary parts, or some real combination of them, and writes a real solution as

$$F_i = A_i \sin(\boldsymbol{k}\cdot\boldsymbol{r} - \omega t + \phi), \tag{7.19}$$

where ϕ is an arbitrary phase angle.

In Fig. 7.1 is illustrated a plane surface which is normal to the direction of $\boldsymbol{k}$. The vector $\boldsymbol{r}$ locates a point in this plane. The term $\boldsymbol{k}\cdot\boldsymbol{r}$ is given by $|\boldsymbol{k}||\boldsymbol{r}|\cos\theta$. But, since $|\boldsymbol{r}|\cos\theta$ is the projection of $\boldsymbol{r}$ on to $\boldsymbol{k}$, this quantity is independent of the location of point $\boldsymbol{r}$, so long as that point lies on the plane. Hence, $\boldsymbol{k}\cdot\boldsymbol{r}$ and $\sin(\boldsymbol{k}\cdot\boldsymbol{r} - \omega t + \phi)$ are constant over the plane at any instant of time. As time increases, this plane must move outward along $\boldsymbol{k}$ with velocity $c = \omega/|\boldsymbol{k}|$ in order to keep $(\boldsymbol{k}\cdot\boldsymbol{r} - \omega t)$ constant over the plane and constant in time. Thus, Eq. (7.19) represents a plane wave solution traveling in the direction of $\boldsymbol{k}$ with velocity c.

We have shown, therefore, that there exists a class of solutions to Maxwell's equations which propagate through space with a velocity equal to the velocity of light. Incidentally, we are not surprised to find that the scalar state functions which we found in previous chapters are acceptable as components of the electric and magnetic fields. We are surprised to find that whereas we assumed before that we could describe light by a scalar function, we must now use two vector functions. This is an indication of difficulties whose discussion we postpone.

7.3 CONSERVATION OF ENERGY IN THE ELECTROMAGNETIC FIELD

If we are to associate an electromagnetic field with light, then we must show that the field transports energy and we must find an expression for the energy transport in terms of the field variables. Consider some fixed volume in space which contains a charge and current distribution, ρ and $\boldsymbol{J}$. We make the following definitions.

u is the density of any energy associated with the electromagnetic field itself. This quantity does not include any kinetic or potential energy of the charge distribution.

$\boldsymbol{S}$ is the flux vector of the field energy, known as the Poynting vector. If $\boldsymbol{n}$ is a unit vector normal to a small area dA, $(\boldsymbol{S} \cdot \boldsymbol{n})\, dA$ is the amount of field energy which flows through dA per unit time.

dW/dt is the rate at which the electromagnetic field does work on the charge and current distributions.

We may now write the conservation of energy in the following equation:

$$-\frac{\partial}{\partial t}\int u\,dV = \int \frac{dW}{dt}\,dV + \int dA\,\boldsymbol{S}\cdot\boldsymbol{n}. \tag{7.20}$$

The left-hand side is obviously the rate at which the total field energy inside the volume is decreasing. Assuming conservation of energy, the energy lost by the field must appear either as an increase in the energy of the charge and current distributions, the first term on the right, or as a flow of energy out of the volume, the second term on the right.

The force density on the distributions is given by Eq. (7.8) and the rate at which this force does work on the distributions is $\boldsymbol{f}\cdot\boldsymbol{v}$, where $\boldsymbol{v}$ is the velocity of the charge distributions. Thus,

$$\frac{dW}{dt} = \boldsymbol{f}\cdot\boldsymbol{v} = (\rho\boldsymbol{E} + \boldsymbol{J}\times\boldsymbol{B})\cdot\boldsymbol{v}.$$

The second term in the dot product is zero, since the current density is in the direction of the charge velocity. Thus, we have

$$\frac{dW}{dt} = \rho\boldsymbol{E}\cdot\boldsymbol{v} = \boldsymbol{E}\cdot\boldsymbol{J}. \tag{7.21}$$

By using Eqs. (7.1) and (7.3), we may write

$$\boldsymbol{E}\cdot\boldsymbol{J} = \boldsymbol{E}\cdot\left(\nabla\times\boldsymbol{H} - \frac{\partial\boldsymbol{D}}{\partial t}\right),$$

$$0 = \boldsymbol{H}\cdot\left(\nabla\times\boldsymbol{E} + \frac{\partial\boldsymbol{B}}{\partial t}\right).$$

Subtracting one equation from the other, we obtain

$$\boldsymbol{H}\cdot(\nabla\times\boldsymbol{E})-\boldsymbol{E}\cdot(\nabla\times\boldsymbol{H})+\boldsymbol{H}\cdot\frac{\partial\boldsymbol{B}}{\partial t}+\boldsymbol{E}\cdot\frac{\partial\boldsymbol{D}}{\partial t}+\boldsymbol{E}\cdot\boldsymbol{J}=0.$$

We then apply vector identity 9 to the first two terms and eliminate $\boldsymbol{B}$ and $\boldsymbol{D}$ by using Eqs. (7.5) and (7.6) and find

$$\nabla\cdot(\boldsymbol{E}\times\boldsymbol{H})+\frac{1}{2}\frac{\partial}{\partial t}[\mu_0 H^2+\varepsilon_0 E^2]+\boldsymbol{E}\cdot\boldsymbol{J}=0.$$

This may be integrated over the volume of space in question to yield

$$-\frac{\partial}{\partial t}\int\tfrac{1}{2}(\mu_0 H^2+\varepsilon_0 E^2)dV=\int\boldsymbol{E}\cdot\boldsymbol{J}\,dV+\int\nabla\cdot(\boldsymbol{E}\times\boldsymbol{H})\,dV.$$

Finally, using vector identity 14, we obtain

$$-\frac{\partial}{\partial t}\int\tfrac{1}{2}(\varepsilon_0 E^2+\mu_0 H^2)\,dV=\int\frac{dW}{dt}\,dV+\int(\boldsymbol{E}\times\boldsymbol{H})\cdot\boldsymbol{n}\,dA. \tag{7.22}$$

If we compare this with Eq. (7.20), it is reasonable to make the identifications

$$u=\tfrac{1}{2}(\mu_0 H^2+\varepsilon_0 E^2), \tag{7.23}$$

$$\boldsymbol{S}=\boldsymbol{E}\times\boldsymbol{H}, \tag{7.24}$$

which are the expressions we were seeking for the energy density of the electromagnetic field and the Poynting vector.

7.4 CONSERVATION OF LINEAR MOMENTUM

In the previous chapters we have discussed the experimental result that the ratio of the energy carried per photon to the linear momentum per photon is the velocity of light. We might then postulate a vector flux of linear momentum given by $\boldsymbol{S}/c$. In some sense, the flux of a quantity is the density of the quantity times the velocity with which the quantity is being carried. It is tempting to associate a momentum density with the field given by

$$\boldsymbol{g}=\frac{1}{c}\left(\frac{\boldsymbol{S}}{c}\right). \tag{7.25}$$

This is indeed correct, but we must prove that the field equations predict this result. We shall also see that the momentum flux is not of the simple form postulated.

In the previous section we were able to define a flux vector $\boldsymbol{S}$ of the scalar quantity u. Linear momentum, $\boldsymbol{p}$, is itself a vector and its flux will be more complicated than that of a simple vector. We define $\boldsymbol{P}$ to be the total momentum density, which is the sum of $\boldsymbol{p}$, the momentum density of the charge and current distributions, and $\boldsymbol{g}$, the momentum density of the fields. Not only is total momentum conserved, but each component is individually conserved. We can therefore talk about a flux vector for each scalar component of linear momentum. Define $\boldsymbol{T}_i$ to be the flux vector of the i-th component of field momentum g_i. Then, as in the preceding section, $\boldsymbol{T}_i \cdot \boldsymbol{n}\, dA$ is the amount of the i-th component of field momentum which flows through dA per unit time. $\boldsymbol{T}_i$, being a vector, has components. Define

$$T_{ij} = j\text{-th component of the flux of the } i\text{-th component of field linear momentum.} \tag{7.26}$$

The nine quantities T_{ij} may be thought of as a matrix which is known as Maxwell's tensor.

The rate at which the momentum $\boldsymbol{p}\, dV$ in volume dV is increasing through interaction with the fields is given by $d\boldsymbol{p}/dt\, dV$. We may now write the equation for conservation of the i-th component of linear momentum over some fixed volume of space:

$$-\frac{\partial}{\partial t}\int g_i\, dV = \int \frac{dp_i}{dt}\, dV + \int \boldsymbol{T}_i \cdot \boldsymbol{n}\, dA. \tag{7.27}$$

A decrease in the field momentum must appear either as an increase in the charge momentum or as a flux of momentum out of the volume. From Newton's second law

$$\frac{d\boldsymbol{p}}{dt} = \boldsymbol{f} = \rho \boldsymbol{E} + \boldsymbol{J} \times \boldsymbol{B}.$$

By use of Eqs. (7.2) and (7.3), this may be written

$$\begin{aligned}\frac{d\boldsymbol{p}}{dt} &= (\nabla \cdot \boldsymbol{D})\boldsymbol{E} + \left(\nabla \times \boldsymbol{H} - \frac{\partial \boldsymbol{D}}{\partial t}\right) \times \boldsymbol{B} \\ &= (\nabla \cdot \boldsymbol{D})\boldsymbol{E} - \mu_0 \boldsymbol{H} \times (\nabla \times \boldsymbol{H}) - \frac{1}{c^2}\left(\frac{\partial}{\partial t}(\boldsymbol{E} \times \boldsymbol{H}) - \boldsymbol{E} \times \frac{\partial \boldsymbol{H}}{\partial t}\right) \\ &= -\frac{\partial}{\partial t}\left(\frac{\boldsymbol{S}}{c^2}\right) + \left[(\nabla \cdot \boldsymbol{D})\boldsymbol{E} + \frac{1}{c^2}\boldsymbol{E} \times \frac{\partial \boldsymbol{H}}{\partial t} - \mu_0 \boldsymbol{H} \times (\nabla \times \boldsymbol{H})\right].\end{aligned}$$

By comparing this with Eqs. (7.25) and (7.27), we see that we must express the i-th component of the vector in brackets as the divergence of some vector. If this is possible, then vector identity 14 will allow us to write the volume integral as a surface integral, as in Eq. (7.27). Adding $\boldsymbol{H}(\nabla \cdot \boldsymbol{B}) = 0$ to the bracket and using Eq. (7.1), we obtain

$$(\nabla \cdot \boldsymbol{D})\boldsymbol{E} + (\nabla \cdot \boldsymbol{B})\boldsymbol{H} - \varepsilon_0 \boldsymbol{E} \times (\nabla \times \boldsymbol{E}) - \mu_0 \boldsymbol{H} \times (\nabla \times \boldsymbol{H}).$$

From vector identity 8,

$$\boldsymbol{H}\times(\nabla\times\boldsymbol{H}) = \tfrac{1}{2}\nabla(H^2) - (\boldsymbol{H}\cdot\nabla)\boldsymbol{H},$$
$$\boldsymbol{E}\times(\nabla\times\boldsymbol{E}) = \tfrac{1}{2}\nabla(E^2) - (\boldsymbol{E}\cdot\nabla)\boldsymbol{E}.$$

This yields for the bracketed term

$$\varepsilon_0[(\nabla\cdot\boldsymbol{E})\boldsymbol{E} + (\boldsymbol{E}\cdot\nabla)\boldsymbol{E}] + \mu_0[(\nabla\cdot\boldsymbol{H})\boldsymbol{H} + (\boldsymbol{H}\cdot\nabla)\boldsymbol{H}] - \tfrac{1}{2}\nabla[\varepsilon_0 E^2 + \mu_0 H^2].$$

Letting x_i ($i = 1, 2, 3$) represent any of the three coordinates x, y, or z, we may express the i-th component of this vector as

$$-\frac{\partial}{\partial x_i}[\tfrac{1}{2}(\varepsilon_0 E^2 + \mu_0 H^2)] + \sum_j \varepsilon_0\left(E_j\frac{\partial}{\partial x_j}E_i + E_i\frac{\partial}{\partial x_j}E_j\right) + \sum \mu_0\left(H_j\frac{\partial}{\partial x_j}H_i + H_i\frac{\partial}{\partial x_j}H_j\right)$$
$$= \sum_j \frac{\partial}{\partial x_j}[\mu_0 H_j H_i + \varepsilon_0 E_j E_i - \tfrac{1}{2}\delta_{ij}(\varepsilon_0 E^2 + \mu_0 H^2)]$$
$$= \nabla\cdot\boldsymbol{T}_i,$$

where

$$(\boldsymbol{T}_i)_j = \mu_0 H_j H_i + \varepsilon_0 E_j E_i - \tfrac{1}{2}\delta_{ij}(\varepsilon_0 E^2 + \mu_0 H^2),$$

which completes the identification. We can thus set

$$\boldsymbol{g} = \frac{\boldsymbol{S}}{c^2}.$$

7.5 PLANE WAVE SOLUTIONS

Returning now to the solutions we have found for the wave equation, we write the electric and magnetic field as

$$\boldsymbol{E} = \boldsymbol{E}_0\, e^{i(\boldsymbol{k}\cdot\boldsymbol{r}-\omega t)},$$
$$\boldsymbol{H} = \boldsymbol{H}_0\, e^{i(\boldsymbol{k}\cdot\boldsymbol{r}-\omega t)}, \tag{7.28}$$

where $\boldsymbol{E}_0$ and $\boldsymbol{H}_0$ are constant vectors. The entire space and time dependence of the field is included in the exponential. The vectors $\boldsymbol{E}_0$ and $\boldsymbol{H}_0$ are not independent of each other, or of $\boldsymbol{k}$, as we shall see. The solutions in Eq. (7.28) are related by Maxwell's equations. We now put Eq. (7.28) into Eqs. (7.9) and (7.11). Using vector identity 7, we obtain

$$\nabla\times\boldsymbol{E} = \nabla(e^{i(\boldsymbol{k}\cdot\boldsymbol{r}-\omega t)})\times\boldsymbol{E}_0 + e^{i(\boldsymbol{k}\cdot\boldsymbol{r}-\omega t)}\,\nabla\times\boldsymbol{E}_0,$$
$$\nabla\times\boldsymbol{H} = \nabla(e^{i(\boldsymbol{k}\cdot\boldsymbol{r}-\omega t)})\times\boldsymbol{H}_0 + e^{i(\boldsymbol{k}\cdot\boldsymbol{r}-\omega t)}\,\nabla\times\boldsymbol{H}_0,$$

where the last terms vanish, since $\boldsymbol{E}_0$ and $\boldsymbol{H}_0$ are constant vectors. Using the fact that

$$\nabla(e^{(i\boldsymbol{k}\cdot\boldsymbol{r}-\omega t)}) = i\boldsymbol{k}\, e^{i(\boldsymbol{k}\cdot\boldsymbol{r}-\omega t)},$$

we obtain

$$\begin{aligned} i\boldsymbol{k} \times \boldsymbol{E}_0 &= i\omega\mu_0\boldsymbol{H}_0, \\ i\boldsymbol{k} \times \boldsymbol{H}_0 &= -i\omega\varepsilon_0\boldsymbol{E}_0, \end{aligned} \tag{7.29}$$

which tell us that $\boldsymbol{E}_0$, $\boldsymbol{H}_0$, and $\boldsymbol{k}$ are mutually perpendicular and that $\boldsymbol{k}$ is in the direction of $\boldsymbol{S} = \boldsymbol{E} \times \boldsymbol{H}$. Thus, the direction of propagation of the wave, $\boldsymbol{k}$, is in the direction of propagation of the energy flux, $\boldsymbol{S}$. Although this is a reasonable result, it must be pointed out that it results from the particular form of the wave we have picked and is not general. Since the vectors are orthogonal, we may replace them by their magnitude. Remembering that $\omega = |\boldsymbol{k}|c$, we obtain

$$|\boldsymbol{E}_0| = \mu_0 c|\boldsymbol{H}_0|, \qquad |\boldsymbol{H}_0| = \varepsilon_0 c|\boldsymbol{E}_0|,$$

or, for the plane wave solutions,

$$\sqrt{\varepsilon_0}|\boldsymbol{E}_0| = \sqrt{\mu_0}|\boldsymbol{H}_0|. \tag{7.30}$$

Thus, for plane waves we may write

$$\boldsymbol{S} = \varepsilon_0 c E_0^2 \boldsymbol{k} = \mu_0 c H_0^2 \boldsymbol{k}, \tag{7.31}$$

$$u = \varepsilon_0 E_0^2 = \mu_0 H_0^2. \tag{7.32}$$

7.6 THE POLARIZATION OF PLANE WAVES

A representation of a wave of the form of Eq. (7.28) is given in Fig. 7.2, where we have taken $\boldsymbol{k}$ to be along the z-axis. The vectors drawn indicate the direction of $\boldsymbol{E}$ and $\boldsymbol{H}$ at any point in a plane, perpendicular to the z-axis, which cuts the z-axis at the point which locates the tail of the vector. Since we have chosen a particular direction for these vectors, Eq. (7.28) becomes

$$\begin{aligned} \boldsymbol{E} &= E_x\, e^{i(kz-\omega t)}\, \boldsymbol{e}_x, \\ \boldsymbol{H} &= H_y\, e^{i(kz-\omega t)}\, \boldsymbol{e}_y. \end{aligned} \tag{7.33}$$

Here and in what follows we plot only the real part of the exponential. Actually, it is not necessary to draw both $\boldsymbol{E}$ and $\boldsymbol{H}$. If the direction of $\boldsymbol{k}$ is given, then the results of the last section uniquely determine the magnitude and direction of $\boldsymbol{H}$ when $\boldsymbol{E}$ is given. From now on we shall draw only the vector $\boldsymbol{E}$. For the wave shown in Fig. 7.2 the vector $\boldsymbol{E}$ points in the x-direction. This will not always be so. The particular wave we have shown is the electromagnetic representation for a $|P_x\rangle$-polarization state. We are thus

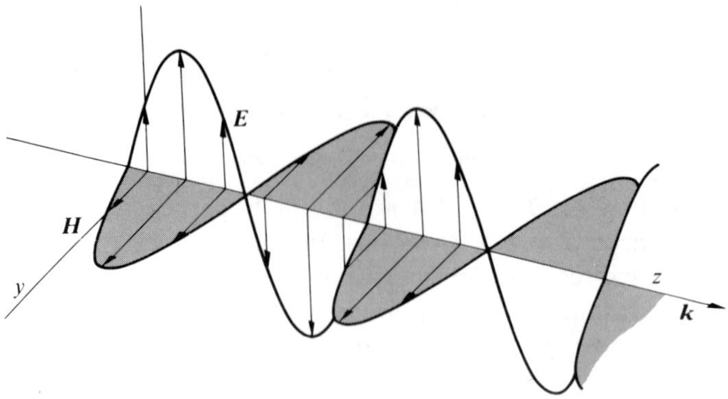

Fig. 7.2. The electromagnetic plane wave.

identifying the line of a P-state with the direction of the electric vector. To obtain a P-state whose line is at angle θ with the x-axis, we would simply rotate Fig. 7.2 through an angle θ about the z-axis. To complete the identification of this wave with a P-state, we will describe how a P-filter may be constructed whose transmission properties obey Malus' law. In Fig. 7.3 the electromagnetic wave, whose $\boldsymbol{E}$ vector is shown, is assumed to be coming out of the page. The series of straight lines represents fine conducting wires stretched in front of the wave. The electric field has been broken into components parallel and perpendicular to these wires. The spacing of the wires is comparable to the wavelength of the light. From Eq. (7.31) the energy flux incident on the wire can be written as the sum of fluxes associated with these two perpendicular components of the electric field:

$$\boldsymbol{S}_0 = \varepsilon_0 c E_0^2 \boldsymbol{k} = \varepsilon_0 c (E_\perp^2 + E_\parallel^2)\boldsymbol{k}. \tag{7.34}$$

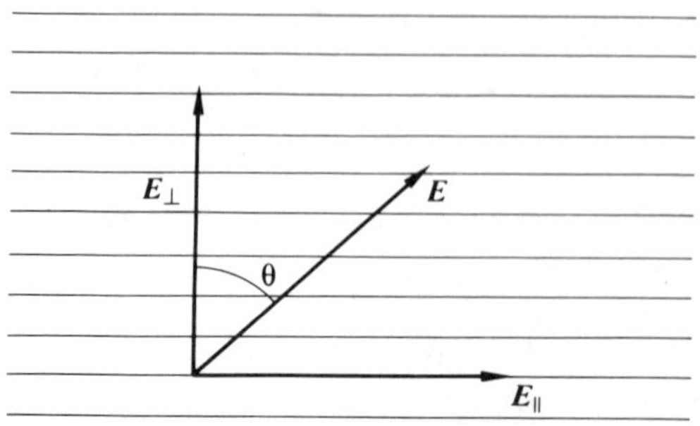

Fig. 7.3. The electric vector of a plane wave incident on a wire grid may be broken into components $\boldsymbol{E}_\perp$ and $\boldsymbol{E}_\parallel$, perpendicular and parallel to the wires.

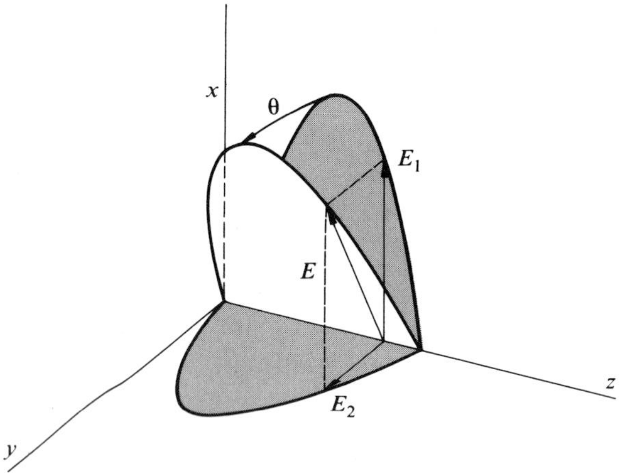

Fig. 7.4. The electric vector for a *P*-state as the sum of electric vectors of orthogonal *P*-states.

The oscillating component of the field parallel to the wires will set up an oscillating current in these wires. The energy of this component of the field will be dissipated in the form of heat owing to the resistance of the wires. The perpendicular component cannot set up a current across the wires, however, and its energy will not be dissipated but will be transmitted through the array. The transmitted intensity, $\boldsymbol{S}$, will therefore be

$$\boldsymbol{S} = \varepsilon_0 c E_{\perp}^2 \boldsymbol{k} = \boldsymbol{S}_0 \cos^2 \theta, \tag{7.35}$$

which is Malus' law. For microwaves whose wavelength is of the order of centimeters the construction of such a *P*-filter is trivial. For light near the visible region one would need to place about 20,000 wires side by side in a space of 1 cm! It has, in fact, been possible to do this! For the details the reader is referred to the literature.*

Since the wave equation is a linear equation, the sum of any two solutions is also a solution. Let us add together two waves, one whose electric vector is in the x-direction and one whose electric vector is in the y-direction. In Fig. 7.4 the vector addition is shown explicitly at only one point but it is clear that the result is a new *P*-state whose line makes an angle θ with the x-axis given by

$$\tan \theta = E_2/E_1. \tag{7.36}$$

This reminds us of the fact that the *P*-state

$$|P\rangle = a|P_x\rangle + b|P_y\rangle,$$

---

* Shurcliff, W. A., and S. S. Ballard, *Polarized Light*, pp. 26 ff, Van Nostrand, Princeton, N.J. (1964).

is a P-state whose line is given by

$$\tan \theta = b/a.$$

We will return to this shortly. Let us now examine the electromagnetic representation of elliptical states. In Fig. 7.4 we have added two waves which are exactly in phase. The two waves reach their maxima and minima at exactly the same points on the axis. Let us now add the two waves in such a way that when one has a maximum, the other is zero. The result of the vector addition is indicated in Fig. 7.5 for the case of equal amplitudes for the x- and y-waves. It is seen that the tip of the $\boldsymbol{E}$-vector lies on a circular spiral. The spiral is that of a left-handed screw. The wave in Fig. 7.5 is the electromagnetic plane wave representation of right circular polarization. As this wave passes through any plane perpendicular to the z-axis, such as the $x'y'$-plane, it is seen that the electric vector in that plane will rotate in a counterclockwise sense, when the wave is traveling toward the observer. By comparing Figs. 7.4 and 7.5 we see that right circularly polarized light may be obtained from a P-state whose line is midway between the x- and y-axes, by retarding $\boldsymbol{E}_y$ by one-quarter of a wavelength. Alternatively, one

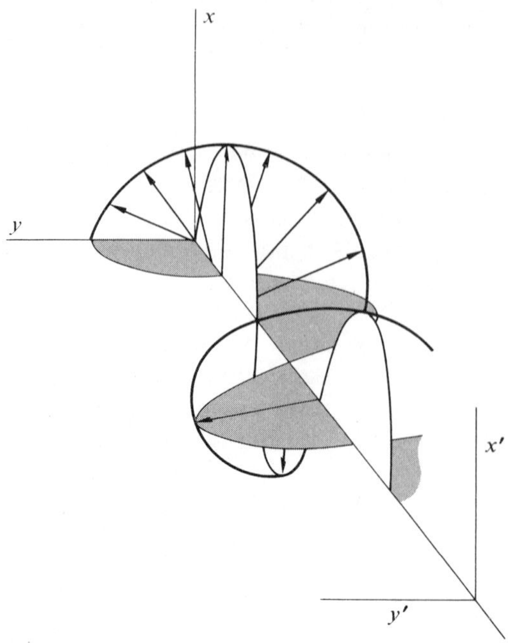

Fig. 7.5. The representation of a right circular polarization state in terms of the electric vector of a plane wave.

may retard $\boldsymbol{E}_y$ by one-eighth of a wavelength while advancing $\boldsymbol{E}_x$ by the same amount. This yields

$$\begin{aligned} \boldsymbol{E}_y &= E_y\, e^{i[k(z+\lambda/8)-\omega t]}\, \boldsymbol{e}_y \\ &= e^{+i\pi/4}\, E_y\, e^{i(kz-\omega t)}\, \boldsymbol{e}_y, \\ \boldsymbol{E}_x &= E_x\, e^{i[k(z-\lambda/8)-\omega t]}\, \boldsymbol{e}_x \\ &= e^{-i\pi/4}\, E_x\, e^{i(kz-\omega t)}\, \boldsymbol{e}_x. \end{aligned} \tag{7.37}$$

Thus, if a P-state whose line is midway between the x- and y-axes may be represented by

$$\begin{aligned} \boldsymbol{E} &= \boldsymbol{E}_1 + \boldsymbol{E}_2, \\ \boldsymbol{E}_1 &= E_0\, e^{i(kz-\omega t)}\, \boldsymbol{e}_x, \\ \boldsymbol{E}_2 &= E_0\, e^{i(kz-\omega t)}\, \boldsymbol{e}_y, \end{aligned}$$

a right circular state is given by

$$\boldsymbol{E} = e^{-i\pi/4}\, \boldsymbol{E}_1 + e^{+i\pi/4}\, \boldsymbol{E}_2.$$

The preceding examples are sufficient to identify the Jones vector associated with a polarization state as

$$|E\rangle = C \begin{pmatrix} E_x \\ E_y \end{pmatrix}, \tag{7.38}$$

where the constant C will be determined by the interpretation of $\langle E|E\rangle$ as an intensity or a probability. The general elliptical state differs from the wave shown in Fig. 7.5 only in that the amplitudes of the two waves need not be equal and that the entire wave may be rotated about the z-axis. As a general elliptical state passes through a plane perpendicular to the z-axis, the electric vector traces out an ellipse, as shown in Fig. 7.6. The direction

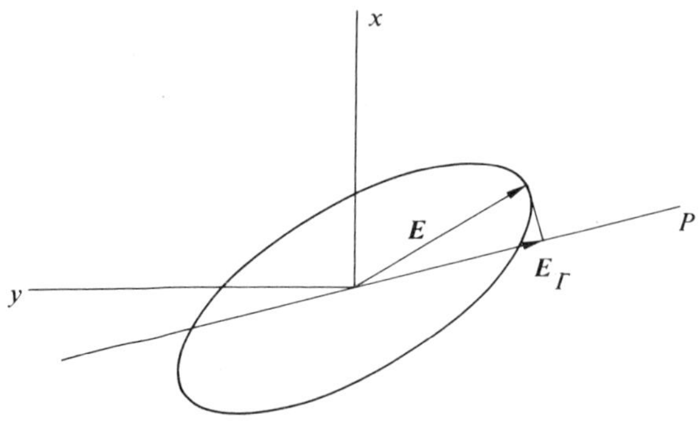

Fig. 7.6. The maximum projection of the electric vector of an elliptical state along the line of a given P-filter.

of rotation is counterclockwise or clockwise as the polarization is right- or left-handed.

If a *P*-filter is placed in this beam, the transmitted electric vector will have the magnitude of the maximum projection of $\boldsymbol{E}$ on the line of the filter. This is shown as $\boldsymbol{E}_T$ in Fig. 7.6. In terms of the electromagnetic representation of states the interpretation of Fig. 3.3 should now be apparent.

7.7 THE LAWS OF REFLECTION AND REFRACTION

Continuing to develop the electromagnetic representation of light, we must show that electromagnetic waves obey the laws of geometric optics, Snell's law, and the law of reflection. In Fig. 7.7 the xy-plane is taken to be the boundary between two dielectric media. Above this plane the index of refraction is given by

$$n_1 = c/v_1 = \sqrt{\frac{\varepsilon_1 \mu_1}{\varepsilon_0 \mu_0}}, \tag{7.39}$$

and below the plane it is given by

$$n_2 = c/v_2 = \sqrt{\frac{\varepsilon_2 \mu_2}{\varepsilon_0 \mu_0}}. \tag{7.40}$$

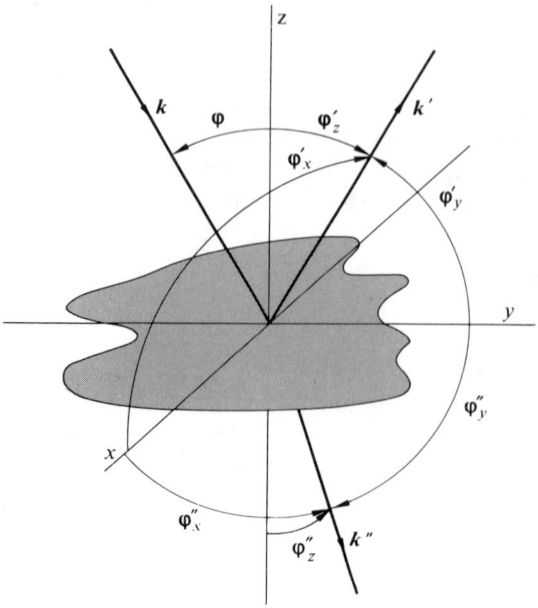

Fig. 7.7. General orientation of the wave vectors $\boldsymbol{k}$, $\boldsymbol{k}'$, and $\boldsymbol{k}''$ for reflection and refraction at a plane interface.

A plane wave is incident having wave vector $\boldsymbol{k}$ which, without loss of generality, is placed in the yz-plane. A reflected wave is taken in the general direction $\boldsymbol{k}'$ whose orientation is given by the ϕ'-angles. A transmitted or refracted wave is located by $\boldsymbol{k}''$ and the angles ϕ''. In what follows the cosines of angles are written γ, e.g., $\cos\phi''_z = \gamma''_z$. Remembering that $|\boldsymbol{k}| = \omega/c$ and $n = c/v$, we write a component of a plane electromagnetic wave as

$$E_i = A_i \exp\left(in\frac{\omega}{c}\frac{\boldsymbol{k}}{|\boldsymbol{k}|}\cdot \boldsymbol{r}\right)\exp(-i\omega t).$$

The constants A_i may be complex, thus allowing for a phase difference between the different components. We do not, therefore, limit ourselves to plane polarized P-states. We thus write for the incident wave

$$\begin{aligned} E_x &= A_x \exp[in_1\omega/c\,(-z\cos\phi + y\sin\phi)]\exp(-i\omega t),\\ E_y &= A_y \exp[in_1\omega/c\,(-z\cos\phi + y\sin\phi)]\exp(-i\omega t); \end{aligned} \tag{7.41}$$

for the refracted wave

$$\begin{aligned} E''_x &= A''_x \exp[in_2\omega''/c(\gamma''_x x + \gamma''_y y + \gamma''_z z)]\exp(-i\omega'' t),\\ E''_y &= A''_y \exp[in_2\omega''/c(\gamma''_x x + \gamma''_y y + \gamma''_z z)]\exp(-i\omega'' t); \end{aligned} \tag{7.42}$$

and for the reflected wave

$$\begin{aligned} E'_x &= A'_x \exp[in_1\omega'/c(\gamma'_x x + \gamma'_y y + \gamma'_z z)]\exp(-i\omega' t),\\ E'_y &= A'_y \exp[in_1\omega'/c(\gamma'_x x + \gamma'_y y + \gamma'_z z)]\exp(-i\omega' t), \end{aligned} \tag{7.43}$$

where we have included the possibility that reflection and refraction might alter ω.

It is well known that the component of the total electric field $\boldsymbol{E}$ which is tangent to the boundary between two simple dielectrics must take on the same value on both sides of the boundary. Thus, for $z = 0$,

$$E_x + E'_x = E''_x, \tag{7.44}$$

$$E_y + E'_y = E''_y. \tag{7.45}$$

Since Eqs. (7.44) and (7.45) are valid for all x and y, they are valid for x and y equal to zero. It is left to the problems to prove that this implies $\omega = \omega' = \omega''$. We may then cancel the time exponentials in Eqs. (7.44) and (7.45) and obtain

$$\begin{aligned} A_x \exp[in_1\omega/cy\sin\phi] + A'_x \exp[in_1\omega/c(\gamma'_x x + \gamma'_y y)]\\ = A''_x \exp[in_2\omega/c(\gamma''_x x + \gamma''_y y)],\\ A_y \exp[in_1\omega/cy\sin\phi] + A'_y \exp[in_1\omega/c(\gamma'_x x + \gamma'_y y)]\\ = A''_y \exp[in_2\omega/c(\gamma''_x x + \gamma''_y y)]. \end{aligned}$$

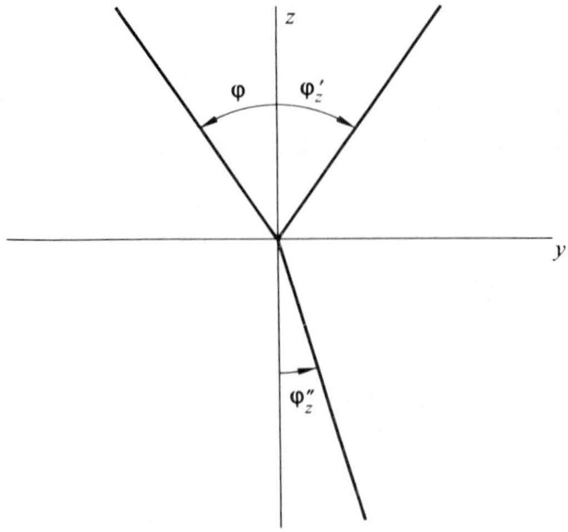

Fig. 7.8. Coplanar geometry for reflection and refraction.

The only way in which these equations may be satisfied for all values of x and y is if the exponentials may be cancelled from both sides. This is possible only if

$$\gamma_x' = \gamma_x'' = 0, \tag{7.46}$$

$$\begin{aligned} n_1 \sin \phi &= n_2 \gamma_y'' = n_1 \gamma_y', \\ n_1 \sin \phi &= n_2 \sin \phi_z'' = n_1 \sin \phi_z'. \end{aligned} \tag{7.47}$$

Equation (7.46) tells us that $\boldsymbol{k}$, $\boldsymbol{k}'$, and $\boldsymbol{k}''$ are coplanar and may be represented as in Fig. 7.8. Equation (7.47) is equivalent to the law of reflection,

$$\phi = \phi_z',$$

and the law of refraction, or Snell's law,

$$n_1 \sin \phi = n_2 \sin \phi_z''.$$

7.8 FRESNEL'S EQUATIONS

In the last section we have determined the relation between the directions of the incident wave and the directions of the reflected and refracted waves. We proceed to derive Fresnel's equations, which relate the intensities of these waves to the incident intensity. These intensities depend on the orientation of the incident electric vector. We have shown that $\boldsymbol{k}$, $\boldsymbol{k}'$, and $\boldsymbol{k}''$ lie in the yz-plane, which will be called the plane of incidence. We distinguish the parallel case where the electric vector is parallel to the plane of incidence and the perpendicular case where the electric vector is perpendicular to this plane. We first consider the parallel case, which is illustrated in Fig. 7.9. We now

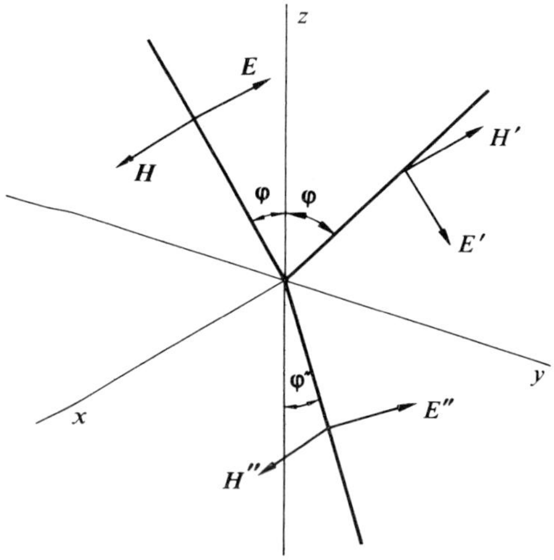

Fig. 7.9. The parallel case. The direction of the electric vector of the incident wave is parallel to the plane of incidence.

impose the condition that the tangential components of both the electric field $\boldsymbol{E}$ and the magnetic field $\boldsymbol{H}$ must be continuous across the boundary between two simple dielectrics. In the parallel case this condition leads to

$$\begin{aligned} E\cos\phi + E'\cos\phi &= E''\cos\phi'', \\ H - H' &= H''. \end{aligned} \tag{7.48}$$

Since for a plane wave $\sqrt{\varepsilon}E = \sqrt{\mu}H$, the last equation may be written

$$\sqrt{\varepsilon_1/\mu_1}\,E - \sqrt{\varepsilon_1/\mu_1}\,E' = \sqrt{\varepsilon_2/\mu_2}\,E''. \tag{7.49}$$

If the two media are simple dielectrics whose magnetic properties do not appreciably differ from those of the vacuum, we may set $\mu_1 = \mu_2 = \mu_0$, which leads to

$$n_1E - n_1E' = n_2E'',$$

and, imposing Snell's law, we obtain

$$\sin\phi''E - \sin\phi''E' = \sin\phi E''. \tag{7.50}$$

Equations (7.48) and (7.50) may be treated as two simultaneous linear equations which may be solved for E' and E'' to yield for the parallel case

$$E_p'' = \frac{2\sin\phi''\cos\phi}{\sin(\phi+\phi'')\cos(\phi-\phi'')}\,E_p, \tag{7.51}$$

$$E_p' = -\frac{\tan(\phi-\phi'')}{\tan(\phi+\phi'')}\,E_p. \tag{7.52}$$

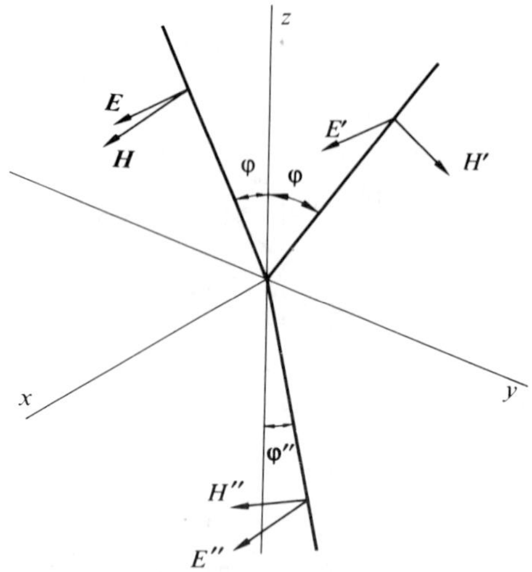

Fig. 7.10. The perpendicular case. The direction of the electric vector of the incident wave is perpendicular to the plane of incidence.

The perpendicular case is illustrated in Fig. 7.10. By methods completely equivalent to those which led to Eqs. (7.51) and (7.52) one may show that in this case

$$E_s'' = \frac{2 \sin \phi'' \cos \phi}{\sin (\phi + \phi'')} E_s, \tag{7.53}$$

$$E_s' = -\frac{\sin (\phi - \phi'')}{\sin (\phi + \phi'')} E_s. \tag{7.54}$$

Equations (7.51) through (7.54) are known as Fresnel's equations. The subscript p has an obvious origin in the parallel case. The subscript s comes from the German, *senkrecht*, meaning perpendicular. The angles ϕ and ϕ'' are, of course, related by Snell's law, $n_1 \sin \phi = n_2 \sin \phi''$. The numerical values given by Fresnel's equations are highly dependent on the ratio n_2/n_1. If this ratio is less than 1, there exists a maximum angle of incidence ϕ_c for which Snell's law admits a solution, known as the critical angle, given by

$$\sin \phi_c = n_2/n_1. \tag{7.55}$$

For angles larger than this there is no transmitted wave.

In Figs. 7.11 and 7.12 the coefficients in Fresnel's equations are plotted as a function of the incident angle ϕ, for $n_2/n_1 = 3/2 > 1$ and $n_2/n_1 = 2/3 < 1$.

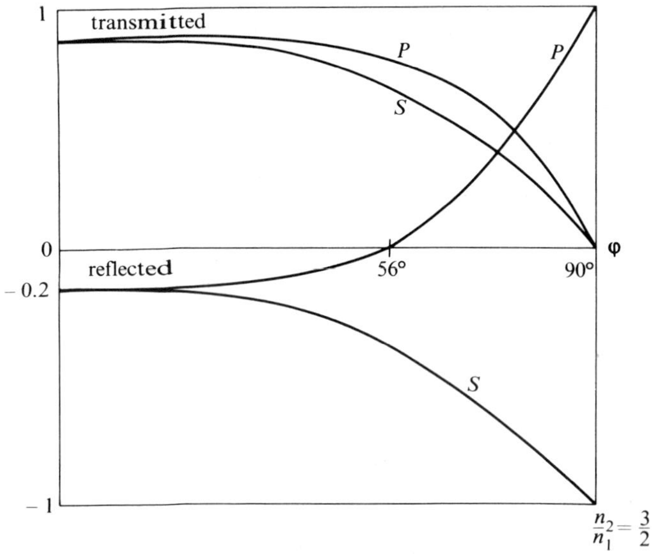

Fig. 7.11. A plot of Fresnel's equations when the incident wave is in the less dense medium. The ratio n_2/n_1 has been taken as 3/2.

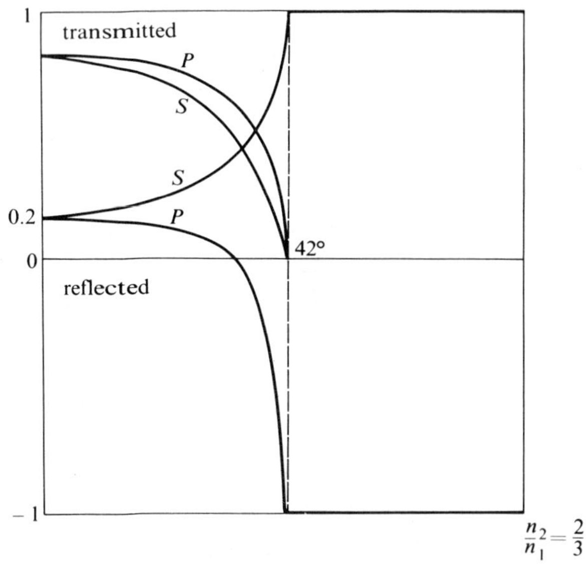

Fig. 7.12. A plot of Fresnel's equations when the incident wave is in the more dense medium. The ratio n_2/n_1 has been taken as 2/3.

7.9 THE COMPLEX FIELD

We have been using complex, rather than real, solutions to Maxwell's equations because of the simplicity this introduces into many calculations. However, it must be remembered that whenever an actual measurement of the electromagnetic fields is made, it is a real number which is obtained. In what follows we are going to show that if $V^R(\mathbf{r}, t)$ is a real scalar function of position and time—for example, one of the components of the electric field—then there is a unique and useful way to associate with it a complex function $V(\mathbf{r}, t)$. We do not assume that the field has a simple monochromatic time variation but that it may be represented as a Fourier superposition of such variations. We also suppress the explicit space dependence and write

$$V^R(t) = \frac{1}{\sqrt{2\pi}} \int_{-\infty}^{+\infty} \mathcal{V}(\omega)\, e^{-i\omega t}\, d\omega. \tag{7.56}$$

Since $V^R(t)$ is real, $V^{R\star} = V^R$. Applying this to Eq. (7.56),

$$\int_{-\infty}^{+\infty} \mathcal{V}^\star(\omega)\, e^{i\omega t}\, d\omega = \int_{-\infty}^{+\infty} \mathcal{V}(\omega)\, e^{-i\omega t}\, d\omega,$$

but, by a change of dummy variable on the right, ω being replaced by $-\omega$, we find

$$\int_{-\infty}^{\infty} \mathcal{V}^\star(\omega)\, e^{i\omega t}\, d\omega = \int_{-\infty}^{\infty} \mathcal{V}(-\omega)\, e^{i\omega t}\, d\omega,$$

or

$$\int_{-\infty}^{\infty} [\mathcal{V}^\star(\omega) - \mathcal{V}(-\omega)]\, e^{i\omega t}\, d\omega = 0.$$

Thus, $\mathcal{V}^\star(\omega) - \mathcal{V}(-\omega)$ is the Fourier transform of zero and is itself zero, giving

$$\mathcal{V}^\star(\omega) = \mathcal{V}(-\omega), \tag{7.57}$$

which is known as the *reality condition.*

We now split Eq. (7.56) into two integrals:

$$V^R(t) = \frac{1}{\sqrt{2\pi}} \int_0^{\infty} \mathcal{V}(\omega)\, e^{-i\omega t}\, d\omega + \frac{1}{\sqrt{2\pi}} \int_{-\infty}^{0} \mathcal{V}(\omega)\, e^{-i\omega t} d\omega.$$

Making use of Eq. (7.57) and again replacing ω by $-\omega$ in the second integral, we obtain

$$V^R(t) = \frac{1}{\sqrt{2\pi}} \int_0^{\infty} \mathcal{V}(\omega)\, e^{-i\omega t}\, d\omega + \frac{1}{\sqrt{2\pi}} \int_0^{\infty} \mathcal{V}^\star(\omega)\, e^{+i\omega t}\, d\omega,$$

which is of the form

$$V^R(t) = \tfrac{1}{2}[V(t) + V^\star(t)],$$

where

$$V(t) = \frac{2}{\sqrt{2\pi}} \int_0^\infty \mathscr{V}(\omega)\, e^{-i\omega t}\, d\omega. \tag{7.58}$$

Equation (7.58) defines the complex function associated with the real function given by Eq. (7.56). If V^R and V^I are the real and imaginary parts of V, it is easily shown that

$$\int_{-\infty}^{\infty} (V^R)^2\, dt = \int_{-\infty}^{\infty} (V^I)^2\, dt = \tfrac{1}{2} \int_{-\infty}^{\infty} VV^\star\, dt \tag{7.59}$$

and

$$\int_{-\infty}^{\infty} V^2\, dt = \int_{-\infty}^{\infty} (V^\star)^2\, dt = \int_{-\infty}^{\infty} V^I V^R\, dt = 0. \tag{7.60}$$

7.10 THE PSEUDOMONOCHROMATIC CASE

In no case is it possible to produce perfectly monochromatic electromagnetic radiation. Of great practical importance, therefore, is the case of nearly monochromatic radiation. This may be called either the pseudomonochromatic or the quasimonochromatic case. By these terms is meant that the Fourier transform of the radiation, $\mathscr{V}(\omega)$, is zero except in some narrow band of frequencies, $\Delta\omega$, small in comparison with the midfrequency of the band, ω_0. Values for $\Delta\omega/\omega_0$ of 10^{-2} to 10^{-3} are easily attained in the range of optical frequencies and much lower values can be obtained from laser sources; 10^{-7} to 10^{-8} are typical for lasers. In practice the pseudomonochromatic case arises from nearly monochromatic sources and/or the use of narrow band transmissions filters with all detectors.

We wish to show that in the pseudomonochromatic case $V(t)$ has a simple monochromatic time dependence, $e^{-i\omega_0 t}$, modulated by a slowly varying envelope function. To see this we multiply Eq. (7.58) by $e^{+i\omega_0 t}$:

$$e^{i\omega_0 t}\, V(t) = \frac{1}{\sqrt{2\pi}} \int_0^\infty \mathscr{V}(\omega)\, e^{-i(\omega-\omega_0)t}\, d\omega.$$

If $\mathscr{V}(\omega)$ is different from zero only if $\omega \simeq \omega_0$, then only those exponentials will contribute to the integral for which $\Delta\omega = \omega - \omega_0 \simeq 0$. Thus, the integral and, hence, $V(t)\, e^{i\omega_0 t}$, are slowly varying and we may write

$$V(t) = A(t)\, e^{i\phi(t)}\, e^{-i\omega_0 t},$$

where $A(t)$ and $\phi(t)$ are slowly varying functions compared to $e^{-i\omega_0 t}$. An illustration of the variation of $V^R(t)$ is given in Fig. 7.13, though the variation

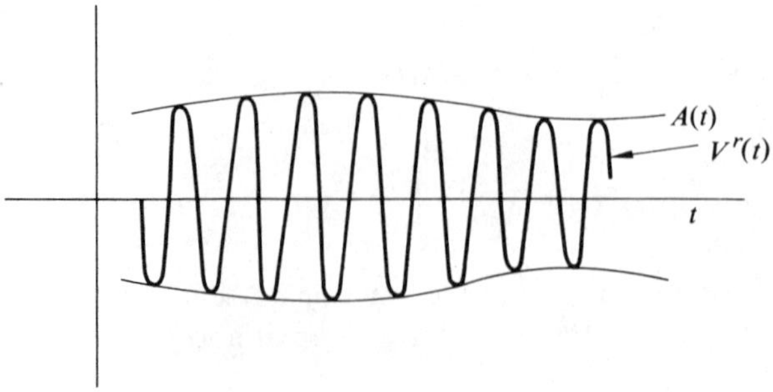

Fig. 7.13. The time variation of a pseudomonochromatic variable $V^R(t)$.

is greatly exaggerated, since $\Delta\omega/\omega_0$ gives a measure of the relative rates of variation of the envelope function $A(t)$ and the carrier $e^{-i\omega_0 t}$. Since this quantity can easily be taken as 10^{-3} or smaller, it would appear that there is little difference between the monochromatic and the pseudomonochromatic cases. On the contrary, we shall see that this difference introduces us to a whole new range of optical phenomena.

7.11 COHERENCE

The concept of coherence was briefly examined in Section 4.7. There the two extremes of completely coherent and completely incoherent superposition were discussed. We now examine in more detail these extremes and also the intermediate cases. For the moment we consider only one field variable or component of the field and postpone the problems of polarization until later. Consider any two points within the electromagnetic field located at r_1 and r_2 and a third point p. As in Fig. 7.14, points r_1 and r_2 have been isolated by the introduction of a screen with small holes at these points. If these holes are small enough, then scalar diffraction theory tells us that the field variable at p will be proportional to the sum of two disturbances, one due to each of the source holes, and may be written

$$V(p, t) = C[V_1(\mathbf{r}_1, t_1) + V_2(\mathbf{r}_2, t_2)]. \tag{7.61}$$

We have explicitly indicated that the disturbance at p at time t is due to the disturbances at r_1 and r_2 at early times t_1 and t_2. Because of the different path lengths, t_1 and t_2 are not necessarily equal. The intensity measured at point p by a detector will be given by $I \propto V^\star V$, with

$$\begin{aligned} I &= V^\star V \\ &= V_1V_1^\star + V_2V_2^\star + V_1V_2^\star + V_1^\star V_2, \end{aligned} \tag{7.62}$$

where the constants of proportion have either been set equal to 1 or absorbed into the V's. The detector will actually measure some time average of $V^\star V$.

A lower bound for this averaging time with modern detectors can be placed at about 10^{-9} sec. This is to be compared with the period of oscillation of a wavelength 6000 Å, which is 2×10^{-15} sec. The average is therefore over a comparatively long time. We are also going to assume that the field is stationary. That is, although there may be short-range time fluctuations in the field variables, an averaging process will yield the same result independent of the origin of the averaging process in time. We identify

$$I_1 = \overline{V_1 V_1^\star}, \qquad I_2 = \overline{V_2 V_2^\star},$$

where the bar indicates a time average of the intensities measured at p if only one of the apertures was open. Following Wolf, we introduce the *complex mutual coherence function* defined by

$$\Gamma_{12}(\tau) = \overline{V_1(\boldsymbol{r}_1, t + \tau) V_2^\star(\boldsymbol{r}_2, t)}. \tag{7.63}$$

Equation (7.62) may be written in terms of this function as

$$I = I_1 + I_2 + 2 \operatorname{Re} \Gamma_{12}(\tau),$$

where Re indicates the real part of the function. When $\boldsymbol{r}_1 = \boldsymbol{r}_2$, $\Gamma_{11}(\tau)$ is known as the *self-coherence function*, which as $\tau \to 0$ becomes simply the intensity. The *complex degree of coherence* is defined as

$$\gamma_{12}(\tau) = \frac{\Gamma_{12}(\tau)}{\sqrt{\Gamma_{11}(0)\Gamma_{22}(0)}} = \frac{\Gamma_{12}(\tau)}{\sqrt{I_1 I_2}}. \tag{7.64}$$

From the definition of a time average

$$\Gamma_{12}(\tau) = \overline{V_1(t + \tau) V_2^\star(t)} = \frac{1}{2T} \int_{-T}^{+T} V_1(t + \tau) V_2^\star(t)\, dt,$$

and from the Schwarz inequality for any two complex functions f and g

$$\left| \int f^\star g\, dt \right|^2 \leqslant \int f^\star f\, dt \int g^\star g\, dt.$$

It follows that

$$0 \leqslant |\gamma_{12}(\tau)| \leqslant 1.$$

In keeping with Section 4.7, the extreme values 0 and 1 are, respectively, associated with completely incoherent and coherent superpositions.

It is of some usefulness to write Γ_{12} explicitly in terms of the real and imaginary parts of V:

$$\begin{aligned}\Gamma_{12}(\tau) = {} & \overline{V_1^R(t + \tau) V_2^R(t)} + \overline{V_1^I(t + \tau) V_2^I(t)} \\ & + i\, \overline{V_1^I(t + \tau) V_2^R(t)} - i\overline{V_1^R(t + \tau) V_2^I(t)}.\end{aligned} \tag{7.65}$$

Now consider

$$\overline{V_1(t+\tau)V_2(t)} = \frac{1}{2T}\int_{-T}^{T} dt \int_0^\infty \int_0^\infty d\omega\, d\omega' \mathscr{V}(\omega)\mathscr{V}(\omega')\, e^{i\omega\tau}\, e^{i(\omega+\omega')t}.$$

If the averaging time, T, is very large, the integral over t yields a delta function, $\delta(\omega + \omega')$. The remaining integrals are therefore nonzero only for $\omega + \omega' = 0$, which implies that either ω or ω' is negative or that both are zero. But $\mathscr{V}(0) = 0$ and the integrations are only over positive ω and ω'. Therefore,

$$\overline{V_1(t+\tau)V_2(t)} = 0,$$

which, if written out in terms of real and imaginary parts, yields

$$\begin{aligned}\overline{V_1^R(t+\tau)V_2^R(t)} &= \overline{V_1^I(t+\tau)V_2^I(t)},\\ \overline{V_1^I(t+\tau)V_2^R(t)} &= -\overline{V_1^R(t+\tau)V_2^I(t)}.\end{aligned} \tag{7.66}$$

Equations (7.65) and (7.66) may be combined to yield

$$\mathrm{Re}\,\Gamma_{12}(\tau) = 2\overline{V_1^R V_2^R} = 2\overline{V_1^I V_2^I}, \tag{7.67}$$

$$\mathrm{Im}\,\Gamma_{12}(\tau) = 2\overline{V_1^I V_2^R} = -2\overline{V_1^R V_2^I}. \tag{7.68}$$

7.12 MEASUREMENT OF γ_{12}

In this section we wish to show how the modulus of the complex degree of coherence, $|\gamma_{12}(\tau)|$, may be measured and how in principle the phase of this function may also be determined. The determination of the phase of γ_{12} is still an important theoretical and experimental problem.

We consider Fig. 7.14 to represent an actual two-slit diffraction experiment with S_0 a pseudomonochromatic source. Since the source is pseudomonochromatic, we may write

$$V_1(t+\tau) = A_1(t+\tau)\, e^{-i\phi_1(t+\tau)}\, e^{-i\omega_0(t+\tau)},$$

$$V_2^\star(t) = A_2(t)\, e^{+i\phi_2(t)}\, e^{+i\omega_0 t},$$

where the functions A_1, A_2, ϕ_1, and ϕ_2 are slowly varying compared to $e^{-i\omega_0 t}$. It follows from Eq. (7.63) that $\Gamma_{12}(\tau)$ and, hence, $\gamma_{12}(\tau)$ will be slowly varying functions of time compared to $e^{-i\omega_0 t}$. It follows that we may write γ_{12} as

$$\gamma_{12}(\tau) = |\gamma_{12}(\tau)|\, e^{-i[\alpha_{12}(\tau)-\omega_0\tau]},$$

where $|\gamma_{12}|$ and α_{12} are slowly varying compared to $\omega_0\tau$. The intensity at p may then be written

$$I = I_1 + I_2 + 2\sqrt{I_1 I_2}\,|\gamma_{12}| \cos\,[\alpha_{12}(\tau) - \omega_0\tau]. \tag{7.69}$$

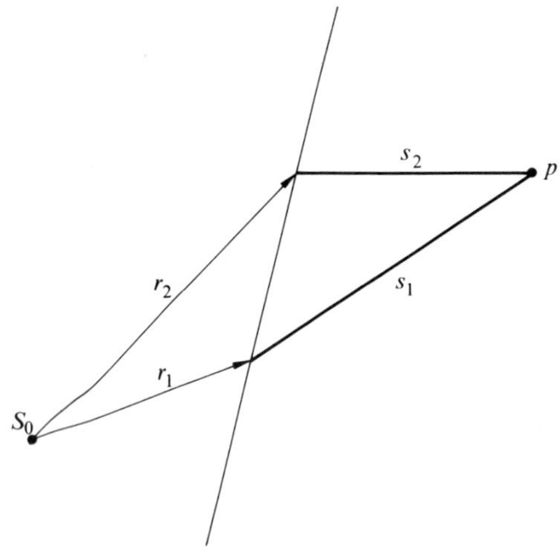

Fig. 7.14. Geometry used in defining the correlation function $\Gamma_{12}(\tau)$.

Now, τ is simply the difference in arrival times at p of the disturbances originating at r_1 and r_2 and is given by

$$\tau = \frac{s_1 - s_2}{c},$$

where s_1 and s_2 are the distances between p and r_1 and r_2. As the point p is moved across the two-slit diffraction pattern, the variation will be due primarily to $\omega_0\tau$ rather than $\alpha_{12}(\tau)$, which is slowly varying. Thus, the pattern will be essentially a cosine variation over several fringes. Using Michelson's definition of the fringe visibility,

$$\text{Visibility} = \frac{I_{\text{max.}} - I_{\text{min.}}}{I_{\text{max.}} + I_{\text{min.}}}, \tag{7.70}$$

we obtain

$$\text{Visibility} = 2\frac{\sqrt{I_1 I_2}}{I_1 + I_2}|\gamma_{12}(\tau)|.$$

If the size of the slits is adjusted so that $I_1 = I_2$, this becomes

$$\text{Visibility} = |\gamma_{12}(\tau)|,$$

which may therefore be measured experimentally. In principle, knowing ω_0 and τ as a function of position, one should be able to measure $\alpha_{12}(\tau)$ also. In practice, however, the variation of $\omega_0\tau$ is so large as to hide the dependence on α_{12}.

7.13 COMPLEX CORRELATION AND INCOHERENT SOURCES

In the preceding section we have seen how a complex correlation function, γ_{12}, can be measured by using a two-slit interference experiment. We now wish to examine how the correlation function may be related to the sources of the electromagnetic field. When the actual time variation of the field is known or can be assumed to be known, as was the case in Chapter 6, where completely coherent illumination was assumed, then the calculation of the correlation function follows directly from the definition of Eqs. (7.63) and (7.64). We now examine the other extreme, an extended source whose individual elements radiate incoherently.

In Fig. 7.15, P_1 and P_2 are the points for which we wish to calculate the function $V_1(t)\,V_2^\star(t)$. We are limiting ourselves to the case $\tau = 0$. If the source is broken into small elements dA, then the variables V_1 and V_2 will be the sum of contributions $v_1\,dA$ and $v_2\,dA$, where $v dA$ is the contribution to V from the area of the source dA. We then write

$$V_1(t) = \int dA\, v_1(t), \qquad V_2(t) = \int dA\, v_2(t).$$

We use a prime to distinguish the dummy variables of integration and $\overline{V_1(t)V_2^\star(t)}$ becomes

$$\overline{V_1(t)V_2^\star(t)} = \int\!\!\int dA\, dA'\, \overline{v_1(t)v_2^{\star\prime}(t)}.$$

If δ represents a delta function which is zero unless dA and dA' are the same, the integral may be broken into two parts:

$$\begin{aligned}\overline{V_1V_2^\star} &= \int\!\!\int dA\, dA'\, \delta\, \overline{v_1v_2^{\star\prime}} + \int\!\!\int dA\, dA'(1-\delta)\, \overline{v_1v_2^{\star\prime}} \\ &= \int dA\, \overline{v_1v_2^\star} + \int\!\!\int dA\, dA'(1-\delta)\, \overline{v_1v_2^{\star\prime}}.\end{aligned}$$

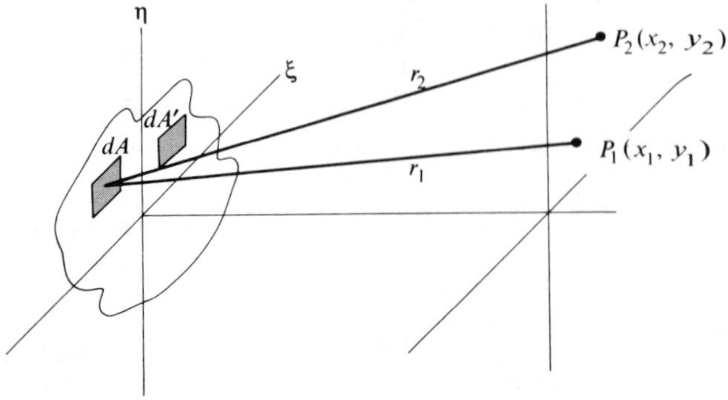

Fig. 7.15. Geometry used for computing the correlation function due to an incoherent source.

The element $\overline{v_1(t)v_2^{\star\prime}(t)}$ is recognized as a correlation function between different parts of the source. Since we are assuming a completely incoherent source, this must vanish through the time average. We are left with

$$\overline{V_1(t)V_2^\star(t)} = \int dA\ \overline{v_1(t)v_2^\star(t)}. \tag{7.71}$$

Restricting ourselves to the pseudomonochromatic case, we put

$$v(t) = a(t)\, e^{i\phi(t)}\, e^{-it\omega_0 t},$$

and Eq. (7.71) becomes

$$\overline{V_1(t)V_2^\star(t)} = \int dA\ \overline{a_1(t)a_2(t)}\ e^{\overline{i[\phi_1(t)-\phi_2(t)]}}. \tag{7.72}$$

Now, since v_1 and v_2 are disturbances which originate from a common source element at dA, they differ, if we assume isotropic radiation, only through different arrival times at the points p_1 and p_2. Thus,

$$\begin{aligned} a_2(t) &= a_1\left(t - \frac{(r_2 - r_1)}{c}\right), \\ \phi_2(t) &= \phi_1\left(t - \frac{(r_2 - r_1)}{c}\right). \end{aligned} \tag{7.73}$$

If the source dA were a perfect monochromatic radiator, Eq. (7.73) would simplify, when r_1 and r_2 are approximately equal, to

$$a_2 = a_1 \tag{7.74}$$

and

$$\phi_1 - \phi_2 = k(r_1 - r_2). \tag{7.75}$$

We wish to examine the extent to which Eqs. (7.74) and (7.75) may be used in Eq. (7.72) for the pseudomonochromatic case.

We have indicated in Section 7.10 that $\Delta\omega/\omega_0$ is a measure of the ratio of the variation rates of the envelope and the carrier of a pseudomonochromatic wave. This follows from taking the frequency, $f = \omega/2\pi$, as the rate of variation of an individual monochromatic wave. The reciprocal of the frequency, or period T, is a reasonable measure of the minimum time in which the wave varies appreciably. Thus, for a pseudomonochromatic wave $T_c = 1/\Delta f_0$, known as the coherency time, may be taken as the length of time over which we may approximate the wave by a monochromatic wave. Associated with the coherency time is the coherency length, $L_c = T_c c$. Over this distance in space a pseudomonochromatic wave may be approximated by a monochromatic wave of wavelength λ_0.

In the example to follow, the source will be taken as a star, the points

P_1 and P_2 will be separated by approximately 10 meters along a line perpendicular to the direction of the star, and the angular diameter of the star as viewed from the earth may be put at about 0.05 seconds of arc. This leads to a maximum variation in $r_1 - r_2$ of a few wavelengths for light in the visible range. Since the coherency length may be taken as a few hundred wavelengths when $\Delta\omega/\omega_0 \simeq 10^{-3}$, we may easily make use of Eqs. (7.74) and (7.75). Equation (7.72) thus becomes

$$\overline{V_1(t)V_2^\star(t)} = \int dA\, a^2\, e^{ik(r_1 - r_2)}. \tag{7.76}$$

Now, as indicated in Fig. 7.15, we limit ourselves to the case where the source is confined to a plane and the element of area dA is located by the coordinates ξ and η. The points P_1 and P_2 are located in the xy-plane, which is taken parallel to the $\xi\eta$-plane. The distances r_1 and r_2 are given by

$$r_1 = [R^2 + (x_1 - \xi)^2 + (y_1 - \eta)^2]^{1/2},$$
$$r_2 = [R^2 + (x_2 - \xi)^2 + (y_2 - \eta)^2]^{1/2}.$$

The distance from the earth to the star is R. These may be approximated as

$$r_1 \simeq R + \frac{(x_1 - \xi)^2 + (y_1 - \eta)^2}{2R},$$

$$r_2 \simeq R + \frac{(x_2 - \xi)^2 + (y_2 - \eta)^2}{2R}.$$

After squaring the indicated terms, $r_1 - r_2$ becomes

$$r_1 - r_2 = \phi - (x_1 - x_2)\frac{\xi}{R} - (y_1 - y_2)\frac{\eta}{R},$$

where ϕ, which is given by

$$\phi = \frac{1}{2R}(x_1^2 - x_2^2 + y_1^2 - y_2^2),$$

does not depend on ξ or η and may be taken through the integral in Eq. (7.76).

If we now make the substitutions

$$\alpha = \frac{x_1 - x_2}{R}, \qquad \beta = \frac{y_1 - y_2}{R},$$

Eq. (7.76) becomes

$$\overline{V_1(t)V_2^\star(t)} = e^{i\phi} \iint d\xi\, d\eta\, a^2\, e^{-ik(\alpha\xi + \beta\eta)}. \tag{7.77}$$

The intensities I_1 and I_2 at points p_1 and p_2 are given by

$$I_1 = I_2 = \iint d\xi \, d\eta \, a^2(\xi, \eta). \tag{7.78}$$

Combining Eqs. (7.78), (7.77), and (7.64), we obtain for the complex degree of coherence between points one and two

$$\gamma_{12}(0) = e^{i\phi} \frac{\iint d\xi \, d\eta \, a^2 \, e^{-ik(\alpha\xi + \beta\eta)}}{\iint d\xi \, d\eta \, a^2}. \tag{7.79}$$

Equation (7.79) is a form of the Van Cittert–Zernike theorem.

7.14 THE MICHELSON STELLAR INTERFEROMETER

In Fig. 7.16 aspects of the last two sections are combined. The source S is taken as a star a distance R from earth. Lens L is the objective of a telescope which would ordinarily produce the Fraunhofer diffraction pattern of its aperture in its focal plane at point p. The lens, however, is masked, leaving two slits separated by a distance d, at points P_1 and P_2. As described in Section 7.12, a pattern of fringes will now be obtained at point p. The visibility of these fringes will be given by $|\gamma_{12}(0)|$. But $\gamma_{12}(0)$ will be determined according to Eq. (7.79) by the intensity distribution $a^2(\xi, \eta)$ across the face of the star. By measuring the visibility of the fringes as a function of the slit separation d, we obtain information about γ_{12} and from this we may infer information about the size of the star.

To evaluate Eq. (7.79) we take the intensity across the disk of the

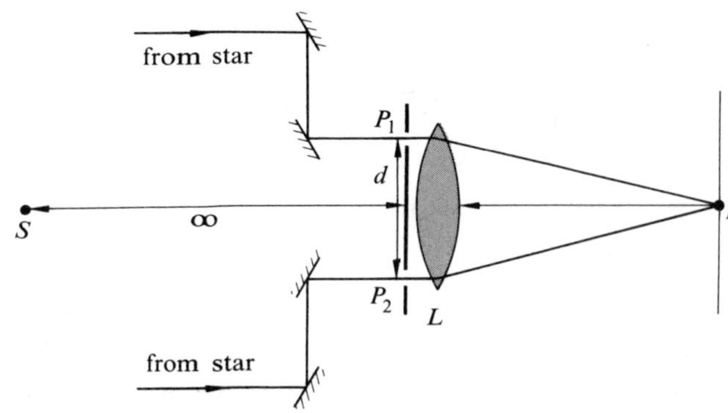

Fig. 7.16. The Michelson stellar interferometer.

star to be constant. By Eq. (7.79) $\gamma_{12}(0)$ should then be given by the proportionality

$$\gamma_{12}(0) \propto \int\int d\xi\, d\eta\, e^{-ik(\alpha\xi+\beta\eta)}, \tag{7.80}$$

where the integration is over the disk of the star of radius ρ. The integral is recognized as the one encountered in Section 6.9 for the Fraunhofer diffraction pattern of a circular aperture. By the result of that section γ_{12} should therefore be proportional to $2J_1(v)/v$, where $v = kd\rho/R$ and $J_1(v)$ is Bessel's function of order one of the first kind. The proportionality may be made an equality by noting that when $d = 0$, we also have $\alpha = \beta = v = 0$. In this case Eq. (7.79) reduces to $|\gamma_{12}| = |\gamma_{11}| = 1$, which is also the value of $2J_1(v)/v$ for $v = 0$. We then can write

$$|\gamma_{12}(0)| = \left|\frac{2J_1(v)}{v}\right|. \tag{7.81}$$

As the separation of the slits is increased, the visibility of the fringes will decrease from a maximum of one, with zero separation, and become zero when the first zero of $J_1(v)$ is reached at $v = 1.22\pi$. Thus, if the fringes vanish for a slit separation d, the angular diameter of the star, $2\rho/R$, is given by

$$2\rho/R = 1.22\lambda/d. \tag{7.82}$$

In practice the experimental arrangement cannot be exactly as discussed. For example, Michelson's measurement of the star Betelgeuse yielded an angular diameter for this star of 0.047 seconds of arc. Assuming an effective wavelength of 5500 Å, the slit separation at minimum visibility is 5.9 meters, which exceeds the diameter of the largest telescope objectives.

In order to overcome this difficulty, Michelson used four small auxiliary mirrors, as shown in Fig. 7.16. The separation of the slits remains fixed and the distance between the two outer mirrors is varied. In this arrangement it is the separation of these outer mirrors which must be inserted for d in Eq. (7.82).

7.15 COHERENCE AND POLARIZATION

Until now we have treated only a single field variable. The field, of course, consists of three such components. If we limit ourselves to waves which are at least approximately plane waves we may take the component along the direction of propagation to be zero and consider only two Cartesian components which are perpendicular to this direction. We take these components to be along the x- and y-axes. We modify our notation to include a subscript indicating the direction of the field component. For example, $V_x^R(r_1, t)$ is the real part of the complex field variable associated with the x-component of the field. There are now four possible correlation functions which may be arranged in a correlation matrix

$$\mathsf{J}_{12}(\tau) = \begin{pmatrix} \Gamma_{xx}(r_1, r_2, \tau) & \Gamma_{xy}(r_1, r_2, \tau) \\ \Gamma_{yx}(r_1, r_2, \tau) & \Gamma_{yy}(r_1, r_2, \tau) \end{pmatrix}, \tag{7.83}$$

where

$$\Gamma_{ij}(r_1, r_2, \tau) = \overline{V_i(r_1, t + \tau)V_j^{\star}(r_2, t)}. \tag{7.84}$$

We will not consider the general case given by Eq. (7.83) but will limit ourselves to $r_1 = r_2$, $\tau = 0$. Since the points r_1 and r_2 are coincident, in what follows we drop the subscripts on J_{12} and write $\mathsf{J}_{12} = \mathsf{J}$.

Just as γ_{12}, as defined by Eq. (7.64), is a measure of the degree of coherence between a single field variable at two different points in space, the quantity

$$\mu_{xy} = \Gamma_{xy}/\sqrt{\Gamma_{xx}\Gamma_{yy}} \tag{7.85}$$

may be taken as a measure of the coherence between two different components of the field at the same point in space. Again appealing to the Schwartz inequality and the definition of Γ_{ij} by Eq. (7.84), it is clear that

$$0 \leqslant \mu_{xy} \leqslant 1. \tag{7.86}$$

In Section 7.5 we have seen that in the plane wave representation of light, the Jones vector may be taken as proportional to the field vector:

$$|S\rangle = \begin{pmatrix} V_x \\ V_y \end{pmatrix}.$$

Within this same representation, it should be apparent that the coherency matrix J is simply the time average of the projection matrix $|S\rangle\langle S|$ associated with the Jones vector, $|S\rangle$.

Thus, the matrix J is an old acquaintance, arrived at from a new point of view. Since most of Chapter 4 was spent in a study of $|S\rangle\langle S|$ and its expansion in terms of the Stokes parameters, we do not need to develop the mathematics again. It will be useful, however, to write the Stokes parameters in terms of the Γ_{ij}. Following exactly the procedures of Chapter 4, we obtain

$$\begin{aligned} P_0 &= \Gamma_{xx} + \Gamma_{yy}, \\ P_1 &= \Gamma_{xx} - \Gamma_{yy}, \\ P_2 &= \Gamma_{xy} + \Gamma_{yx}, \\ P_3 &= i(\Gamma_{xy} - \Gamma_{yx}), \end{aligned} \tag{7.87}$$

which may easily be inverted to yield the Γ_{ij} as functions of the Stokes parameters:

$$\begin{aligned}\Gamma_{xx} &= \tfrac{1}{2}(P_0 + P_1),\\ \Gamma_{yy} &= \tfrac{1}{2}(P_0 - P_1),\\ \Gamma_{xy} &= \tfrac{1}{2}(P_2 - iP_3).\end{aligned} \tag{7.88}$$

Since the measurement of the Stokes parameters was discussed, independently of any representation in Chapter 4, we need not examine the measurement of the Γ_{ij}.

It remains, then, to compare in some way the theory of partial polarization with that of partial coherence. In Chapter 4 the degree of polarization was defined as

$$\mathscr{P} = \left[\frac{P_1^2 + P_2^2 + P_3^2}{P_0^2}\right]^{1/2}. \tag{7.89}$$

Using Eq. (7.87), we may write this in terms of the Γ_{ij}:

$$1 - \mathscr{P}^2 = \frac{4\,\mathrm{Det}\,(\mathrm{J})}{(\Gamma_{xx} + \Gamma_{yy})^2}. \tag{7.90}$$

Using Eq. (7.85), we obtain

$$1 - \mathscr{P}^2 = \frac{\Gamma_{xx}\Gamma_{yy}}{[\frac{1}{2}(\Gamma_{xx} + \Gamma_{yy})]^2}\,[1 - |\mu_{xy}|^2]. \tag{7.91}$$

Now, Γ_{xx} and Γ_{yy} are real positive scalars and it is easily shown by performing the square that

$$\Gamma_{xx}\Gamma_{yy} \leqslant [\tfrac{1}{2}(\Gamma_{xx} + \Gamma_{yy})]^2.$$

It follows immediately that

$$|\mu_{xy}| \leqslant \mathscr{P}. \tag{7.92}$$

Therefore, the degree of correlation between orthogonal components of the electromagnetic field does not exceed the degree of polarization of the wave. We now show that the strict equality holds in Eq. (7.92) if the orientation of the axes x and y are picked correctly.

As was seen in Section 4.7, the degree of polarization, $\mathscr{P}$, is independent of the choice of axes x and y. This is not true of μ_{xy}. The results of Section 4.5, however, indicate that we may pick a set of axes for which $|V_x|^2$ and $|V_y|^2$ are equal and therefore for which $\Gamma_{xx} = \Gamma_{yy}$ and $|\Gamma_{xy}| = |\Gamma_{yx}|$. If these conditions are placed in Eq. (7.91), one finds

$$|\mu_{xy}| = \mathscr{P}. \tag{7.93}$$

Thus, the maximum value of $|\mu_{xy}|$ which will be measured is equal to the degree of polarization of the state in question.

7.16 INTENSITY FLUCTUATIONS

It has been seen that a correlation function $\Gamma_{12}(r_1, r_2, \tau)$ can be measured by allowing the disturbances at points r_1 and r_2 to form a diffraction pattern whose intensity is measured. In some ways it would be more appealing if separate measurements could be made directly at r_1 and r_2 which could then be compared to yield information about Γ_{12}. If we substitute the word *detect* for the word *measure*, this is possible at the lower frequencies in the electromagnetic spectrum, e.g., radio and television frequencies. At these frequencies antennae and preamplifiers can detect and transmit the instantaneous value of the field variable to a central receiver where correlation may be performed. At optical frequencies this is impossible. In this range we can detect only intensities $\overline{V^\star V}$, rather than the instantaneous field $V^R(t)$. However, in some cases the slow intensity variations as well as the rapid variation of the field variables yield information of $\Gamma_{12}(\tau)$.

We are thus interested in the relation between the correlation of field variables

$$\Gamma_{12}(\tau) = \overline{V_1^\star(t+\tau)V_2(t)}$$

and the correlation of intensities

$$\overline{I_1 I_2} = \overline{V_1^\star(t+\tau)V_1(t+\tau)\, V_2^\star(t)V_2(t)}.$$

The relation between these measurables depends on a more detailed knowledge of the manner in which the field variable fluctuates than we have so far considered.

Whether or not we may actually measure $V^R(t)$ as a function of time, our mathematics presume the existence of this function. Within this mathematical framework we may thus ask for the probability that within a sufficiently small time interval, dt, the field variable takes on a particular value, $V^R(t)$. We define the function $p(V^R)\, dV^R$ as the probability that within the interval from t to $t + dt$ the field variable has a value between V^R and $V^R + dV^R$.

The functional form of $p(V^R)$ will depend on the nature of the source and its coherency properties. We shall consider only one case, that of a thermal incoherent source. In this case we take the distribution $p(V^R)$ to be a normal Gaussian distribution given by Eq. (7.94) and Fig. 7.17:

$$p(V^R) = \frac{1}{\sigma\sqrt{2\pi}}\, e^{-V^{R2}/2\sigma^2}. \tag{7.94}$$

The parameter σ is known as the standard deviation and is a measure of the probable deviation of the variable from its average, which is zero.

The choice of this distribution results from considering V^R to be a sum of many contributions from independently (randomly) radiating parts of an

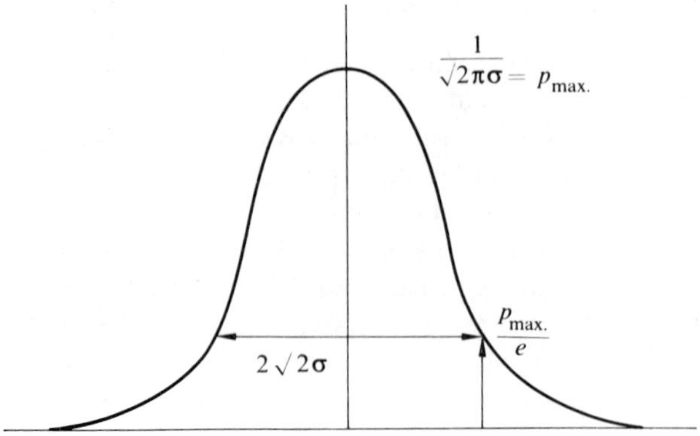

Fig. 7.17. The Gaussian distribution.

incoherent source. Without attempting to specify necessary and sufficient conditions, we know from the central limit theorem of statistics that the distribution of the sum of a large number of randomly varying quantities will approach a Gaussian distribution function, independent of the distribution functions for the individual quantities.

The standard deviation of the distribution may be determined by the conditions, Eq. (7.59),

$$\tfrac{1}{2}\bar{I} = \overline{V^{R2}} = \int_{-\infty}^{\infty} V^{R^2} p(V^R)\, dV^R,$$

which may be evaluated, using the results of Appendix VII, to give

$$\tfrac{1}{2}\bar{I} = \sigma^2. \tag{7.95}$$

Now, consider two field variables V_1^i and V_2^j, which could be any two of $V^R(r_1)$, $V^R(r_2)$, $V^I(r_1)$, and $V^I(r_2)$, which are assumed to have Gaussian distributions but have a nonvanishing correlation $\overline{V_1^i V_2^j}/\overline{V_1^i}\,\overline{V_2^j} = \rho$. The appropriate bivariant Gaussian distribution

$$p(V_1^i, V_2^j, \sigma_1, \sigma_2, \rho)\, dV_1^i\, dV_2^j,$$

which is the probability that V_1^i will lie between V_1^i and $V_1^i + dV_1^i$ and V_2^j simultaneously between V_2^j and $V_2^j + dV_2^j$, is given by

$$p = \frac{1}{2\pi\sqrt{1-\rho^2}}\frac{1}{\sigma_1\sigma_2}\exp\left\{-\frac{1}{2}\left[\left(\frac{V_1^i}{\sigma_1}\right)^2 + \left(\frac{V_2^j}{\sigma_2}\right)^2 - \frac{2\rho}{\sigma_1\sigma_2} V_1^i V_2^j\right]\right\}. \tag{7.96}$$

Using this distribution, we may determine the relation between $\overline{V_1^{i2} V_2^{j2}}$ and $(\overline{V_1^i V_2^j})^2$ for the Gaussian variables V_1^i and V_2^j:

$$\overline{V_1^{i2} V_2^{j2}} = \int\!\!\int dV_1^i\, dV_2^j\, V_1^{i2} V_2^{j2} p(V_1^i, V_1^j, \sigma_1, \sigma_2, \rho).$$

The integration may be found by use of Appendix VII, and we obtain

$$\overline{V_1^{i^2} V_2^{j^2}} = \overline{V_1^{i^2}}\ \overline{V_2^{j^2}} + 2(\overline{V_1^i V_2^j})^2. \tag{7.97}$$

Using Eq. (7.97), we may now examine the correlation between the intensities:

$$\overline{I_1(t+\tau) I_2(t)} = \overline{V_1(t+\tau) V_1^\star(t+\tau) V_2(t) V_2^\star(t)}. \tag{7.98}$$

If the right-hand side of Eq. (7.98) is written out in terms of the real and imaginary parts of V_1 and V_2, we obtain

$$\overline{I_1(t+\tau) I_2(t)} = \overline{V_1^{R^2} V_2^{R^2}} + \overline{V_1^{R^2} V_2^{I^2}} + \overline{V_1^{I^2} V_2^{R^2}} + \overline{V_1^{I^2} V_2^{I^2}}. \tag{7.99}$$

By use of Eqs. (7.97),

$$\begin{aligned}\overline{I_1 I_2} = {} & \overline{V_1^{R^2}}\ \overline{V_2^{R^2}} + \overline{V_1^{I^2}}\ \overline{V_2^{I^2}} + \overline{V_1^{I^2}}\ \overline{V_2^{R^2}} + \overline{V_2^{I^2}}\ \overline{V_1^{R^2}} \\ & + 2[(\overline{V_1^R V_2^I})^2 + (\overline{V_1^I V_2^R})^2 + (\overline{V_1^R V_2^R})^2 + (\overline{V_1^I V_2^I})^2].\end{aligned} \tag{7.100}$$

By use of Eqs. (7.59), (7.67), and (7.68), Eq. (7.100) becomes

$$\begin{aligned}\overline{I_1(t+\tau) I_2(t)} &= \overline{I_1}\ \overline{I_2} + [\mathrm{Re}\,(\Gamma_{12})]^2 + [\mathrm{Im}\,(\Gamma_{12})]^2 \\ &= \overline{I_1}\ \overline{I_2} + |\Gamma_{12}(\tau)|^2 \\ &= \overline{I_1}\ \overline{I_2}[1 + |\gamma_{12}(\tau)|^2].\end{aligned} \tag{7.101}$$

Thus, measurement of an intensity correlation yields a knowledge of $|\gamma_{12}(\tau)|$. A measurement made in this way may be put to the same uses as when $|\gamma_{12}|$ is obtained by measuring $\overline{V_1(t) V_2^\star(t)}$. In some ways the intensity correlation is the easier method. Hanbury-Brown and Twiss have used intensity correlation for the determination of the angular diameter of stars. Figure 7.18 is a schematic representation of their apparatus. Since high-quality imaging of the star is not required, the mirrors A_1 and A_2 were searchlight mirrors which focused the stellar image on photoelectric detectors P_1 and P_2. The output of these detectors was fed through amplifiers B to a multiplier C and then to an integrator M which performed the time average. The actual electronics involved was, of course, more complicated than this simple block diagram indicates. The time delay Δt which is introduced between P_2 and B_2 compensates for the difference in arrival time at the two mirrors of the light from the star. It should not be confused with the τ of $\gamma_{12}(\tau)$. What is measured in this experiment is $|\gamma_{12}(0)|$. Because it is the slowly varying intensities that are measured and compared rather than the rapid amplitude and phase variation of $V(t)$, the intensity correlation technique is less affected by slight variation in path length, allowing longer base-lines. It is also less

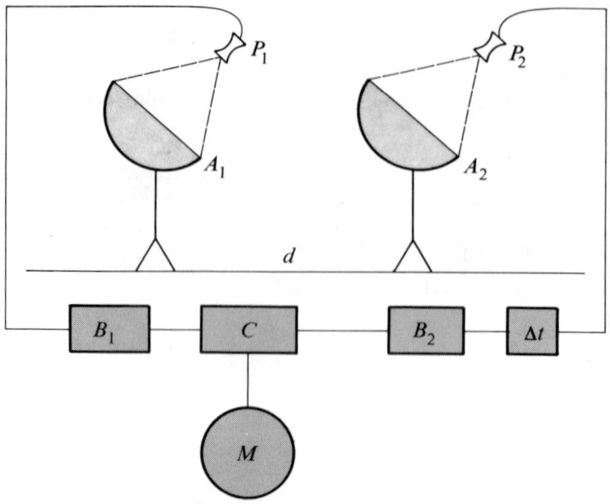

Fig. 7.18. Block diagram of the stellar intensity interferometer. (From Hanbury-Brown, R., and R. Q. Twiss, *Proc. Roy. Soc. A***248**, 201, 1958).

influenced by atmospheric turbulence and "seeing" than is Michelson's method. In Fig. 7.19 are shown the results of measurements made on the star Sirius A. The curve is the theoretical result for an angular diameter of 0.0069 seconds of arc.

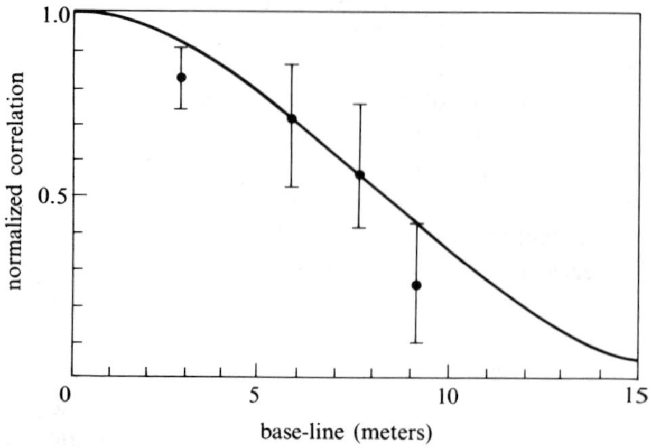

Fig. 7.19. The correlation function as determined experimentally for the star Sirius A. (From Hanbury-Brown, R., and R. Q. Twiss, *Proc. Roy. Soc. A***248**, 235, 1958).

PROBLEMS

7.1 In the presence of a conducting medium in which the current density and the electric field are related by $\boldsymbol{J} = g\boldsymbol{E}$, where the conductivity g is a constant, show that the wave equation becomes

$$\left\{\nabla^2 - \frac{1}{c^2}\frac{\partial^2}{\partial t^2} - g\mu\frac{\partial}{\partial t}\right\}\boldsymbol{E}(\boldsymbol{r}, t) = 0.$$

7.2 In terms of the electromagnetic plane wave representation of polarization states, show that

$$\frac{1}{\sqrt{2}}\{e^{+i\pi/4}\,|P_x\rangle + e^{-i\pi/4}\,|P_y\rangle\}$$

represents an $|R\rangle$ state if the time dependence $e^{+i\omega t}$ is chosen in Eq. (7.17) instead of $e^{-i\omega t}$.

7.3 The Poynting vector for a plane electromagnetic wave is in the direction of propagation of the wave. This conforms to our physical intuition. Consider a long straight wire carrying current I. Show that the Poynting vector is perpendicular to the wire and points in toward the center of the wire. Show that $\boldsymbol{S}$ vanishes as the resistance of the wire goes to zero, keeping I constant. Interpret this result physically.

7.4 A large radar antenna emits a narrow beam carrying a total average power of 10^6 watts. Show that there is a reaction force on the antenna of 3.3×10^{-3} newtons.

7.5 Consider an electromagnetic wave traveling in the z-direction:

$$\boldsymbol{E} = \boldsymbol{E}_0 e^{i(kz-\omega t)}, \quad \boldsymbol{H} = \boldsymbol{H}_0 e^{i(kz-\omega t)}.$$

Show that if $\boldsymbol{E}_0$ and $\boldsymbol{H}_0$ are not constants but are functions of the coordinates x and y, the fields E and $\boldsymbol{H}$ must have nonzero z-components.

7.6 Show that for normal incidence, $\phi = \phi' = \phi'' = 0$, about 4% of the incident intensity of a wave is reflected at a glass-air interface when the index of refraction of the glass is $n = 1.5$.

7.7 When $\phi + \phi'' = 90°$ in Fig. 7.8, the reflected wave is completely polarized. If the incident beam is completely unpolarized, show that the ratio of the reflected intensity to the incident intensity in this case is

$$\frac{1}{2}\left(\frac{1-n^2}{1+n^2}\right)^2,$$

where n is the index of refraction of the glass. The angle of incidence in this case is known as the polarizing angle. When $n = 1.5$, show that the above equation yields about 7.4% reflected totally polarized light of the incident light.

7.8 Derive Eqs. (7.53) and (7.54).

7.9 Verify Eqs. (7.59) and (7.60).

7.10 If a real field variable is given by $V^r(t) = A \cos \omega_0 t$, show that Eqs. (7.56) and (7.58) result in

$$\mathscr{V}(\omega) = \frac{A}{2\sqrt{2\pi}} [2\pi\delta(\omega + \omega_0) + 2\pi\delta(\omega - \omega_0)],$$
$$V(t) = Ae^{i\omega_0 t}.$$

7.11 Explain from diffraction theory why the constant C in Eq. (7.61) must be a pure imaginary.

7.12 For the complex signal $V(x, t) = Ae^{i(kx - \omega t)}$ show that

$$\Gamma_{x_1 x_2}(\tau) = A^2 e^{i[k(x_1 - x_2) - \omega\tau]}.$$

7.13 Consider a source of light emitting incoherently in the range $\lambda = 5000$ Å $\pm$ 1/2 Å. Let this source be used with the Mach-Zehnder interferometer of Chapter 2. Using simple arguments based on coherence length, say how many wavelengths may one beam be retarded over the other and still produce the results given in Chapter 2. Show that this retardation is equivalent to inserting a plate of glass ($n = 1.5$) in one beam whose thickness is of the order of 1 mm.

7.14 Prove that $Ae^{i\omega_1 t} + Be^{i\omega_2 t} = 0$ implies that $\omega_1 = \omega_2$. Show in general that

$$\sum_{n=1}^{N} A_n e^{i\omega_n t} = 0$$

implies

$$\omega_n = \omega_m.$$

These results were used in deriving the laws of reflection and refraction.

7.15 If $V^R(t) = A \sin \omega t$, show that the probability that the function at some particular time takes on a value between V^R and $V^R + dV^R$ is given by

$$p(V^R)dV^R = \frac{1}{\pi} \frac{dV^R}{\sqrt{A^2 - V^{R^2}}}.$$

7.16 An inebriated mathematician starts from a lamp-post, which we locate at $x = 0$, and takes equal steps one after another toward the right ($+x$) or the left ($-x$) with equal probability. After taking a large number of steps, N, of equal unit length his most probable x-coordinate will be zero, as he is just as likely to wander to the right as to the left. However, his most probable distance from the lamp-post, $|x|$, is given by $x^2 = N\,|x| = \sqrt{N}$. To show this, let $x(N)$ be his probable distance from the origin after N steps and explain why

$$[X(N+1)]^2 = [(X(N)+1)^2 + (X(N)-1)^2]/2.$$

Using finite induction, prove that $x(N) = \sqrt{N}$. Discuss how this result might be applied to the incoherent addition of P-states whose lines are parallel.

7.17 As a very crude model of a star, we picture a uniformly radiating core of radius R surrounded by a partially absorbing atmosphere of thickness T, with $T \ll R$. Use this model to explain why the rim of a stellar disk is not as bright as the central portions. How would this phenomenon, known as limb darkening, affect the determination of stellar diameters?

CHAPTER 8

THE MOMENTUM SPACE REPRESENTATION

Elements of a mathematical language have been developed in preceding chapters with which the quantum mechanical description of physical reality can be stated. This description involves a state function or state vector from which all measurable aspects of a system can be determined. In Chapters 2, 3, and 4 only polarization states were considered. In the last three chapters the space and time dependence of the state vectors was taken up. In this chapter the meaning of the state function is reexamined. In particular we shall see that a state vector which is a function of position in real space cannot rigorously be defined for the photon and that our description of light must be extended to include a more abstract momentum space. This is not to imply that the quantum mechanical formulation we have developed is invalid. Rather, this accurately reflects the difficulties, mentioned in Chapter 2, of locating a photon—for example, in one arm or the other of an interferometer.

8.1 THE TIME DEPENDENCE OF THE STATE FUNCTION

Consider a state function which is a function of position and whose bracket product is normalized to unity. The equation

$$\langle \boldsymbol{r} | \boldsymbol{r} \rangle = \int d\boldsymbol{r} \psi^\star(\boldsymbol{r}) \psi(\boldsymbol{r}) = 1 \tag{8.1}$$

is interpreted to mean that the probability of finding the particle, somewhere, is identically one. It is the strict validity of this last statement that we wish to reexamine.

Let us consider a "free" photon, one which is in a region of space devoid of other matter, and for which the probability of interacting with other matter is therefore zero over some finite time. Thus, the probability of finding this photon somewhere within this region must be constant over this time. If the probability is constant, its time derivative must vanish. If $|\boldsymbol{r}\rangle$ is the state function of the photon, as in Eq. (8.1), we must have

$$\frac{\partial}{\partial t}\langle \boldsymbol{r}|\boldsymbol{r}\rangle = \frac{\partial}{\partial t}\int d\boldsymbol{r}\psi^\star\psi = 0, \tag{8.2}$$

which yields

$$\begin{aligned}\frac{\partial}{\partial t}\langle \boldsymbol{r}|\boldsymbol{r}\rangle &= \int d\boldsymbol{r}\left(\frac{\partial\psi^\star}{\partial t}\psi + \psi^\star\frac{\partial\psi}{\partial t}\right)\\ &= 2\,\mathrm{Re}\int d\boldsymbol{r}\,\frac{\partial\psi}{\partial t}\psi^\star.\end{aligned}$$

Now assume that no relationship necessarily exists between ψ and its time derivative $\partial\psi/\partial t$. If this is so, then one could imagine at some particular time the possibility that $\psi = \partial\psi/\partial t$, which would imply that

$$\psi^\star\psi = \psi^\star\partial\psi/\partial t$$

would be real and positive. Thus,

$$\int d\boldsymbol{r}\,\frac{\partial\psi}{\partial t}\psi^\star > 0,$$

which is contrary to Eq. (8.2).

Thus, if our probability interpretation of the bracket product is correct, there must exist some relationship between ψ and $\partial\psi/\partial t$ at each instant of time. The condition in its weakest form would simply deny that ψ and $\partial\psi/\partial t$ could ever be equal. The condition, however, must be strong enough to guarantee that $\psi^\star\partial\psi/\partial t$ cannot be everywhere positive.

As an example of such a relationship, let us consider a particle such as an electron moving in empty space. As was pointed out in Chapter 5, the operator formalism is not limited to light and we should be able to write eigenvalue equations for electron state functions. For a classical particle of mass m the kinetic energy and the linear momentum are given by

$$E = \tfrac{1}{2}mv^2, \qquad \boldsymbol{p} = m\boldsymbol{v},$$

which leads to the relation

$$E = p^2/2m. \tag{8.3}$$

To obtain the quantum mechanical equation for such a particle, we simply reinterpret this as an operator equation. We thus make the replacements

$$\begin{aligned}E &\to i\hbar\partial/\partial t,\\ \boldsymbol{p} &\to -i\hbar\nabla,\end{aligned}$$

and obtain

$$-\frac{\hbar^2}{2m}\nabla^2\psi = i\hbar\frac{\partial\psi}{\partial t}, \tag{8.4}$$

which is the Schrödinger equation for a free particle of mass m. We see that in this case there is a relation between ψ and $\partial\psi/\partial t$. If we know the state function as a function of position, we may compute its time derivative by using Eq. (8.4).

Let us try the same procedure for the photon. The difference is that Eq. (8.3) does not hold for the photon. For the photon the relation between energy and momentum is given by

$$|\boldsymbol{p}| = E/c.$$

The absolute value sign may be removed by squaring both sides, giving

$$E^2 = p^2c^2. \tag{8.5}$$

If this is interpreted as an operator equation, we obtain

$$-\hbar^2\frac{\partial^2}{\partial t^2}\psi = -\hbar^2c^2\nabla^2\psi$$

or

$$\nabla^2\psi - \frac{1}{c^2}\frac{\partial^2\psi}{\partial t^2} = 0, \tag{8.6}$$

which, not surprisingly, is the scalar wave equation obeyed by each component of the electric and magnetic fields. Although this yields a relation between ψ and $\partial^2\psi/\partial t^2$, it imposes no condition on $\partial\psi/\partial t$. It is well known that a unique solution of a second-order differential equation can be obtained only after auxiliary conditions are imposed on *both* the function and its first derivative.

The state functions for light that we have been using which are solutions of Eq. (8.6) may not be used with rigor to compute the probability of finding a photon in the volume dV. This interpretation for the photon which was used in earlier chapters must be viewed as an approximation. Its success in the study of diffraction indicates, however, that the approximation is a good one if the volume dV is not too small.

The difficulties pointed out here do not indicate that we must search for another equation to replace Eq. (8.6). Rather, the results obtained reflect the difficulty of locating individual photons in space. In the following sections we shall see how $\partial\psi/\partial t$ may be determined if we abandon a description of the photon in real space.

8.2 MOMENTUM REPRESENTATION OF CLASSICAL PARTICLES

In the following sections we are going to see that a good state function may be obtained for light if we allow the state function to be a function of momentum rather than position. To perform the transition from real space to momentum space as clearly as possible, we will examine how it can be done for a classical particle. We confine the discussion to two-dimensional motion without a loss of generality. The motion of a particle is usually considered as given if its position is specified as a function of time. A particle thus traces out some curve in space with the coordinates given by

$$x = x(t), \qquad y = y(t),$$

as shown in Fig. 8.1. Since the particle is moving, it possesses at each point of its path a velocity and a momentum whose direction is tangent to the path and whose components are given by

$$p_x = m\frac{d}{dt}x(t), \qquad p_y = m\frac{d}{dt}y(t).$$

We now consider a two-dimensional momentum space whose axes are labeled p_x and p_y. At each point of the motion in real space the components of the momentum give us a pair of numbers which determine a point in momentum space. Thus, the particle traces out a path in momentum space, as shown in Fig. 8.2.

Since the particular path shown starts from the origin in momentum space, it represents a particle which starts from rest in real space at t_0. The point where the path crosses the p_x-axis in momentum space corresponds to the point A in real space where the particle reaches its maximum y-value and, hence, has zero y-velocity. In general, the shapes of the two paths in real space and in momentum space are not apparently or simply related.

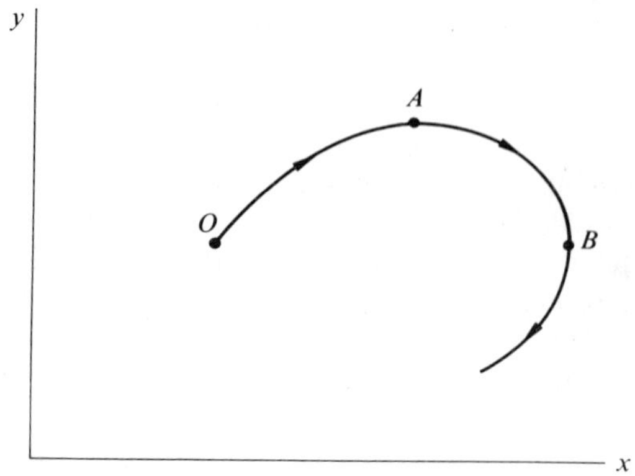

Fig. 8.1. The path of a particle in real space (two-dimensional).

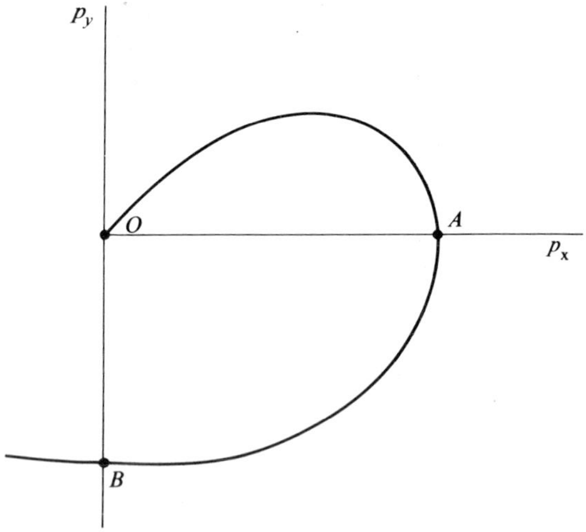

Fig. 8.2. The path of the particle of Fig. 8.1 in momentum space.

For example, a particle which moves with constant velocity in real space is represented by a single point in momentum space.

We have seen how the momentum space path is determined by the path in real space. It is easy to turn this procedure around. If we are given the path in momentum space as a function of time, we may obtain the path in real space through

$$x(t) = x_0 + \frac{1}{m}\int_{t_0}^{t} p_x(t)dt,$$

$$y(t) = y_0 + \frac{1}{m}\int_{t_0}^{t} p_y(t)dt.$$

Thus, the whole story of the particle's motion in momentum space is known if we know its position in real space as a function of time and the motion in real space is known if we know the position of the particle in momentum space as a function of time and also know the initial position of the particle, x_0 and y_0. In fact the entire content of classical mechanics can be cast in the formalism of either real space or momentum space. However, at an elementary level the momentum space representation offers no particular advantage.

Consider, however, not a single classical particle but a collection of them—say, the 10^{23} particles which make up a mole of an ideal gas. In this case one has little interest in the actual location of the individual particles in real space. The measurable properties of the system, the pressure and temperature, specific heats, etc., are determined by the distribution of velocities of the molecules. It is the gas considered as a distribution of points in momentum space that is physically meaningful. One is more interested in the gas density in momentum space, the number of molecules of mass m with momentum between $m\boldsymbol{v}$ and $m\boldsymbol{v} + m d\boldsymbol{v}$, than in the density in real space, the number of particles within a real volume element $d\boldsymbol{r}$.

When the study of classical particles is replaced by the study of photons, the use of momentum space is absolutely necessary since, as we have seen, the location of photons in real space cannot be always rigorously defined and, hence, no density function in real space can always be rigorously given. Therefore, we proceed to rewrite the theory of the electromagnetic field in momentum space and then impose upon it the state vector interpretation.

8.3 MAXWELL'S EQUATIONS IN MOMENTUM SPACE

Let us assume a solution to Maxwell's equations which consists of two vector functions of position, $\boldsymbol{E}(\boldsymbol{r})$, $\boldsymbol{H}(\boldsymbol{r})$. These functions may be written as Fourier superpositions of plane waves of definite linear momentum $\hbar\boldsymbol{k}$ as

$$\boldsymbol{E}(\boldsymbol{r}) = \frac{1}{(2\pi)^{3/2}} \int d\boldsymbol{k}\, \mathscr{E}(\boldsymbol{k}) e^{i\boldsymbol{k}\cdot\boldsymbol{r}},$$

$$\boldsymbol{H}(\boldsymbol{r}) = \frac{1}{(2\pi)^{3/2}} \int d\boldsymbol{k}\, \mathscr{H}(\boldsymbol{k}) e^{i\boldsymbol{k}\cdot\boldsymbol{r}}. \tag{8.7}$$

If $\boldsymbol{E}(\boldsymbol{r})$ and $\boldsymbol{H}(\boldsymbol{r})$ obey Maxwell's equations, we will determine the equations which $\mathscr{E}$ and $\mathscr{H}$ obey. Since the momentum of a photon is given by $\boldsymbol{p} = \hbar\boldsymbol{k}$, a description using a space whose coordinates are k_x, k_y, and k_z will be called either the K-space representation or the momentum space representation. Consider the Maxwell equation, Eq. (7.3),

$$\nabla \times \boldsymbol{E}(\boldsymbol{r}) = -\frac{\partial \boldsymbol{B}(\boldsymbol{r})}{\partial t} = -\mu \frac{\partial \boldsymbol{H}(\boldsymbol{r})}{\partial t}.$$

By use of Eq. (8.7) the left-hand side may be written

$$\nabla \times \boldsymbol{E} = \frac{1}{(2\pi)^{3/2}} \int d\boldsymbol{k} \nabla \times [\mathscr{E}(\boldsymbol{k}) e^{i\boldsymbol{k}\cdot\boldsymbol{r}}],$$

where we have taken the vector operator through the integral sign, since it operated on the spatial coordinates and the integration is over the k-space coordinates. By use of identity 7 of Appendix V, the integrand may be written

$$\nabla \times [\mathscr{E}(\boldsymbol{k}) e^{i\boldsymbol{k}\cdot\boldsymbol{r}}] = \nabla e^{i\boldsymbol{k}\cdot\boldsymbol{r}} \times \mathscr{E}(\boldsymbol{k}) + [\nabla \times \mathscr{E}(\boldsymbol{k})] e^{i\boldsymbol{k}\cdot\boldsymbol{r}}.$$

The second term vanishes because $\mathscr{E}(\boldsymbol{k})$ is not a function of the spatial coordinates. After taking the gradient in the first term, we obtain

$$\nabla \times [\mathscr{E}(\boldsymbol{k}) e^{i\boldsymbol{k}\cdot\boldsymbol{r}}] = i[\boldsymbol{k} \times \mathscr{E}(\boldsymbol{k})] e^{i\boldsymbol{k}\cdot\boldsymbol{r}}.$$

Finally, the Maxwell equation, Eq. (7.3), becomes

$$\frac{1}{(2\pi)^{3/2}} \int dk[ik \times \mathscr{E} e^{ik\cdot r}] = -\frac{\mu}{(2\pi)^{3/2}} \int dk \frac{\partial \mathscr{H}}{\partial t} e^{ik\cdot r}$$

or

$$\int dk\left[ik \times \mathscr{E} + \mu \frac{\partial \mathscr{H}}{\partial t}\right] e^{ik\cdot r} = 0.$$

Thus, the quantity in brackets is the inverse Fourier transform of zero and is itself zero. We then have

$$ik \times \mathscr{E}(k) = -\mu \frac{\partial \mathscr{H}(k)}{\partial t}.$$

By similar arguments we may obtain the *K*-space representation for all of Maxwell's equations. These are listed below beside their corresponding forms in real space. The dot signifies partial differentiation in respect to time.

$$ik \times \mathscr{E} = -\mu\dot{\mathscr{H}}, \qquad \nabla \times E = -\mu \dot{H}, \tag{8.8}$$

$$ik \times \mathscr{H} = \varepsilon\dot{\mathscr{E}}, \qquad \nabla \times H = \varepsilon \dot{E}, \tag{8.9}$$

$$k \cdot \mathscr{E} = 0, \qquad \nabla \cdot E = 0, \tag{8.10}$$

$$k \cdot \mathscr{H} = 0, \qquad \nabla \cdot H = 0. \tag{8.11}$$

It must be remembered that $E(r)$ and $H(r)$ are real quantities. Since $\mathscr{E}$ and $\mathscr{H}$ are related to these real quantities through a Fourier transform, $\mathscr{E}$ and $\mathscr{H}$ must obey a reality condition, as was shown in Section 7.9. Thus, in addition to Eqs. (8.8) through (8.11) $\mathscr{E}$ and $\mathscr{H}$ obey

$$\mathscr{E}^\star(k) = \mathscr{E}(-k), \tag{8.12}$$

$$\mathscr{H}^\star(k) = \mathscr{H}(-k). \tag{8.13}$$

8.4 THE FIELD VARIABLES IN *K*-SPACE

Just as Maxwell's equations couple E and H in real space, Eqs. (8.8) through (8.11) couple $\mathscr{E}$ and $\mathscr{H}$ in *K*-space. In real space we were able to decouple Maxwell's equations and obtain wave equations. We now decouple Eqs. (8.8) through (8.11) in *K*-space. Starting with Eq. (8.8), we vector multiply both sides by k, yielding

$$ik \times (k \times \mathscr{E}) = -\mu k \times \frac{\partial \mathscr{H}}{\partial t}.$$

Expanding the left-hand side using vector identity 2 and taking $\boldsymbol{k}$ through the derivative on the right, we obtain

$$i[\boldsymbol{k}(\boldsymbol{k}\cdot\mathscr{E}) - \mathscr{E}(\boldsymbol{k}\cdot\boldsymbol{k})] = -\mu\frac{\partial}{\partial t}[\boldsymbol{k}\times\mathscr{H}].$$

The first term on the left is zero by Eq. (8.10) and the right-hand side may be replaced by using Eq. (8.9). We then obtain

$$ik^2\mathscr{E} = \frac{\mu\varepsilon}{i}\frac{\partial^2\mathscr{E}}{\partial t^2}$$

or

$$\left(\frac{\partial^2}{\partial t^2} + k^2c^2\right)\mathscr{E}(\boldsymbol{k}) = 0, \tag{8.14}$$

where $c = (\varepsilon\mu)^{-1/2}$. One can likewise show that $\mathscr{H}$ obeys

$$\left(\frac{\partial^2}{\partial t^2} + k^2c^2\right)\mathscr{H}(\boldsymbol{k}) = 0. \tag{8.15}$$

In each case we are interested in solutions $f(\boldsymbol{k})$ of the equation

$$\left(\frac{\partial^2}{\partial t^2} + k^2c^2\right)f(\boldsymbol{k}, t) = 0,$$

which may be factored to yield

$$\left(\frac{\partial}{\partial t} + ikc\right)\left(\frac{\partial}{\partial t} - ikc\right)f(\boldsymbol{k}, t) = 0, \tag{8.16}$$

and we seek solutions $f(\boldsymbol{k})$ which obey one of the following equations:

$$\left(\frac{\partial}{\partial t} + ikc\right)f(\boldsymbol{k}, t) = 0, \tag{8.17}$$

$$\left(\frac{\partial}{\partial t} - ikc\right)f(\boldsymbol{k}, t) = 0. \tag{8.18}$$

A general solution of Eq. (8.16) will then be a linear combination of independent solutions of Eqs. (8.17) and (8.18).

Now, it is readily shown that if $f(\boldsymbol{k}, t)$ is a solution of either Eq. (8.17) or Eq. (8.18), the function $f(\boldsymbol{k}, -t)$ is a solution for the other. It is possible to set up a one-to-one correspondence between the solutions of these two equations in still another way. It is left to the exercises to show that if $f(\boldsymbol{k}, t)$ is a solution to either equation, its complex conjugate $f^*(\boldsymbol{k}, t)$ is a solution of the other.

Since $|-\boldsymbol{k}| = |\boldsymbol{k}| = k$, we point out that if $f(\boldsymbol{k})$ is a solution to either equation $f(-k)$ is also a solution to the same equation. In what follows we explicitly use the symbol $f(k, t)$ to indicate a solution to Eq. (8.17). From the foregoing discussion symbols such as $f^\star(k, t)$, $f(k, -t)$, etc., denote solutions to Eq. (8.18).

We may now write a general solution for $\mathscr{E}(k)$ as

$$\mathscr{E}(k) = Af_1(k, t) + Bf_2^\star(-k, t),$$

where A and B are independent of time but may be functions of $k = |k|$. We may impose the reality conditions, Eq. (8.12), on this form by setting $f_1 = f_2$ and $B = A$, giving

$$\mathscr{E}(k) = A[f(k) + f^\star(-k)]. \tag{8.19}$$

The reality condition which leads to the form of Eq. (8.19) also imposes a special form on $\dot{\mathscr{E}}(k)$. By direct differentiation of Eq. (8.19), we obtain

$$\dot{\mathscr{E}}(k) = A\left[\frac{\partial f(k)}{\partial t} + \frac{\partial f^\star(-k)}{\partial t}\right].$$

Since f and $f^\star$ are, respectively, solutions of Eqs. (8.17) and (8.18), we may use these equations to obtain

$$\dot{\mathscr{E}}(k) = -ikcA[f(k) - f^\star(-k)]. \tag{8.20}$$

Combining Eqs. (8.19) and (8.20), we may solve for $f(k)$ in terms of $\mathscr{E}$ and $\dot{\mathscr{E}}$:

$$f(k) = \frac{1}{2A}\left[\mathscr{E}(k) - \frac{1}{ikc}\dot{\mathscr{E}}(k)\right]. \tag{8.21}$$

If we take the scalar product of this equation with k, we obtain

$$k \cdot f = \frac{1}{2A}\left[k \cdot \mathscr{E} - \frac{1}{ikc} k \cdot \dot{\mathscr{E}}\right].$$

The first term on the right is zero by Eq. (8.10). The second term also vanishes, since, by Eq. (8.9),

$$k \cdot \dot{\mathscr{E}} = \frac{i}{\varepsilon} k \cdot (k \times \mathscr{H})$$

and $k \cdot (k \times \mathscr{H})$ vanishes identically. We thus obtain an important condition on the function $f(k)$ known as the *tranversality condition*:

$$k \cdot f(k) = 0. \tag{8.22}$$

8.5 THE STATE FUNCTION IN K-SPACE

Culminating in Eqs. (8.19) and (8.20) we have related a function $f(k, t)$ in momentum space with the Fourier transform $\mathscr{E}(k, t)$ and, hence, with the real electric field $\boldsymbol{E}(\boldsymbol{r}, t)$. Furthermore, we have done this in such a way that the reality condition on the field is automatically satisfied. Let us reconsider Eq. (8.17):

$$\frac{\partial \boldsymbol{f}}{\partial t} = -ikc\boldsymbol{f}. \tag{8.23}$$

We see that $\boldsymbol{f}(\boldsymbol{k}, t)$ is a candidate for the state function in k-space, since Eq. (8.23) gives a direct relation between $\boldsymbol{f}$ and its first time derivative. To complete this identification we multiply both sides of Eq. (8.23) by $i\hbar$ and obtain

$$i\hbar \frac{\partial \boldsymbol{f}}{\partial t} = \hbar kc\boldsymbol{f}. \tag{8.24}$$

Recalling the energy eigenvalue equation,

$$i\hbar \frac{\partial}{\partial t} |S\rangle = E|S\rangle, \tag{8.25}$$

and the fact that for the photon the energy is given by $E = k\hbar c$, the identification between Eq. (8.24) and Eq. (8.25) is complete.

Before the function $\boldsymbol{f}(\boldsymbol{k})$ may be uniquely related to a particular real field $\boldsymbol{E}(\boldsymbol{r})$, it is necessary to evaluate the constant A in Eq. (8.19). From Eq. (8.24) the energy operator in real space $i\hbar(\partial/\partial t)$ may be replaced in K-space by the multiplicative operator $k\hbar c$. Thus, the expectation value of the energy evaluated in the k-space representation should be given by

$$\langle S|\hbar kc|S\rangle = \int d\boldsymbol{k}\boldsymbol{f}^{\star}(k\hbar c)\boldsymbol{f},$$

and this should correspond to the classical energy of the corresponding electromagnetic field obtained from Eq. (7.23)

$$W = \tfrac{1}{2} \int d\boldsymbol{r}[\varepsilon \boldsymbol{E} \cdot \boldsymbol{E} + \mu \boldsymbol{H} \cdot \boldsymbol{H}].$$

We thus demand that

$$\int d\boldsymbol{k}\boldsymbol{f}^{\star}(\hbar kc)\boldsymbol{f} = \tfrac{1}{2} \int d\boldsymbol{r}[\varepsilon \boldsymbol{E}(\boldsymbol{r}) \cdot \boldsymbol{E}(\boldsymbol{r}) + \mu \boldsymbol{H}(\boldsymbol{r}) \cdot \boldsymbol{H}(\boldsymbol{r})]. \tag{8.26}$$

Using Eq. (8.7), we insert the Fourier transforms into the classical energy integral and obtain

$$W = \frac{1}{2(2\pi)^3} \int d\boldsymbol{r} \left\{ \varepsilon \int d\boldsymbol{k}\mathscr{E}(\boldsymbol{k})e^{i\boldsymbol{k}\cdot\boldsymbol{r}} \cdot \int d\boldsymbol{k}'\mathscr{E}(\boldsymbol{k}')e^{i\boldsymbol{k}'\cdot\boldsymbol{r}} \right.$$
$$\left. + \mu \int d\boldsymbol{k}\mathscr{H}(\boldsymbol{k})e^{i\boldsymbol{k}\cdot\boldsymbol{r}} \cdot \int d\boldsymbol{k}'\mathscr{H}(\boldsymbol{k}')e^{i\boldsymbol{k}'\cdot\boldsymbol{r}} \right\}$$

or

$$W = \frac{1}{2}\frac{1}{(2\pi)^3}\iint d\boldsymbol{k}d\boldsymbol{k}'[\varepsilon\mathscr{E}(\boldsymbol{k})\cdot\mathscr{E}(\boldsymbol{k}') + \mu\mathscr{H}(\boldsymbol{k})\cdot\mathscr{H}(\boldsymbol{k}')] \times [e^{i(\boldsymbol{k}+\boldsymbol{k}')\cdot\boldsymbol{r}}]d\boldsymbol{r}.$$

From Appendix IV the integral over the spatial coordinates may be interpreted as a delta function, leading to the form

$$W = \frac{1}{2}\frac{1}{(2\pi)^3}\iint d\boldsymbol{k}d\boldsymbol{k}'[\varepsilon\mathscr{E}(\boldsymbol{k})\cdot\mathscr{E}(\boldsymbol{k}') + \mu\mathscr{H}(\boldsymbol{k})\cdot\mathscr{H}(\boldsymbol{k}')](2\pi)^3\delta(\boldsymbol{k}+\boldsymbol{k}').$$

After integration over the variable $d\boldsymbol{k}'$, we obtain

$$W = \tfrac{1}{2}\int d\boldsymbol{k}[\varepsilon\mathscr{E}(\boldsymbol{k})\cdot\mathscr{E}(-\boldsymbol{k}) + \mu\mathscr{H}(\boldsymbol{k})\cdot\mathscr{H}(-\boldsymbol{k})].$$

Now, from Eq. (8.9),

$$\dot{\mathscr{E}}(\boldsymbol{k})\cdot\dot{\mathscr{E}}(-\boldsymbol{k}) = \left[\frac{i}{\varepsilon}\Big((\boldsymbol{k}\times\mathscr{H}(\boldsymbol{k})\Big)\right]\cdot\left[\frac{i}{\varepsilon}\Big(-\boldsymbol{k}\times\mathscr{H}(-\boldsymbol{k})\Big)\right],$$

and, using vector identity 3, we obtain

$$\dot{\mathscr{E}}(\boldsymbol{k})\cdot\dot{\mathscr{E}}(-\boldsymbol{k}) = \frac{1}{\varepsilon^2}k^2\mathscr{H}(\boldsymbol{k})\cdot\mathscr{H}(-\boldsymbol{k}).$$

Thus,

$$W = \tfrac{1}{2}\int d\boldsymbol{k}\left[\varepsilon\mathscr{E}(\boldsymbol{k})\cdot\mathscr{E}(-\boldsymbol{k}) + \frac{\varepsilon^2\mu}{k^2}\dot{\mathscr{E}}(\boldsymbol{k})\cdot\dot{\mathscr{E}}(-\boldsymbol{k})\right],$$

and, using Eqs. (8.19) and (8.20), we may write $\mathscr{E}$ and $\dot{\mathscr{E}}$ in terms of $\boldsymbol{f}(\boldsymbol{k})$, yielding

$$W = \varepsilon\int d\boldsymbol{k}A^2[\boldsymbol{f}^\star(\boldsymbol{k})\cdot\boldsymbol{f}(\boldsymbol{k}) + \boldsymbol{f}^\star(-\boldsymbol{k})\cdot\boldsymbol{f}(-\boldsymbol{k})].$$

Making the change of dummy variable $\boldsymbol{k}\to-\boldsymbol{k}$ in the last integral, we can show that

$$\int d\boldsymbol{k}A^2\boldsymbol{f}^\star(\boldsymbol{k})\cdot\boldsymbol{f}(\boldsymbol{k}) = \int d\boldsymbol{k}A^2\boldsymbol{f}^\star(-\boldsymbol{k})\cdot\boldsymbol{f}(-\boldsymbol{k}),$$

and, therefore,

$$W = 2\varepsilon\int d\boldsymbol{k}A^2\boldsymbol{f}^\star(\boldsymbol{k})\cdot\boldsymbol{f}(\boldsymbol{k}).$$

It is seen that we can satisfy Eq. (8.26) by setting

$$A = \left(\frac{k\hbar c}{2\varepsilon}\right)^{1/2}. \qquad (8.27)$$

Treating $f(k)$ as the state function of the photon in momentum space leads to the correct correspondence between the classical energy and the expectation value of the energy when the normalization given by Eq. (8.27) is used. The normalization of Eq. (8.27) would be meaningless, however, if it did not lead to a correct correspondence between other classical properties of the electromagnetic field and their expectation values in momentum space. In particular, we consider the expectation value of linear momentum.

If $f(k)$ is the state function of the photon, then

$$f^{\star}(k) \cdot f(k) dk$$

should be the probability that the photon will be found in the volume element of k-space, dk, and possess linear momentum $\hbar k$. The expectation value for the i-th component of linear momentum should, therefore, be of the form

$$\langle S|p_i|S\rangle = \int dk \hbar k_i f^{\star}(k) \cdot f(k).$$

This expectation value must correspond to the linear momentum of the classical field, which is given by

$$p = \frac{1}{c^2} \int dr E(r) \times H(r). \tag{8.28}$$

The integral in Eq. (8.28) may be transformed in a manner similar to that used for the energy integral. The details of the transformation are given in Appendix VI. The results of this transformation yield for each component p_i, of the linear momentum p,

$$\begin{aligned} p_i &= \frac{1}{c^2} \int dr [E \times H]_i \\ &= \int dk f^{\star}(k) \cdot \hbar k_i f(k). \end{aligned} \tag{8.29}$$

Thus, one can, with the normalization of Eq. (8.27), identify the linear momentum operator in K-space as $\hbar k$.

Let us summarize. We have seen that in momentum space we may define an acceptable state function $f(k, t)$, as given by Eq. (8.21). This function obeys a Schrödinger equation and admits a probability interpretation for the bracket product. The energy and linear momentum operators for this state function have been identified. The remaining problem of angular momentum and its operators is taken up in the next chapter.

PROBLEMS

8.1 Show that the motion in momentum space of a classical simple harmonic oscillator is an ellipse centered on the origin. Show that the area of the ellipse is proportional to the total energy of the oscillator.

8.2 Consider a very large number of classical particles, each of mass m, which move with constant velocities along the x-axis. At $t = 0$ those particles located by x-coordinates with $X \geqslant x \geqslant -X$ and with velocities satisfying $V \geqslant v \geqslant -V$ are represented in momentum space by points lying within a rectangle of area $4mVX$. Show that at some later time the area containing these points in momentum space has become distorted into a parallelogram but that the area containing them is still $4mVX$.

8.3 By transforming Maxwell's equations to momentum space obtain Eqs. (8.9), (8.10), and (8.11).

8.4 If $\boldsymbol{f}(\boldsymbol{k}, t)$ is a solution of either Eq. (8.17) or Eq. (8.18), show that $\boldsymbol{f}(-\boldsymbol{k}, t)$ is a solution of the same equation and $\boldsymbol{f}^{\star}(\boldsymbol{k}, t)$ is a solution of the other.

8.5 Show that $\mathscr{H}(\boldsymbol{k}, t)$ is given in terms of $\boldsymbol{f}(\boldsymbol{k}, t)$ by

$$\mathscr{H}(\boldsymbol{k}, t) = \sqrt{\frac{\varepsilon \hbar c}{2k}}\, c\boldsymbol{k} \times [\boldsymbol{f}(\boldsymbol{k}, t) - \boldsymbol{f}^{\star}(-\boldsymbol{k}, t)].$$

8.6 If a particle of mass m which is described by a scalar state function $\Psi(\boldsymbol{r})$ possesses a potential energy $V(\boldsymbol{r})$ as well as kinetic energy, explain why you would expect $\Psi(\boldsymbol{r})$ to obey a Schrödinger equation of the form

$$-\frac{\hbar^2}{2m}\nabla^2\Psi + V\Psi = i\hbar\frac{\partial\Psi}{\partial t}\cdot$$

8.7 Consider the plane wave electric field defined by

$$\boldsymbol{E}(\boldsymbol{r}, t) = E_0 \cos(kz - \omega t)\boldsymbol{e}_x.$$

For this field show by the inverse of Eq. (8.7) that

$$\mathscr{E}(\boldsymbol{k}, t) = \tfrac{1}{2}E_0\delta(k_x)\delta(k_y)[\delta(k - k_z)e^{-i\omega t} + \delta(k + k_z)e^{i\omega t}]\boldsymbol{e}_x,$$

and, hence,

$$\boldsymbol{f}(\boldsymbol{k}, t) = E_0/2A\ \delta(k_x)\delta(k_y)\delta(k - k_z)e^{-i\omega t}\boldsymbol{e}_x,$$

where A is given by Eq. (8.27). What difficulty is encountered in attempting to interpret these functions as single-photon functions?

8.8 Show that if the electric field is of the form

$$\boldsymbol{E}(\boldsymbol{r}, t) = \boldsymbol{E}_0(\boldsymbol{r}) \cos \omega t,$$

the state function is of the form

$$\boldsymbol{f}(\boldsymbol{k}, t) = \boldsymbol{f}_0(\boldsymbol{k})e^{-i\omega t}.$$

Show that the converse is not necessarily true.

8.9 It is often convenient to demand that a photon be found within some given finite (although perhaps large) volume of space. A simple example is a cubic box of side L, with one corner at the origin. As a formal mathematical device one may then stack these boxes side by side to fill all of space, with one photon in each box.

Since the electric and magnetic fields are identical at equivalent points in different boxes, these fields are periodic in space, that is,

$$\boldsymbol{E}(\boldsymbol{r}) = \boldsymbol{E}(\boldsymbol{r} + \boldsymbol{L}),$$

where

$$\boldsymbol{L} = L(n_x\boldsymbol{e}_x + n_y\boldsymbol{e}_y + n_z\boldsymbol{e}_z),$$

with n_x, n_y, and n_z being positive or negative integers or zero. Show that this condition implies

$$\boldsymbol{E}(\boldsymbol{r})[e^{i\boldsymbol{k}\cdot\boldsymbol{L}} - 1] = 0,$$
$$\mathscr{E}(\boldsymbol{k})[e^{i\boldsymbol{k}\cdot\boldsymbol{L}} - 1] = 0,$$

and, hence, that $\mathscr{E}(\boldsymbol{k})$ is zero except for those values of $\boldsymbol{k}$ of the form

$$\boldsymbol{k}_m = \frac{2\pi}{L}(m_x\boldsymbol{e}_x + m_y\boldsymbol{e}_y + m_z\boldsymbol{e}_z),$$

where the m_i are integers or zero.

8.10 Proceeding from the results of the previous problem, show that if Eq. (8.7) is to remain meaningful, $\mathscr{E}$ must take on the form

$$\mathscr{E}(\boldsymbol{k}, t) = \Sigma\mathscr{E}_{\boldsymbol{k}_m}(t)\delta(\boldsymbol{k} - \boldsymbol{k}_m),$$

leading to

$$\boldsymbol{E}(\boldsymbol{r}, t) = \frac{1}{(2\pi)^{3/2}}\Sigma\mathscr{E}_{\boldsymbol{k}_m}(t)e^{i\boldsymbol{k}_m\cdot\boldsymbol{r}},$$

where the sums are over all allowed values of $\boldsymbol{k}_m$ and $\mathscr{E}_{\boldsymbol{k}_m}(t)$ is given by

$$\mathscr{E}_{\boldsymbol{k}_m}(t) = \frac{(2\pi)^{3/2}}{L^3}\int d\boldsymbol{r}\boldsymbol{E}(\boldsymbol{r}, t)e^{-i\boldsymbol{k}_m\cdot\boldsymbol{r}},$$

where the integral is taken over the volume of the box of side L. Thus, the continuous function $\mathscr{E}(\boldsymbol{k}, t)$ is replaced by the discrete set of functions $\mathscr{E}_{\boldsymbol{k}_m}(t)$, one function for each allowed value of $\boldsymbol{k}_m$. Likewise, the continuous function $\boldsymbol{f}(\boldsymbol{k}, t)$ must be replaced by a set of discrete functions $\boldsymbol{f}_{\boldsymbol{k}_m}(t)$ given by

$$\boldsymbol{f}_{\boldsymbol{k}_m}(t) = \tfrac{1}{2}\sqrt{\frac{2\varepsilon}{k_m\hbar c}}\left[\mathscr{E}_{k_m}(t) - \frac{1}{ik_mc}\dot{\mathscr{E}}_{\boldsymbol{k}_m}(t)\right].$$

8.11 Let the state function $\boldsymbol{f}(\boldsymbol{k}, t)$ represent a monochromatic superposition of the form $\boldsymbol{f}(\boldsymbol{k}, t) = \boldsymbol{f}(\boldsymbol{k})e^{-i\omega t}$. If we assume that the energy content of the electromagnetic field over all space is equal to the energy, $\hbar\omega$, of one photon, show that Eq. (8.26) reduces to

$$\hbar\omega = \hbar\omega\int d\boldsymbol{k}\boldsymbol{f}^{\star}(\boldsymbol{k})\boldsymbol{f}(\boldsymbol{k}),$$

and that the normalization $\int d\boldsymbol{k}\boldsymbol{f}^{\star}\boldsymbol{f} = 1$ of the state function in momentum space corresponds to one photon in real space.

8.12 Follow through the derivation leading from Eq. (8.26) to Eq. (8.27) in the case outlined in Problems 8.9 and 8.10, and show that in the monochromatic case, with

$$\mathscr{E}_{\boldsymbol{k}_m}(t) = \mathscr{E}_{\boldsymbol{k}_m} e^{-i\omega t},$$

the assumption of one photon of energy $\hbar\omega$ within each box of side L leads to

$$\hbar\omega = \hbar\omega \left(\frac{L}{2\pi}\right)^3 \Sigma \boldsymbol{f}_{\boldsymbol{k}_m} \cdot \boldsymbol{f}_{\boldsymbol{k}_m},$$

and, hence, that the normalization

$$\Sigma \boldsymbol{f}_{\boldsymbol{k}_m} \cdot \boldsymbol{f}_{\boldsymbol{k}_m} = \left(\frac{2\pi}{L}\right)^3$$

in momentum space corresponds to one photon per box in real space. This procedure, for obvious reasons, is known as box normalization.

8.13 Consider the wave of Problem 8.7,

$$\boldsymbol{E}(\boldsymbol{r}, t) = E_0 \cos(kz - \omega t)\boldsymbol{e}_x,$$

from the viewpoint of box normalization, where k must now be of the form $k = 2\pi m/L$ and as a vector is given by $\boldsymbol{k} = 2\pi m/L\boldsymbol{e}_z$. Show that there are only two nonzero $\mathscr{E}_{\boldsymbol{k}_m}(t)$, given by

$$\mathscr{E}_{\pm \boldsymbol{k}} = \frac{E_0}{2}(2\pi)^{3/2} e^{\mp i\omega t},$$

and, hence, only two nonzero $\boldsymbol{f}_{\boldsymbol{k}_m}(t)$:

$$\boldsymbol{f}_{\pm k}(t) = \frac{E_0}{2}(2\pi)^{3/2} \sqrt{\frac{\hbar\omega}{2\varepsilon}}\, e^{\mp i\omega t}.$$

Show that box normalization leads to

$$E_0 = \sqrt{\frac{\hbar\omega}{\varepsilon L^3}}.$$

CHAPTER 9

ANGULAR MOMENTUM

Following the introduction of the photon in Chapter 2, its particle nature was justified by the absorption in a detector of discrete bundles of energy within a very small region of space and within a very short interval of time. Intimately associated with this absorption of energy were other measurable quantities. In particular, linear and angular momentum have been discussed. Of these, angular momentum, within the Stokes parameter formalism, was used to introduce the entire scheme of quantum mechanics.

Perhaps it has been noticed that since that point the topic of angular momentum has been scrupulously avoided. In neither the electromagnetic representation of the classical field nor the quantum description in momentum space was it discussed, although both energy and linear momenta were considered. The reason is that angular momentum cannot be adequately considered until both the classical and quantum representations are available. In this chapter we shall examine the angular momentum of light from both viewpoints.

9.1 ANGULAR MOMENTUM OF THE ELECTROMAGNETIC FIELD

To repeat what has been given before, if a classical particle possesses linear momentum $\boldsymbol{p}$, then, in respect to an origin from which the position $\boldsymbol{r}$ is measured, its angular momentum is given by $\boldsymbol{r} \times \boldsymbol{p}$. Thus, if a physical entity, e.g., the electromagnetic field, possesses a linear momentum density $\boldsymbol{g}$, the angular momentum density, computed in respect to the origin, would be $\boldsymbol{r} \times \boldsymbol{g}$. The total angular momentum of the electromagnetic field is then

$$\boldsymbol{J} = \int d\boldsymbol{r}\, \boldsymbol{r} \times \boldsymbol{g}. \tag{9.1}$$

From Eq. (7.25) this can be written

$$\boldsymbol{J} = \int d\boldsymbol{r}\, \boldsymbol{r} \times \left(\frac{\boldsymbol{E} \times \boldsymbol{H}}{c^2}\right). \tag{9.2}$$

For the plane wave states which were considered in Chapter 7, $\boldsymbol{E}$ and $\boldsymbol{H}$ were mutually perpendicular and their cross product was parallel to the direction of propagation $\boldsymbol{k}$. Thus, the linear momentum density for such waves is also in the direction of propagation. It follows that for such waves $\boldsymbol{r} \times \boldsymbol{g}$ must be perpendicular to the direction of propagation, and the total angular momentum of a plane wave can have no component of angular momentum parallel to the direction of propagation.

This leads to an apparent contradiction in our development. We know that circularly polarized states carry angular momentum whose direction is along the direction of propagation. And yet, in Chapter 7 we represented these states by superpositions of plane waves which cannot carry angular momentum whose direction is along the direction of propagation. Obviously, there is a contradiction.

One way out of this dilemma is to realize that perfect plane waves which fill all of space do not exist. Hence, we should not worry about that which cannot be observed. This reasoning has some merit, perhaps, but it must be remembered that there is nothing within the classical framework which prohibits us from approximating a plane wave as closely as we wish. We proceed to examine the angular momentum of approximately plane waves.

Recall that the plane wave representation of right and left circularly polarized light is given in terms of the electric intensity vector by

$$\boldsymbol{E} = E_0[e^{\mp i\pi/4}\boldsymbol{e}_x + e^{\pm i\pi/4}\boldsymbol{e}_y]e^{i(kz-\omega t)}. \tag{9.3}$$

Since for plane waves $\boldsymbol{k} \times \boldsymbol{E} \propto \boldsymbol{H}$ and $\sqrt{\varepsilon}|\boldsymbol{E}| = \sqrt{\mu}|\boldsymbol{H}|$, the magnetic vector for this wave is given by

$$\begin{aligned}\boldsymbol{H} &= H_0[-e^{\pm i\pi/4}\boldsymbol{e}_x + e^{\mp i\pi/4}\boldsymbol{e}_y]e^{i(kz-\omega t)}\\ &= \mp i\sqrt{\frac{\varepsilon}{\mu}}\,\boldsymbol{E}.\end{aligned} \tag{9.4}$$

E_0 and H_0 in Eqs. (9.3) and (9.4) are constants. This wave propagates in the z-direction and uniformly fills all of space. We will now adjust this solution so that $\boldsymbol{E}$ and $\boldsymbol{H}$ are nonzero only within some finite, although perhaps large, region around the z-axis. We accomplish this by allowing E_0 to be a function of x and y but not of z. Furthermore, for simplicity we assume symmetry about the z-axis so that the dependence is only a function of the distance from the z-axis, $r = \sqrt{x^2 + y^2}$. We also assume that the function is slowly varying in r, in a sense to be made more precise, and goes to zero outside some finite cylinder of radius R_0 about the z-axis. We thus have

$$\boldsymbol{E} = E_0(x, y)[e^{\mp i\pi/4}\boldsymbol{e}_x + e^{\pm i\pi/4}\boldsymbol{e}_y]e^{i(kz-\omega t)}, \tag{9.5}$$

$$\boldsymbol{H} = \mp i\sqrt{\frac{\varepsilon}{\mu}}\,\boldsymbol{E}. \tag{9.6}$$

We would now have a realistic model of a light beam of finite cross section if it were not for the fact that the function defined by Eqs. (9.5) and (9.6) do not satisfy Maxwell's equations. To see this we recall Eq. (7.11):

$$\nabla \times \boldsymbol{H} = \varepsilon \frac{\partial \boldsymbol{E}}{\partial t}.$$

Using the time dependence of Eq. (9.5) and replacing $\boldsymbol{H}$ by $\boldsymbol{E}$ through Eq. (9.6), we obtain

$$\nabla \times \boldsymbol{E} = \pm k\boldsymbol{E}. \tag{9.7}$$

From Eq. (9.5), $\boldsymbol{E}$, on the right-hand side of Eq. (9.7), has no z-component. By direct computation we find the z-component of the left-hand side of Eq. (9.7) to be non-zero and given by

$$(\nabla \times \boldsymbol{E})_z = \left[e^{\pm i\pi/4}\frac{\partial E_0}{\partial x} - e^{\mp i\pi/4}\frac{\partial E_0}{\partial y}\right]e^{i(kz-\omega t)}. \tag{9.8}$$

We thus see that the fields given by Eqs. (9.5) and (9.6) do not satisfy Maxwell's equations because they lack z-components. We attempt a second approximate solution by explicitly including nonzero components E_z and H_z. It is first convenient to rewrite Eqs. (9.5) and (9.6) in cylindrical coordinates, as shown in Fig. 9.1. The direction of propagation is kept along the z-axis. The unit vectors $\boldsymbol{e}_r$ and $\boldsymbol{e}_\theta$ are shown in Fig. 9.2, and it is easily shown that they are related to the Cartesian unit vectors $\boldsymbol{e}_x$ and $\boldsymbol{e}_y$ through

$$\begin{aligned} \boldsymbol{e}_x &= \cos\theta\boldsymbol{e}_r - \sin\theta\boldsymbol{e}_\theta, \\ \boldsymbol{e}_y &= \sin\theta\boldsymbol{e}_r + \cos\theta\boldsymbol{e}_\theta. \end{aligned} \tag{9.9}$$

The combinations

$$e^{\mp i\pi/4}\boldsymbol{e}_x + e^{\pm i\pi/4}\boldsymbol{e}_y = e^{\mp i\pi/4}[\boldsymbol{e}_x \pm i\boldsymbol{e}_y]$$

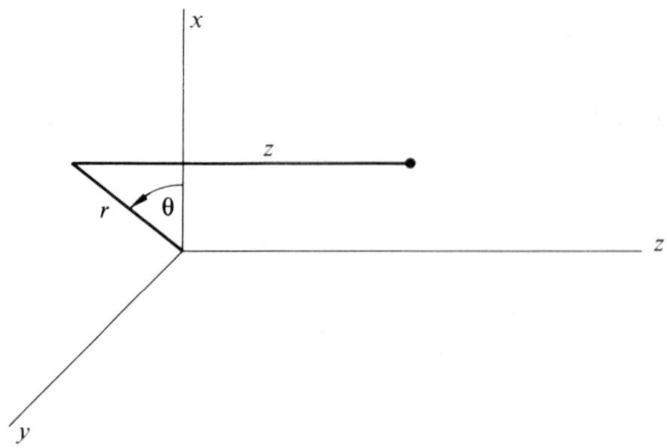

Fig. 9.1. System of cylindrical coordinates.

then become

$$e^{\mp i\pi/4}[\boldsymbol{e}_x \pm i\boldsymbol{e}_y] = e^{\mp i\pi/4}\{(\cos\theta \pm i\sin\theta)\boldsymbol{e}_r \pm i(\cos\theta \pm i\sin\theta)\boldsymbol{e}_\theta\}$$
$$= e^{\mp i\pi/4}\{\boldsymbol{e}_r \pm i\boldsymbol{e}_\theta\}e^{\pm i\theta}. \tag{9.10}$$

Our second approximate solution is then written

$$\boldsymbol{E}' = [E_0(r)(\boldsymbol{e}_r \pm i\boldsymbol{e}_\theta) + f(r)\boldsymbol{e}_z]e^{i\phi} \tag{9.11}$$

$$\boldsymbol{H}' = \mp i\sqrt{\frac{\varepsilon}{\mu}}\,\boldsymbol{E}', \tag{9.12}$$

where

$$\phi = (kz - \omega t \pm \theta \mp \pi/4). \tag{9.13}$$

Because of the proportionality between $\boldsymbol{E}'$ and $\boldsymbol{H}'$ given by Eq. (9.12) and the form of the time dependence, Maxwell's equations reduce to the two independent equations:

$$\nabla \times \boldsymbol{E}' = \pm k\boldsymbol{E}', \tag{9.14}$$

$$\nabla \cdot \boldsymbol{E}' = 0. \tag{9.15}$$

We impose Eq. (9.15) first. Writing the divergence in cylindrical coordinates, we obtain

$$\nabla \cdot \boldsymbol{E}' = \frac{1}{r}\frac{\partial}{\partial r}[rE_r'] + \frac{1}{r}\frac{\partial}{\partial\theta}E_\theta' + \frac{\partial}{\partial z}E_z', \tag{9.16}$$

where, by Eq. (9.11),

$$E_r' = E_0(r)e^{i\phi},$$
$$E_\theta' = \pm iE_0(r)e^{i\phi}, \qquad (9.17$$
$$E_z' = f(r)e^{i\phi}.$$

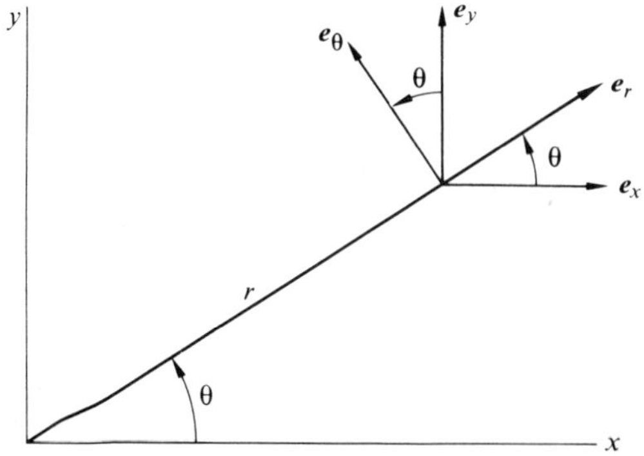

Fig. 9.2. The relation between the unit vectors of cylindrical and Cartesian coordinates.

Performing the differentiations, one finds that Eq. (9.15) will be satisfied if and only if

$$\frac{\partial E_0(r)}{\partial r} + ikf(r) = 0. \tag{9.18}$$

Thus, the wave will have nonvanishing components in the direction of propagation, i.e. $f(r) \neq 0$, only in those regions where $E_0(r)$ is not constant.

Returning to Eq. (9.14), we write out the left-hand side in cylindrical coordinates below:

$$\nabla \times \boldsymbol{E}' = \left(\frac{1}{r}\frac{\partial E'_z}{\partial \theta} - \frac{\partial E'_\theta}{\partial z}\right)\boldsymbol{e}_r + \left(\frac{\partial E'_r}{\partial z} - \frac{\partial E'_z}{\partial r}\right)\boldsymbol{e}_\theta + \left(\frac{1}{r}\frac{\partial}{\partial r}(rE'_\theta) - \frac{1}{r}\frac{\partial E'_r}{\partial \theta}\right)\boldsymbol{e}_z. \tag{9.19}$$

Making use of Eq. (9.17), we obtain

$$(\nabla \times \boldsymbol{E}')_r = \pm kE'_r\left(1 \pm \frac{i}{krE_0(r)} f(r)\right),$$

$$(\nabla \times \boldsymbol{E}')_\theta = \pm kE'_\theta\left(1 + \frac{i}{kE_0(r)}\frac{\partial f(r)}{\partial r}\right),$$

$$(\nabla \times \boldsymbol{E}')_z = \pm kE'_z.$$

Thus, the z-component of Eq. (9.14) is satisfied identically but there are errors in the r and θ components, given by

$$\pm\frac{i}{krE_0} f(r), \quad \frac{i}{kE_0}\frac{\partial f(r)}{dr},$$

which, by use of Eq. (9.18), may be written

$$\mp\frac{1}{rk^2E_0}\frac{\partial E_0}{\partial r}, \quad -\frac{1}{k^2E_0}\frac{\partial^2 E_0}{\partial r^2}. \tag{9.20}$$

$E_0(r)$ was specified as a slowly varying function. We now make this more precise. By slowly varying we mean that the derivatives of $E_0(r)$ are so small that the error terms above are very small compared to 1 and may be neglected. Examples of acceptable functional forms for E_0 will be given later. Thus, $\boldsymbol{E}'$ and $\boldsymbol{H}'$ are a second approximation to a solution of Maxwell's equations which represent a circularly polarized wave contained within a finite cylinder about the z-axis.

In Fig. 9.3 we plot an acceptable function $\boldsymbol{E}_0(x, y) = \boldsymbol{E}_0(r)$. We have explicitly made $\boldsymbol{E}_0$ constant over a large central region of the wave and confined the variation of the function from this constant value to zero to lie within a "skin" of thickness δ which lies a distance R_0 from the axis. From

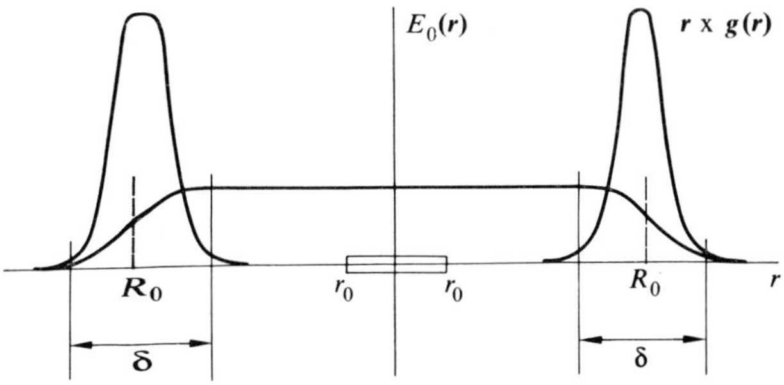

Fig. 9.3. The electric field amplitude and the angular momentum density across a cylindrical beam.

Eq. (9.18) the electric and magnetic fields can have a nonzero z-component only within the skin region of this wave. Having z-components within this region implies the possibility of a nonzero z-component of angular momentum within this region. Since the wave is identically zero outside the skin and constant inside the skin region, the skin region is the only one in which the z-component of angular momentum does not vanish.

Without specifying the exact form of $E_0(r)$, it is, of course, impossible to calculate the total angular momentum of the wave. However, it is possible to calculate the ratio of the total z-component of angular momentum, per unit length, of this wave to the total energy per unit length.

This calculation is most readily carried out in terms of the real components of the fields. These may easily be obtained by referring to Eqs. (9.11), (9.12), and (9.18), and are listed below. $E_0(r)$ is assumed to be a real function.

Field component	Real part of component
E_r'	$E_0 \cos\phi$
E_θ'	$\mp E_0 \sin\phi$
E_z'	$-\dfrac{1}{k}\dfrac{\partial E_0}{\partial r}\sin\phi$
H_r'	$\pm E_0\sqrt{\dfrac{\varepsilon}{\mu}}\sin\phi$
H_θ'	$E_0\sqrt{\dfrac{\varepsilon}{\mu}}\cos\phi$
H_z'	$\pm\dfrac{1}{k}\dfrac{\partial E_0}{\partial r}\sqrt{\dfrac{\varepsilon}{\mu}}\cos\phi.$

(9.21)

Written in cylindrical coordinates, the z-component of angular momentum, as given by Eq. (9.2), is

$$J_z = \iiint \frac{1}{c^2}[E'_z H'_r - E'_r H'_z] r^2 dr d\theta dz.$$

When the real parts of the fields are inserted, this becomes

$$J_z = \mp \frac{1}{\omega} \iiint \frac{\partial}{\partial r}(\tfrac{1}{2}\varepsilon E_0^2) r^2 dr d\theta dz.$$

If the integration over r is performed by parts, we obtain

$$\int_0^\infty \frac{\partial}{\partial r}(\tfrac{1}{2}\varepsilon E_0^2) r^2 dr = \tfrac{1}{2}\varepsilon E_0^2 r^2 \Big|_0^\infty - \int_0^\infty \varepsilon E_0^2 r dr.$$

The first term vanishes at $r = 0$ and $E_0(r)$ is assumed to go to zero for r larger than R_0. We then have

$$J_z = \pm \frac{1}{\omega} \iiint \varepsilon E_0^2 r dr d\theta dz, \tag{9.22}$$

where the plus and minus refer to right and left circular states. In a completely analogous manner the energy content of the field is computed by use of Eq. (7.23) and the real field components from Eqs. (9.21) and is found to be

$$W = \iiint \varepsilon E_0^2 \left[1 + \frac{1}{2}\frac{1}{k^2 E_0^2}\left(\frac{\partial E_0}{\partial r}\right)^2\right] r dr d\theta dz. \tag{9.23}$$

Just as the error terms given by Eq. (9.20) were assumed to be small enough compared to unity to be ignored, so we also ignore the term

$$\frac{1}{2}\frac{1}{k^2 E_0^2}\left(\frac{\partial E_0}{\partial r}\right)^2,$$

assuming that it is also small compared to unity. A specific example in which both of these approximations are obviously justified is given in the problems. Finally, the ratio of the energy content of the wave to the component of angular momentum along the direction of propagation is

$$\frac{W}{J_z} = \pm\omega, \tag{9.24}$$

which is consistent with the quantum viewpoint of photons each carrying energy $\hbar\omega$ and angular momentum $\pm\hbar$.

Perhaps there now appears to be no dilemma, for, as we have seen, when a more realistic approximation to a plane circularly polarized wave is used, we find the correct angular momentum content. However, let us plot the angular momentum density predicted by our results, rather than consider the total angular momentum content of the wave. As we have mentioned, the z-component of angular momentum can be nonzero only in those regions in which the fields themselves have nonzero z-components. Once again referring to Eq. (9.18), we see that the field can have nonzero z-components only in those regions where $E_0(r)$ varies. Thus, the total angular momentum content of the wave lies within the skin region where E_0 is nonconstant.

Now consider the following thought experiment. A small circular absorbing disk, which is free to rotate about the z-axis, is placed in the center of the beam. The radius, r_0, of the disk is indicated in Fig. 9.3 with $r_0 \lll R_0$. Assuming that the beam is a circular polarization state, should we expect the small disk to rotate? There are two possible answers: "yes" and "no". No, it will not rotate, for although it absorbs energy from the beam, the classical angular momentum density is zero in the central region and, hence, no angular momentum can be taken up by the disk. Yes, it will rotate, for each photon in the beam is in the same circular state and carries angular momentum in the direction of propagation, plus or minus $\hbar$ as the beam is right or left circularly polarized. Once again there seems to be a contradiction.

Let us show that there is no contradiction in expecting the disk to rotate. By placing the small disk in the beam we are attempting to measure, or observe, some property of the beam. In this case the property is angular momentum. In any measurement there must be an interaction between the measuring instrument and the observed system. In many cases this interaction is small enough to be ignored but in this case it has a gross effect, for we are removing from the beam the central section which is absorbed by the disk. This is indicated in Fig. 9.4, where we have plotted the

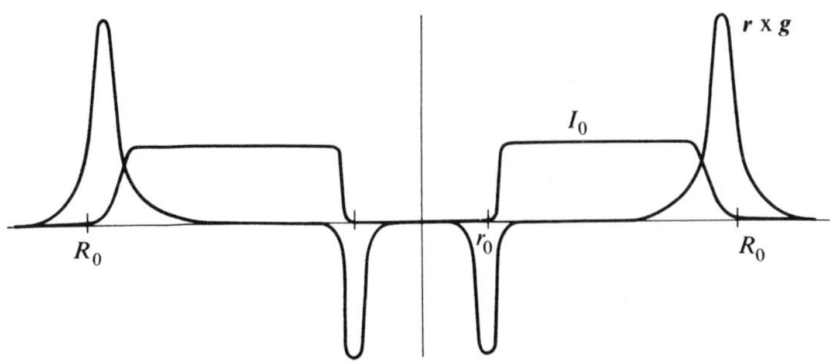

Fig. 9.4. The intensity and the angular momentum density across a cylindrical beam after an absorbing disk has been placed in the center of the beam.

intensity across the beam just behind the disk. The variation of the fields at radius r_0 behind the disk must be much too rapid to apply quantitatively the previous results for approximate plane waves. However, we can attempt to apply them in a qualitative manner. Because the fields are varying as a function of r in this region, Eq. (9.18) tells us that we can expect to find components in this region along the direction of propagation. If the fields have nonzero components in the z-direction, we expect to find a nonzero component of angular momentum along the direction of propagation. In the region around r_0 the fields are increasing as a function of r, whereas at R_0 they are decreasing. Thus, the derivatives of these functions with respect to r will have opposite signs in these two regions. It follows from Eq. (9.18) that the z-components of the fields and, hence, the z-component of the angular momentum density will also have opposite signs in these two regions. This is indicated in Fig. 9.4.

Thus, the inner skin of the wave has picked up an angular momentum opposite to that of the outer skin. In order to have total angular momentum conserved, the disk which absorbed this inner part of the wave must rotate with angular momentum opposite to that of the inner skin or in the direction of the incident beam's angular momentum.

Several important points result from our discussion. First, and perhaps most interesting, is that a classical quantity associated with the electromagnetic field does not necessarily indicate the value of that quantity which will be measured. This is clear from our example. The angular momentum density of the wave was zero at the center, yet when we attempted to measure it there, the classical fields adjusted themselves and produced a nonzero measurement which must be in accord with the quantum result $W/J_z = \pm\omega$. In the preceding thought experiment we have neglected the niceties of diffraction theory. This was done simply because the exact diffraction problem is of great complexity. However, in every classical calculation on realistic fields, the ratio of energy content to angular momentum content for the entire wave is consistent with the quantum relation $E = \hbar\omega$.

The second point follows directly from the last statement. The classical field concept and the quantum viewpoint give consistent results if we consider large volumes of the classical fields. The classical field does not include the concept of the photon as a localizable particle. Hence, apparent contradictions arise when we compare these theories within very small volumes or at particular points in space. For this reason a discription of angular momentum in K-space proves fruitful. In K-space we suppress the explicit space dependence of the state function.

Before discussing the angular momentum of the photon in momentum space, we wish to examine in the following sections some fundamental aspects of angular momentum which are found in all systems whatever the nature of their state function might be.

9.2 THE ANGULAR MOMENTUM OPERATORS

The role of angular momentum in quantum mechanics is of such importance that before studying the angular momentum of the photon we shall present some very general results for the angular momentum of any system.

Quantum mechanics associates with each observable an operator whose eigenvalues yield the measurable values of the observable. We begin by examining the angular momentum operator associated with a particle which possesses a scalar state function. Classically, as we have seen, the angular momentum, $\boldsymbol{L}$, of a point particle is given in terms of its linear momentum, $\boldsymbol{p}$, by

$$\boldsymbol{L} = \boldsymbol{r} \times \boldsymbol{p},$$

where $\boldsymbol{r}$ is the location of the particle, as measured from some particular origin. This may be translated into the language of quantum mechanics by replacing $\boldsymbol{p}$ by the operator $p_{\text{op.}} = -i\hbar\nabla$, and we obtain

$$L_{\text{op.}} = -i\hbar \boldsymbol{r} \times \nabla. \tag{9.25}$$

This, of course, represents three operators, one for each component of angular momentum. In Cartesian coordinates these operators are given by

$$\begin{aligned} L_x &= -i\hbar\left(y\frac{\partial}{\partial z} - z\frac{\partial}{\partial y}\right), \\ L_y &= -i\hbar\left(z\frac{\partial}{\partial x} - x\frac{\partial}{\partial z}\right), \\ L_z &= -i\hbar\left(x\frac{\partial}{\partial y} - y\frac{\partial}{\partial x}\right). \end{aligned} \tag{9.26}$$

In addition to the component operators, we may also define an operator for the magnitude of the total angular momentum by

$$L^2 = L \cdot L. \tag{9.27}$$

One may, of course, set up eigenvalue equations for these operators immediately. However, by so doing a great deal of generality is lost. Consider the commutation relations which are obeyed by the four operators given above. It is left to the problems to show that

$$\begin{gathered} [L_x, L_y] = i\hbar L_z, \quad [L_y, L_z] = i\hbar L_x, \\ [L_z, L_x] = i\hbar L_y, \\ [L^2, L_x] = [L^2, L_y] = [L^2, L_z] = 0. \end{gathered} \tag{9.28}$$

Now, it happens that Eqs. (9.28) are much more general than Eqs. (9.26). By this is meant that there exist operators which satisfy the latter equations which cannot be written in the form of Eqs. (9.26). In fact, we shall see that the angular momentum operators for the photon satisfy Eqs. (9.28), yet we have seen that the photon does not possess a scalar state function and, hence, the operators of Eqs. (9.26) do not apply to the photon. To distinguish from the particular form of Eq. (9.26), we therefore define general Hermitian angular momentum operators by their commutation rules:

$$[J_i, J_j] = i\hbar J_k, \qquad [J^2, J_i] = 0, \qquad J_i = J_i^+. \tag{9.29}$$

In the first of these equations, i, j, and k are assumed to be in cyclic order. Since we imply no particular representation for these operators, the operator J^2 must be interpreted to mean

$$J^2 = J_x^2 + J_y^2 + J_z^2. \tag{9.30}$$

The following section is devoted to exhibiting the eigenvalues of these operators. This exhibition is largely an algebraic tour de force. Thus, let us postpone the algebra and examine the physics which is involved.

Equation (9.29) tells us that the operators associated with the various components of angular momentum do not commute among themselves. Referring to the results of Section 5.2, we may not, therefore, define states which are simultaneously eigenstates of two such components. Physically we may not simultaneously measure two different components of angular momentum with arbitrary accuracy. However, Eq. (9.29) also tells us that the operators associated with the individual components commute with the operator associated with the total angular momentum, J^2. Thus, for any system we may prepare states which are simultaneously pure states in respect to total angular momentum and the component of angular momentum along some one direction. Without loss of generality we may take this direction to be the z-direction. Then, as we shall see in the next section, Eqs. (9.29) lead directly to the eigenvalue equations

$$J^2|S\rangle = j(j+1)\hbar^2|S\rangle, \tag{9.31}$$

$$J_z|S\rangle = m_j\hbar|S\rangle, \tag{9.32}$$

with j taking on integer or half-integer values, $j = \frac{1}{2}, 1, 1\frac{1}{2}, 2, \ldots$, and for a given value of j, m_j may take on only the $(2j+1)$-values

$$m_j = j, j-1, j-2, \ldots, -j+1, -j.$$

In conclusion, the quantum mechanical specification of the total angular momentum of a system is given by the two "quantum numbers" j and m_j, which may be integers or half-integers. The term "quantum number" is used for any convenient label for the eigenstates of a system. For example, one may refer to a system as having angular momentum $j = 1$, meaning that according to the eigenvalues of Eq. (9.31) the magnitude of the angular momentum of the system is

$$\sqrt{j(j+1)\hbar^2} = \sqrt{2}\hbar.$$

9.3 ANGULAR MOMENTUM EIGENVALUES

We now demonstrate that the commutation rules given by Eqs. (9.29) are sufficient to determine the eigenvalues of the operators J^2 and J_z, as given in Eqs. (9.31) and (9.32).

We write the eigenstates of these operators as $|\lambda, m\rangle$, where the quantum numbers or labels of these states are related to the eigenvalues of the operators as shown below:

$$J^2|\lambda, m\rangle = \lambda^2\hbar^2|\lambda, m\rangle, \tag{9.33}$$

$$J_z|\lambda, m\rangle = m\hbar|\lambda, m\rangle. \tag{9.34}$$

We can assume a normalization of these states such that

$$\langle\lambda, m|\lambda, m\rangle = 1.$$

We first show that $\lambda^2 \geqslant m^2$. Let $|S\rangle$ be any eigenstate $|\lambda, m\rangle$. Using Eq. (9.30), we may write

$$\langle S|J^2|S\rangle = \langle S|J_x^2|S\rangle + \langle S|J_y^2|S\rangle + \langle S|J_z^2|S\rangle,$$

which, by Eqs. (9.33) and (9.34), becomes

$$\hbar^2\lambda^2 = \langle S|J_x^\dagger J_x|S\rangle + \langle S|J_y^\dagger J_y|S\rangle + \hbar^2 m^2,$$

where we have made explicit use of the Hermiticity of the operators to write

$$\langle S|J_i^2|S\rangle = \langle S|J_i^\dagger J_i|S\rangle.$$

The remaining bracket products must be real positive numbers, since they are of the form $\langle S'|S'\rangle$ with $|S'\rangle = J_x|S\rangle$ or $|S'\rangle = J_y|S\rangle$. It is thus apparent that $\lambda^2 \geqslant m^2$.

We next show that, given an acceptable eigenvalue m_j, we may create a sequence of acceptable eigenvalues whose differences are integers. The proof requires the introduction of two new operators, which are defined by

$$\begin{aligned} J_+ &= J_x + iJ_y, \\ J_- &= J_x - iJ_y. \end{aligned} \tag{9.35}$$

It is easy to show that these operators satisfy the commutation relations

$$[J^2, J_\pm] = 0, \tag{9.36}$$

$$[J_z, J_\pm] = \pm\hbar J_\pm, \tag{9.37}$$

and obey the identities

$$J_+ = J_-^\dagger, \tag{9.38}$$

$$J_+J_- = J^2 - J_z^2 + \hbar J_z, \tag{9.39}$$

$$J_-J_+ = J^2 - J_z^2 - \hbar J_z. \tag{9.40}$$

If one writes out the commutation relation Eq. (9.37) in full, one obtains

$$J_z J_\pm |\lambda, m\rangle - J_\pm J_z |\lambda, m\rangle = \pm\hbar J_\pm |\lambda, m\rangle$$

or

$$J_z(J_\pm |\lambda, m\rangle) = \hbar(m \pm 1)(J_\pm |\lambda, m\rangle).$$

Thus, the kets $J_\pm |\lambda, m\rangle$ are eigenvectors of J_z whose eigenvalues are $\hbar(m \pm 1)$. Therefore, the kets $J_\pm |\lambda, m\rangle$ can differ from the eigenstates $|\lambda, m \pm 1\rangle$ by at most a constant of proportionality. Thus, if we have any eigenvector, we may create a sequence of eigenvectors

$$|\lambda, m\rangle, |\lambda, m + 1\rangle, |\lambda, m + 2\rangle, \ldots$$

with increasing values of m, by successive application of the operator J_+, and a sequence with decreasing values of m,

$$|\lambda, m\rangle, |\lambda, m - 1\rangle, |\lambda, m - 2\rangle, \ldots,$$

by application of J_-. However, this cannot be carried on indefinitely or the condition $\lambda^2 \geqslant m^2$ would be violated. Thus, there must exist both a maximum and minimum m that can be reached. Let us call these maximum and minimum values of m, j, and j'. For these states we must have

$$J_+|\lambda, j\rangle = 0, \qquad |\lambda, j\rangle \neq 0, \tag{9.41}$$

$$J_-|\lambda, j'\rangle = 0, \qquad |\lambda, j'\rangle \neq 0, \tag{9.42}$$

or the sequence would not terminate. We therefore have a sequence of eigenstates $|\lambda, m\rangle$ where m varies in integral steps from j' to j, with the difference between any two values of m being an integer. The difference between j and j' is therefore integral.

We now show that the sum of j and j' as well as their difference is integral. By applying J_- to Eq. (9.41) and $J_\pm$ to Eq. (9.42) and using Eqs. (9.39) and (9.40), we obtain

$$J_-J_+|\lambda, j\rangle = 0 = \lambda^2\hbar^2 - j^2\hbar^2 - j\hbar^2, \tag{9.43}$$

$$J_+J_-|\lambda, j'\rangle = 0 = \lambda^2\hbar^2 - j'^2\hbar^2 + j'\hbar^2. \tag{9.44}$$

Solving Eq. (9.43) yields

$$\lambda^2 = j(j + 1), \tag{9.45}$$

and, subtracting Eq. (9.43) from (9.44), we find

$$j + j' = 0,$$

which is, of course, an integer. Thus, both the sum and difference of j and j' are integers. This is possible if and only if j and j' are themselves either integers or half-integers.

Therefore, the allowed eigenvalues of the operator J^2 are

$$\hbar^2 j(j+1),$$

where j is integral or half-integral and for each value of j the allowed values of m are the $(2j+1)$-values

$$j, j-1, j-2, -j+1, -j.$$

9.4 ORBITAL ANGULAR MOMENTUM

In Chapter 2 the distinction was made between orbital angular momentum and intrinsic or spin angular momentum. In this section we examine the manner in which the orbital angular momentum operator transforms to a K-space representation.

As we have seen, it is natural to write the orbital angular momentum operators, for a particle which possesses a scalar state function, as

$$L_{\text{op.}} = -i\hbar \boldsymbol{r} \times \nabla_R,$$

where

$$\nabla_R = \boldsymbol{e}_x \frac{\partial}{\partial x} + \boldsymbol{e}_y \frac{\partial}{\partial y} + \boldsymbol{e}_z \frac{\partial}{\partial z}\cdot \tag{9.46}$$

The subscript, R, on the gradient indicates nothing new but is used here to distinguish this operator from

$$\nabla_K = \boldsymbol{e}_{K_x} \frac{\partial}{\partial k_x} + \boldsymbol{e}_{K_y} \frac{\partial}{\partial k_y} + \boldsymbol{e}_{K_z} \frac{\partial}{\partial k_z}, \tag{9.47}$$

which may be thought of as a gradient in momentum space. We shall make use of the operators ∇_K later.

Assuming that a particle has a scalar state function $\psi(\boldsymbol{r})$, the expectation value of its orbital angular momentum would be given by

$$\langle L \rangle = -i\hbar \int d\boldsymbol{r} \psi^\star(r)[\boldsymbol{r} \times \nabla_R]\psi(r). \tag{9.48}$$

We obtain a description of the particle in k-space by writing $\psi(\boldsymbol{r})$ in terms of its Fourier transform:

$$\psi(\boldsymbol{r}) = \frac{1}{(2\pi)^{3/2}} \int d\boldsymbol{k} e^{i\boldsymbol{k}\cdot\boldsymbol{r}} \phi(\boldsymbol{k}), \tag{9.49}$$

where $\phi(\boldsymbol{k})$ is given by

$$\phi(\boldsymbol{k}) = \frac{1}{(2\pi)^{3/2}} \int d\boldsymbol{r} e^{-i\boldsymbol{k}\cdot\boldsymbol{r}} \psi(\boldsymbol{r})$$

Placing Eq. (9.49) in Eq. (9.48), we obtain

$$\langle L\rangle = \frac{-i\hbar}{(2\pi)^3}\iiint dkdk'dr\phi^\star(k')e^{-ik'\cdot r}[r \times \nabla_R]\phi(k)e^{ik\cdot r}. \qquad (9.50)$$

Using the definitions of Eqs. (9.46) and (9.47), one may verify directly that

$$e^{-ik'\cdot r}[r \times \nabla_R]e^{ik\cdot r} = e^{ik\cdot r}[k \times \nabla_{K'}]e^{-ik'\cdot r}.$$

Thus, Eq. (9.50) becomes

$$\langle L\rangle = \frac{-i\hbar}{(2\pi)^3}\iiint dkdk'dr\phi^\star(k')\phi(k)(k \times \nabla_{K'})e^{i(k-k')\cdot r}.$$

Making use of the fact that the expectation value must be real and $\langle L\rangle = \langle L\rangle^\star$, we may rewrite this as

$$\langle L\rangle = \frac{i\hbar}{(2\pi)^3}\iint dkdk'\phi^\star(k)\phi(k')(k \times \nabla_{K'})\int dre^{i(k'-k)\cdot r}$$

$$= i\hbar\int dk\phi^\star(k)k \times \int dk'\phi(k')\nabla_{K'}\delta(k' - k).$$

The integral involving the delta function is simply the three-dimensional generalization of Eq. (IV.11) of Appendix IV and yields

$$\langle L\rangle = -i\hbar\int dk\phi^\star(k)(k \times \nabla_K)\phi(k).$$

If, then, $\phi(k)$ is taken as the state function of the particle in momentum space, the orbital angular momentum operator in momentum space is seen to be

$$L_{\text{op.}} = -i\hbar[k \times \nabla_K],$$

which, of course, represents three operators, one for each component. For example,

$$L_x = -i\hbar\left[k_y\frac{\partial}{\partial k_z} - k_z\frac{\partial}{\partial k_y}\right].$$

It is interesting to note that these operators obey the general commutation rules for angular momentum operators:

$$[L_i, L_j] = i\hbar L_k,$$

$$[L^2, L_i] = 0.$$

Finally, the operators discussed in this section do correspond to our ideas of orbital angular momentum. The quantum mechanical "particle" which is described by a simple scalar function of position and time corresponds to the classical point particle. The quantity $\psi^\star\psi dr$ tells us the probability of finding or not finding the particle within the volume element dr.

There is no information contained in $\psi(\boldsymbol{r})$ about any possible internal structure of the particle. The operator $-i\hbar\nabla_R$ yields information about the linear momentum of this simple particle. The operators $-i\hbar\boldsymbol{r} \times \nabla_R$ and $-i\hbar\boldsymbol{k} \times \nabla_K$ specifically refer to the angular momentum of the particle in respect to a given origin.

This is not to say that the classical expression for angular momentum, $\boldsymbol{L} = \boldsymbol{r} \times \boldsymbol{p}$, generalizes to nothing but orbital angular momentum operators in quantum mechanics. When the classical momentum, $\boldsymbol{p}$, has a greater complexity than that of a simple particle, e.g., in the electromagnetic field, we shall find that the angular momentum operators contain more than orbital terms.

9.5 THE ANGULAR MOMENTUM OF THE PHOTON

In Chapter 8 it was seen that an acceptable state function, $\boldsymbol{f}(\boldsymbol{k})$, could be found for the photon in momentum space. We now wish to find the proper and correct angular momentum operators which can be associated with the momentum space representation of the electromagnetic field. We begin with the classical angular momentum of the electromagnetic field.

The linear momentum density of the field is given by

$$g = \frac{1}{c^2} \boldsymbol{E}(\boldsymbol{r}) \times \boldsymbol{H}(\boldsymbol{r}).$$

We may associate with this an angular momentum density of the field

$$\frac{1}{c^2} \boldsymbol{r} \times (\boldsymbol{E} \times \boldsymbol{H}),$$

and for the angular momentum of the total field we have

$$\boldsymbol{J} = \frac{1}{c^2} \int d\boldsymbol{r}\, \boldsymbol{r} \times (\boldsymbol{E} \times \boldsymbol{H}).$$

Proceeding along the lines of Chapter 8, we write $\boldsymbol{E}$ and $\boldsymbol{H}$ in terms of their Fourier transforms, $\mathscr{E}(\boldsymbol{k})$ and $\mathscr{H}(\boldsymbol{k})$. $\mathscr{E}(\boldsymbol{k})$ and $\mathscr{H}(\boldsymbol{k})$ are then expressed in terms of the state function, $\boldsymbol{f}(\boldsymbol{k})$. After a great deal of manipulation, which is given in Appendix VI, one obtains for the α-component of $\boldsymbol{J}$:

$$\langle J_\alpha \rangle = \sum_{\beta=1}^{3} \int d\boldsymbol{k}\, f_\beta^\star [-i\hbar \boldsymbol{k} \times \nabla_K]_\alpha f_\beta + \int d\boldsymbol{k} [-i\hbar \boldsymbol{f}^\star \times \boldsymbol{f}]_\alpha. \qquad (9.51)$$

The first term may be interpreted in terms of the orbital angular momentum operator in K-space and is in the form of an expectation value. Here the expectation value includes both a summation over the discrete components of the state function and an integration over the continuous variable $\boldsymbol{k}$. The summation over the components indicates that each component of the state function can contribute to each component of the orbital angular momentum.

As an example of orbital momentum, we might consider the approximate plane wave of Section 9.1 displaced in such a way that it does not pass through the origin. It would then possess a nonzero orbital momentum about the origin in much the same way as the ball discussed in Section 2.7.

The second integral in Eq. (9.51) is something new and is not in the form of a bracket product or expectation value. Let us show that it can easily be cast in this form.

Consider any two vectors, whose components may be complex, which we call $\boldsymbol{A}$ and $\boldsymbol{B}$. By the ordinary rules of vector algebra we may compute the components of the vector cross product $\boldsymbol{A}^\star \times \boldsymbol{B}$. Now, if we think of these vectors as kets $|A\rangle$ and $|B\rangle$, $\boldsymbol{A}^\star \times \boldsymbol{B}$ may be written as an operator bracket product.

We introduce the three matrices Σ_i, given by

$$\Sigma_x = \begin{bmatrix} 0 & 0 & 0 \\ 0 & 0 & 1 \\ 0 & -1 & 0 \end{bmatrix},$$

$$\Sigma_y = \begin{bmatrix} 0 & 0 & 1 \\ 0 & 0 & 0 \\ -1 & 0 & 0 \end{bmatrix}, \tag{9.52}$$

$$\Sigma_z = \begin{bmatrix} 0 & 1 & 0 \\ -1 & 0 & 0 \\ 0 & 0 & 0 \end{bmatrix}.$$

By direct computation, one may show that the α-component of the cross product $\boldsymbol{A}^\star \times \boldsymbol{B}$ is correctly given by

$$(\boldsymbol{A}^\star \times \boldsymbol{B})_\alpha = \langle A|\Sigma_\alpha|B\rangle.$$

If we further define a set of matrices, S_α, by

$$S_\alpha = -i\hbar\Sigma_\alpha, \tag{9.53}$$

the second integral in Eq. (9.51) may be written

$$\int dk[-i\hbar \boldsymbol{f}^\star \times \boldsymbol{f}]_\alpha = \langle f|S_\alpha|f\rangle, \tag{9.54}$$

where $|f\rangle$ is the ket representation of the vector $\boldsymbol{f}(\boldsymbol{k})$ and the bracket product is understood to include an integration over the continuous variable $\boldsymbol{k}$ as well as the usual matrix multiplication.

It is the term given in Eq. (9.54) which represents the intrinsic or spin angular momentum of the photon. If we define the matrix

$$S^2 = S_x^2 + S_y^2 + S_z^2 = 2\hbar^2 \begin{bmatrix} 1 & 0 & 0 \\ 0 & 1 & 0 \\ 0 & 0 & 1 \end{bmatrix}, \tag{9.55}$$

we find that the set of matrices S_x, S_y, S_z, and S^2 is Hermitian and obeys the usual commutation rules for angular momentum operators:

$$[S_i, S_j] = i\hbar S_K,$$

$$[S_i, S^2] = 0.$$

Once again, these commutation relations allow us to find simultaneous eigenstates of the angular momentum operator S^2 and any one component. As before, we pick the z-component.

Since the operator S^2 is proportional to the identity matrix, any ket

$$|\chi\rangle = \begin{pmatrix} a \\ b \\ c \end{pmatrix}$$

is an eigenstate of it, with eigenvalue $2\hbar^2$. Thus, by Eq. (9.31), the quantum number for the spin angular momentum of the photon must be $j = 1$. By Eq. (9.32) the projection of this angular momentum along the z-axis may take on the values $+\hbar$, 0, $-\hbar$. The last result could also be obtained by solving the eigenvalue equation

$$S_z|\chi\rangle = m|\chi\rangle.$$

Explicit solution of this equation also yields the eigenvectors associated with these states:

$$|\chi_0\rangle = \begin{pmatrix} 0 \\ 0 \\ 1 \end{pmatrix} \qquad |\chi_\pm\rangle = \frac{1}{\sqrt{2}} \begin{pmatrix} e^{\mp i\pi/4} \\ e^{\pm i\pi/4} \\ 0 \end{pmatrix}. \tag{9.56}$$

The states $|\chi_0\rangle$, $|\chi_+\rangle$, and $|\chi_-\rangle$ are, respectively, associated with the eigenvalues 0, $+\hbar$, and $-\hbar$. These eigenvectors have been normalized to satisfy

$$\langle\chi_i|\chi_j\rangle = \delta_{ij}.$$

The states $|\chi_\pm\rangle$ come as something of a happy surprise. They are simply the ket vectors for right and left circularly polarized states when $\boldsymbol{k}$ is along the z-direction, with a third component, which has been made explicitly zero. The appearance of a third state, $|\chi_0\rangle$, is unexpected. The appearance of the eigenstate $|\chi_0\rangle$ with eigenvalue zero would ordinarily indicate the existence of an observable state with zero projection of spin angular momentum along the z-direction. This state is not observed experimentally. Recall that all observable states must satisfy the transversality condition, Eq. (8.22). Now, for the waves which we have been considering, we have always picked the direction of propagation to be the z-direction. If we continue this practice, then the vector $\boldsymbol{k}$ is of the form

$$\boldsymbol{k} = \begin{bmatrix} 0 \\ 0 \\ k \end{bmatrix} = |k\rangle,$$

and, in accordance with the transversality condition, we find

$$\langle k|\chi_{\pm}\rangle = 0.$$

However, $\langle k|\chi_0\rangle$ is not zero. Thus, it is the transversality condition which forbids the experimental observation of the state $|\chi_0\rangle$ in this case. More will be said about this in the next section.

9.6 THE PURE STATES OF THE FREE PHOTON

The three eigenstates of the operator S_z—$|\chi_0\rangle$, $|\chi_+\rangle$, and $|\chi_-\rangle$—are three independent, orthonormal, three-dimensional vectors and, hence, form a basis in K-space. They may therefore be used to expand any vector in that space. In particular, if $f(\boldsymbol{k})$ is any photon state function, it may be written

$$f(\boldsymbol{k}) = f_+(\boldsymbol{k})|\chi_+\rangle + f_0(\boldsymbol{k})|\chi_0\rangle + f_-(\boldsymbol{k})|\chi_-\rangle. \tag{9.57}$$

Now, the orbital angular momentum operators $L = -i\hbar[\boldsymbol{k} \times \nabla_k]$ and L^2 operate only on the coefficients $f_0(\boldsymbol{k})$, $f_+(\boldsymbol{k})$, and $f_-(\boldsymbol{k})$, and not on the unit vectors $|\chi_0\rangle$, $|\chi_+\rangle$, and $|\chi_-\rangle$. The spin operators S_z and S^2 operate only on these unit vectors. It follows that if one operates on a state function with both an orbital and a spin angular momentum operator, the order of operation has no effect. Thus, the orbital and spin operators commute. Since these operators commute, it follows immediately that the total angular momentum operators,

$$J_z = L_z + S_z,$$
$$J^2 = (L + S)^2,$$

commute with both the spin and orbital angular momentum operators.

We therefore have a set of six operators J^2, J_z, L^2, L_z, S^2, and S_z which mutually commute. It should follow then that one can find state functions of the photon which are simultaneously eigenstates of these six operators. Experimentally, this would mean that we could prepare pure photon states for which we could simultaneously measure the six associated physical quantities. This should be true regardless of which particular direction we have picked for the z-axis.

As we have seen in the last section, the transversality condition does not allow this much generality. If $\boldsymbol{k}$ is not along the z-axis, $\boldsymbol{k} \cdot \boldsymbol{f} = 0$ would not be satisfied for either of the observable pure states of spin angular momentum,

$$f_+(\boldsymbol{k})|\chi_+\rangle,$$
$$f_-(\boldsymbol{k})|\chi_-\rangle.$$

Thus, if the z-direction, along which we wish to measure the components of angular momentum, is not parallel to $\boldsymbol{k}$, the state of the photon must at best be a state of the form

$$|S\rangle = f_+(\boldsymbol{k})|\chi_+\rangle + f_-(\boldsymbol{k})|\chi_-\rangle.$$

In this case it is only the total angular momentum J^2 and its component J_z which may be measured, and the splitting of this momentum into spin and orbital parts is not well defined.

We also note that the operators for linear momentum and angular momentum do not commute. Thus, we may not prepare photon state functions which are simultaneously pure states of linear momentum and total angular momentum. This is also seen from the function $f(\boldsymbol{k})$ which determines the orbital angular momentum. This in general is a continuous function over *all* values of $\boldsymbol{k}$ in momentum space. The energy operator, being a derivative in respect to time, does commute with both momentum operators, however. Thus, we may prepare pure photon states

$$|S\rangle = |\hbar\omega, J^2, J_z\rangle \tag{9.58}$$

which are pure in respect to energy, total angular momentum, and its z-component.

Alternatively, we may prepare states

$$|S\rangle = |\hbar\omega, \boldsymbol{k}, \alpha\rangle \tag{9.59}$$

which are pure in respect to energy, linear momentum, and, of course, polarization. Either type of pure state may be expressed as a coherent superposition of the other type.

Only in the special case where the z-axis is taken to be along the direction of propagation is there a simple relation between these two types of states. In this case the pure polarization states $|R\rangle$ and $|L\rangle$ are also pure states of spin angular momentum with $S_z = \pm\hbar$.

PROBLEMS

9.1 Verify the relations for the operators $J_\pm$ given by Eqs. (9.36) through (9.40).

9.2 Verify that operators defined by Eq. (9.26) obey the commutation relations, Eq. (9.28).

9.3 Verify that the operators defined by Eq. (9.53) obey the commutation relations, Eq. (9.29).

9.4 By direct computation find the eigenvectors and eigenvalues of the matrices defined by Eq. (9.53).

9.5 Prove that if the sum and difference of two real numbers are integers, the numbers themselves are integers or half-integers.

9.6 Show that if E_0 in Eq. (9.17) is allowed to be a function of r and θ, the z-component of Eq. (9.14) may still be satisfied identically although additional error terms are introduced into the r- and θ-components.

9.7 In Chapter 5 the linear momentum operator was introduced by considering a small translation of the system. Consider a system which is described by a function $f(x, y)$. If this system is rotated through a small angle $d\theta$ about the z-axis, show that the new function $f'(x, y)$ can be obtained by a rotation operator

$$f'(x, y) = \mathsf{R} f(x, y),$$

where

$$\mathsf{R} = \mathsf{I} - \frac{i\,d\theta}{\hbar}\mathsf{L}_z,$$

with L_z as given by Eq. (9.36). By the discussion of Chapter 5 give a condition for the constancy of the z-component of angular momentum of the system.

9.8 Show that the commutation rules, Eq. (9.28), may be written in the compact form

$$\boldsymbol{L} \times \boldsymbol{L} = i\hbar \boldsymbol{L}.$$

9.9 Show that in spherical coordinates (see Fig. 6.2) the angular momentum operators take on the forms

$$L_x = i\hbar\left(\cot\theta\cos\phi\frac{\partial}{\partial\phi} + \sin\phi\frac{\partial}{\partial\theta}\right),$$

$$L_y = i\hbar\left(\cot\theta\sin\phi\frac{\partial}{\partial\phi} - \cos\phi\frac{\partial}{\partial\theta}\right),$$

$$L_z = -i\hbar\frac{\partial}{\partial\phi},$$

$$L^2 = -\hbar^2\left(\frac{1}{\sin\theta}\frac{\partial}{\partial\theta}\left(\sin\theta\frac{\partial}{\partial\theta}\right) + \frac{1}{\sin^2\theta}\frac{\partial^2}{\partial\phi^2}\right).$$

9.10 Verify that the following are eigenfunctions of the orbital angular momentum operators with $j = 1$:

$$|1, \pm 1\rangle = \sqrt{\frac{3}{8\pi}}\sin\theta e^{\pm i\phi} = \sqrt{\frac{3}{8\pi}}\frac{(x \pm iy)}{r},$$

$$|1, 0\rangle = \sqrt{\frac{3}{4\pi}}\cos\theta = \sqrt{\frac{3}{4\pi}}\frac{z}{r}.$$

9.11 For the eigenfunctions given in Problem 9.10 verify that

$$L_+|1, 1\rangle = 0, \quad L_-|1, -1\rangle = 0,$$

and that $L_\pm|1, 0\rangle$ is proportional to $|1, \pm 1\rangle$.

9.12 In terms of the matrices given by Eq. (9.52), we may define

$$\Sigma_\pm = \Sigma_x \pm i\Sigma_y.$$

Show that

$$\Sigma_+|\chi_+\rangle = \Sigma_-|\chi_-\rangle = 0$$

and that $\Sigma_\pm|\chi_0\rangle$ is proportional to $|\chi_\pm\rangle$.

9.13 In relation to the development given in Section 9.1, consider the function

$$E_0(r) = \frac{1}{e^{a(r/R_0 - R_0/r)} + 1}.$$

Show that if a is large, $a = 100$, the function $E_0(r)$ is essentially constant for $r < R_0$ and nearly equal to zero for $r > R_0$, and varies appreciably only in a region near $R_0 = r$ whose width is of the order of R_0/a. For $1 \text{ cm} \leqslant R_0 \leqslant 10 \text{ cm}$ and $\lambda = 5000$ Å show that the error terms

$$\frac{1}{rk^2E_0}\frac{\partial E_0}{\partial r}, \quad \frac{1}{k^2E_0}\frac{\partial^2 E_0}{\partial r^2}, \quad \frac{1}{k^2E_0^2}\left(\frac{\partial E_0}{\partial r}\right)^2,$$

are of order $(a/kR_0)^2$ or smaller and may be neglected.

9.14 Consider a system for which the appropriate angular momentum operators J_x, J_y, and J_z are of the form

$$J_i = 2\sqrt{\frac{5}{3}}\,\hbar K_i,$$

where the square of the K_i operators, K^2, is an identity operator. Prove that if a matrix representation can be found for the operators J_i, they must be 9×9-matrices or larger.

APPENDIX I

MATRIX ALGEBRA

For those who may be unfamiliar with matrices, the mastery of the rudiments of matrix algebra presented here is sufficient for comprehending the mathematical proofs in the text and the solutions of problems. An elementary knowledge of the method of solving simultaneous linear equations is assumed on the part of the student.

I.1 MATRICES AND THEIR PROPERTIES

Consider the set of linear equations

$$\begin{aligned} y_1 &= a_{11}x_1 + a_{12}x_2 + a_{13}x_3, \\ y_2 &= a_{21}x_1 + a_{22}x_2 + a_{23}x_3, \\ y_3 &= a_{31}x_1 + a_{32}x_2 + a_{33}x_3, \end{aligned} \tag{AI.1}$$

where the a_{ij} are the coefficients of the unknown variables x_j $(i, j = 1, 2, 3)$. This set of equations may be written as the array

$$\begin{bmatrix} y_1 \\ y_2 \\ y_3 \end{bmatrix} = \begin{bmatrix} a_{11} & a_{12} & a_{13} \\ a_{21} & a_{22} & a_{23} \\ a_{31} & a_{32} & a_{33} \end{bmatrix} \begin{bmatrix} x_1 \\ x_2 \\ x_3 \end{bmatrix}. \tag{AI.2}$$

Now, any array of numbers a_{ij} set out in m rows and n columns is called a matrix of order m by n and the order is generally indicated as $m \times n$. In Eq. (AI.2) the array of the a_{ij} forms a 3×3-matrix (which is read three *by* three not three times three). The entire array of the a_{ij} will be denoted by the symbol A. Each of the a_{ij} is called an element of the matrix A or simply a matrix element. A matrix may consist of a single column or a single row: the x_j and y_j, for example, are the matrix elements of 3×1-matrices. Hereafter we shall frequently refer to a $1 \times n$-matrix as a row vector and an $n \times 1$-matrix as a column vector. Column vectors will be designated by the symbol $|\,)$ and row vectors by $(\,|$. This shorthand allows us to write the column vectors in Eq. (AI.2) as

$$|y) = \begin{bmatrix} y_1 \\ y_2 \\ y_3 \end{bmatrix}, \qquad |x) = \begin{bmatrix} x_1 \\ x_2 \\ x_3 \end{bmatrix},$$

and Eqs. (AI.1) and (AI.2) can be succinctly written as

$$|y) = \mathsf{A}|x), \tag{AI.3}$$

in which the column vector $|y)$ is interpreted as the result of the operation of the matrix A on the column vector $|x)$.

If Eqs. (AI.3) and (AI.1) are to express the same relations among the variables, there must be some rule by which Eq. (AI.1) can be extracted from Eq. (AI.2). That is, the matrix elements of A, the a_{ij}, must multiply the elements x_j to obtain the y_j, as specified in Eq. (AI.1). It is easy to see that the matrix elements of the vector $|y)$ are obtained from

$$y_i = \sum_{j=1}^{3} a_{ij}x_j,$$

and this prescribes how the matrix elements of A must multiply those of $|x)$ for Eq. (AI.2) to convey the same information as Eq. (AI.1).

The correspondence between Eqs. (AI.1) and (AI.2) has been indicated, but now some other properties of matrices must be specified before we can proceed further.

a) Two matrices A and B are equal if and only if they are of the same order and if their corresponding matrix elements are equal: i.e., for any i and j, $a_{ij} = b_{ij}$.

b) The sum or difference of two matrices A and B (whose orders are equal) is defined as a matrix C such that the elements of C are given by

$$c_{ij} = a_{ij} \pm b_{ij}.$$

c) If A is any matrix and a is any scalar, then the matrix $\mathsf{A}' = a\mathsf{A}$ is that obtained by multiplying each matrix element of A by a.

d) The matrix elements c_{ij} of the product $\mathsf{C} = \mathsf{AB}$ are defined by the relation $c_{ij} = \sum_k a_{ik}b_{kj}$. This implies that the number of columns in A must equal the number of rows of B.

e) In general, $\mathsf{AB} \neq \mathsf{BA}$; that is, matrix products do not commute.

f) Matrix multiplication is associative: $\mathsf{A(BC)} = \mathsf{(AB)C}$.

g) Matrix multiplication is distributive: $\mathsf{A(B + C)} = \mathsf{AB} + \mathsf{AC}$.

As examples of matrix addition and multiplication the following are given:

$$\mathsf{A} = \begin{bmatrix} 2 & 1 & 0 \\ 3 & 5 & 4 \\ 1 & 3 & 0 \end{bmatrix}, \qquad \mathsf{B} = \begin{bmatrix} 3 & 2 & 2 \\ 1 & -1 & 0 \\ 0 & 3 & 1 \end{bmatrix},$$

$$\mathsf{A} + \mathsf{B} = \begin{bmatrix} 5 & 3 & 2 \\ 4 & 4 & 4 \\ 1 & 6 & 1 \end{bmatrix},$$

$$\mathsf{AB} = \begin{bmatrix} 6+1+0 & 4-1+0 & 4+0+0 \\ 9+5+0 & 6-5+12 & 6+0+4 \\ 3+3+0 & 2-3+0 & 2+0+0 \end{bmatrix} = \begin{bmatrix} 7 & 3 & 4 \\ 14 & 13 & 10 \\ 6 & -1 & 2 \end{bmatrix}.$$

I.2 DEFINITIONS

a) A *diagonal* matrix is one in which the matrix elements $a_{ij} = 0$ for all $i \neq j$.

b) The *unit matrix* or *identity matrix* I is a diagonal matrix in which $a_{jj} = 1$ for all j.

c) A matrix which has the same number of rows and columns is called a *square matrix.*

d) A *singular* matrix is a square matrix whose determinant is zero. If the determinant is not zero, the matrix is non-singular.

e) The matrix elements of $\mathsf{A}^\star$, the complex conjugate of A, are the complex conjugates of the elements of A.

f) If $\mathsf{A} = \mathsf{A}^\star$, A is a *real matrix.*

g) The interchange of rows and columns of any matrix results in a new matrix $\tilde{\mathsf{A}}$ called the *transpose* of A. The elements $\tilde{a}_{ij}$ of the transpose are related to those of A by $\tilde{a}_{ij} = a_{ji}$.

h) The matrix $\mathsf{A}^\dagger = \tilde{\mathsf{A}}^\star$ is called the *ajdoint* of A.

i) A matrix is *Hermitian* if $\mathsf{A}^\dagger = \mathsf{A}$.

j) A matrix is *unitary* if $\mathsf{A}^\dagger\mathsf{A} = \mathsf{I}$.

k) A real matrix is *orthogonal* if $\mathsf{A}\tilde{\mathsf{A}} = \mathsf{I}$.

l) The *trace* of a square matrix A is the sum of the diagonal elements of A and is written Tr (A). The trace of the product of two matrices is equal to the product of the traces, Tr (AB) = Tr (A) Tr (B).

I.3 THE INVERSE MATRIX

We represented a set of linear equations in Eq. (AI.1). If the determinant of A (written as Det A or $|\mathsf{A}|$) is nonzero, it is possible to solve for $|x)$ in terms of $|y)$. That is,

$$|x) = \mathsf{B}|y). \tag{AI.4}$$

Given the matrix A, the problem is to determine $\boldsymbol{B}$. The relation that A and B must satisfy can be obtained by substituting for $|x)$ from Eq. (AI.4) into Eq. (AI.3):

$$|y) = \mathsf{AB}|y),$$

which means that $\mathsf{AB} = \mathsf{I}$.

We will show how B can be constructed for any non-singular matrix A. Moreover, it turns out that B is unique and is called the inverse of A, which is designated as $\mathsf{B} = \mathsf{A}^{-1}$. We will use the 3×3-matrix A as given in Eq. (AI.2) as an example. The method presented can be extended to $n \times n$-matrices but is quite tedious for $n > 3$. Recall that the cofactor of a matrix element a_{ij} in A is $(-1)^{i+j}$ times the determinant of the matrix obtained when the i-th row and j-th column are deleted from A. Designating the cofactor of a_{ij} as $|A_{ij}|$, the determinant of the matrix A can be written in terms of its cofactors as

$$|\mathsf{A}| = \sum_{j=1}^{3} a_{ij}|A_{ij}| \text{ for fixed } i,$$

$$|\mathsf{A}| = \sum_{i=1}^{3} a_{ij}|A_{ij}| \text{ for fixed } j.$$

We now construct from A the matrix $\hat{\mathsf{A}}$, such that the cofactors of the elements of the i-th row of A are the elements in the j-th column of $\hat{\mathsf{A}}$. That is,

$$\hat{\mathsf{A}} = \begin{pmatrix} |A_{11}| & |A_{21}| & |A_{31}| \\ |A_{12}| & |A_{22}| & |A_{32}| \\ |A_{13}| & |A_{23}| & |A_{33}| \end{pmatrix}.$$

Let us see what happens when we take the product $\mathsf{A}\hat{\mathsf{A}}$. The matrix elements c_{ij} of the product will be

$$(\mathsf{A}\hat{\mathsf{A}})_{ij} = \sum_{k=1}^{3} a_{ik}|A_{jk}|$$

for fixed i. For $i = j$,

$$(\mathsf{A}\hat{\mathsf{A}})_{ii} = \sum_{k=1}^{3} a_{ik}|A_{ik}| = |\mathsf{A}|,$$

and

$$(\mathsf{AA})_{11} = (\mathsf{A\hat{A}})_{22} = (\mathsf{A\hat{A}})_{33} = |\mathsf{A}|.$$

If $j \neq i$, the sum involves the product of the elements of one row with the cofactors of another row, which is equivalent to the expansion of a determinant which has two identical rows. Consequently,

$$(\mathsf{A\hat{A}})_{ij} = \sum_{k=1}^{3} a_{ik}|A_{jk}| = 0 \text{ for } i \neq j.$$

Therefore, the product $\mathsf{A\hat{A}}$ is

$$\mathsf{A\hat{A}} = \begin{pmatrix} |\mathsf{A}| & 0 & 0 \\ 0 & |\mathsf{A}| & 0 \\ 0 & 0 & |\mathsf{A}| \end{pmatrix} = |\mathsf{A}|\mathsf{I}.$$

Since $\mathsf{AA}^{-1} = \mathsf{I}$, we can say that

$$\mathsf{A}^{-1} = \mathsf{\hat{A}}/|\mathsf{A}|. \tag{AI.5}$$

The inverse of a matrix is unique, because if we have another matrix, C, such that $\mathsf{AC} = \mathsf{I}$, then

$$\mathsf{C} = \mathsf{IC} = (\mathsf{A}^{-1}\mathsf{A})\mathsf{C} = \mathsf{A}^{-1}(\mathsf{AC}) = \mathsf{A}^{-1}.$$

Finally, Eq. (AI.5) is equally valid for real matrices or for matrices with complex elements.

As an example of the above, the determinant of the matrix A given in Section AI.1 is -20 and the inverse of A is found to be

$$\mathsf{A}^{-1} = -1/20 \begin{bmatrix} -12 & 0 & 4 \\ 4 & 0 & -8 \\ 4 & -5 & 7 \end{bmatrix}.$$

I.4 THE MATRIX REPRESENTATION OF VECTORS

In ordinary vector notation any vector is generally expressed in the form

$$\boldsymbol{A} = A_1\boldsymbol{e}_1 + A_2\boldsymbol{e}_2 + A_3\boldsymbol{e}_3,$$

where $\boldsymbol{e}_1$, $\boldsymbol{e}_2$, and $\boldsymbol{e}_3$ are unit basis vectors specifying the direction of the axes of a Cartesian coordinate system, and A_1, A_2, and A_3 are the components of the vector $\boldsymbol{A}$ along these axes. As we know, the scalar or dot product of the vector A with the vector $\boldsymbol{B} = B_1\boldsymbol{e}_1 + B_2\boldsymbol{e}_2 + B_3\boldsymbol{e}_3$ is

$$\boldsymbol{A} \cdot \boldsymbol{B} = A_1B_1 + A_2B_2 + A_3B_3.$$

The manner in which A and B have been written above is really a convention.

So long as the order of the components is specified and maintained, the vectors A and B could both be written as column vectors

$$|A) = \begin{pmatrix} A_1 \\ A_2 \\ A_3 \end{pmatrix}, \qquad |B) = \begin{pmatrix} B_1 \\ B_2 \\ B_3 \end{pmatrix}.$$

The usual operations of addition of vectors and the multiplication of a vector by a scalar can be performed. By definition,

$$a(\boldsymbol{A} + \boldsymbol{B}) = a(A_1 + B_1)\boldsymbol{e}_1 + a(A_2 + B_2)\boldsymbol{e}_2 + a(A_3 + B_3)\boldsymbol{e}_3,$$

which in our notation becomes

$$a[|A) + |B)] = \begin{pmatrix} a(A_1 + B_1) \\ a(A_2 + B_2) \\ a(A_3 + B_3) \end{pmatrix}.$$

What of the scalar product? If we are to represent vectors by column vectors, the scalar product can be taken in the following way. Since $|A)$ and $|B)$ are column vectors, their transposes are given by

$$(A| = (A_1, A_2, A_3), \qquad (B| = (B_1, B_2, B_3),$$

and it is seen that the scalar product may be written as a matrix multiplication

$$\boldsymbol{A} \cdot \boldsymbol{B} = (A|B) = (A_1, A_2, A_3) \begin{pmatrix} B_1 \\ B_2 \\ B_3 \end{pmatrix} = A_1B_1 + A_2B_2 + A_3B_3,$$

where the multiplication $(A|B)$ is the matrix product as defined in Section AI.1.

We have shown the column and row representation of the vectors $\boldsymbol{A}$ and $\boldsymbol{B}$, but nothing was said about the representation of the basis vectors, $\boldsymbol{e}_i$. If we write

$$|A) = \begin{pmatrix} A_1 \\ A_2 \\ A_3 \end{pmatrix} = A_1 \begin{pmatrix} 1 \\ 0 \\ 0 \end{pmatrix} + A_2 \begin{pmatrix} 0 \\ 1 \\ 0 \end{pmatrix} + A_3 \begin{pmatrix} 0 \\ 0 \\ 1 \end{pmatrix},$$

the column vector representation of the $\boldsymbol{e}_i$ can be seen to be

$$|e_1) = \begin{pmatrix} 1 \\ 0 \\ 0 \end{pmatrix}, \qquad |e_2) = \begin{pmatrix} 0 \\ 1 \\ 0 \end{pmatrix}, \qquad |e_3) = \begin{pmatrix} 0 \\ 0 \\ 1 \end{pmatrix},$$

and $|A)$ can be expressed as

$$|A) = \sum_{i=1}^{3} A_i|e_i) = A_1|e_1) + A_2|e_2) + A_3|e_3).$$

With this representation, the scalar product can also be written

$$(B|A) = \left[\sum_{j=1}^{3} B_j(e_j|\right]\left[\sum_{i=1}^{3} A_i|e_i)\right] = \sum_{ij} B_jA_i(e_j|e_i)$$
$$= \sum_{i=1}^{3} A_iB_i,$$

since

$$(e_j|e_i) = \delta_{ij}, \qquad \delta_{ij} = \begin{cases} 1 \text{ for } i = j, \\ 0 \text{ for } i \neq j. \end{cases}$$

APPENDIX II

JONES AND MUELLER MATRICES

Certain matrices are tabulated in this appendix which appear within the context of the Jones and Mueller calculuses. The selection has been made on the basis of pertinence to the main text and is not meant to be complete.

II.1 TRANSFORMATION MATRICES

Any two perpendicular directions x and y may be used to determine basis states $|P_x\rangle$ and $|P_y\rangle$ for the Jones and Mueller calculuses. It is convenient to be able to transform from one such basis to another basis $|P_{x'}\rangle$, $|P_{y'}\rangle$. The primed axes will be taken as rotated through angle θ counterclockwise from the unprimed axis. The matrices given below, multiplying a Jones or Stokes vector from the left, perform the desired transformation. The same matrices also transform a matrix M, in either the Jones or Mueller calculus, from one basis to the other through the similarity transformation $\mathsf{M}' = \mathsf{U}\mathsf{M}\mathsf{U}^\dagger$:

$$\begin{bmatrix} \cos\theta & \sin\theta \\ -\sin\theta & \cos\theta \end{bmatrix},$$

$$\begin{bmatrix} 1 & 0 & 0 & 0 \\ 0 & \cos 2\theta & \sin 2\theta & 0 \\ 0 & -\sin 2\theta & \cos 2\theta & 0 \\ 0 & 0 & 0 & 1 \end{bmatrix}.$$

Within the framework of the Jones calculus, frequent use can be made of the basis $|R\rangle$, $|L\rangle$. The matrix which transforms a Jones vector from the basis $|P_x\rangle$, $|P_y\rangle$ to the basis $|R\rangle$, $|L\rangle$ is given below. Through a similarity transformation this matrix also transforms any Jones matrix:

$$\frac{1}{\sqrt{2}}\begin{bmatrix} e^{i\pi/4} & e^{i\pi/4} \\ e^{i\pi/4} & e^{i\pi/4} \end{bmatrix} \quad \text{or} \quad \frac{1}{\sqrt{2}}e^{i\pi/4}\begin{bmatrix} 1 & -i \\ -i & 1 \end{bmatrix}.$$

In the following, Jones matrices will be given in both a $|P_x\rangle$, $|P_y\rangle$- and $|R\rangle$, $|L\rangle$-basis. These representations will be distinguished by subscripts on the matrix, such as M_{XY} or M_{RL}. All Mueller matrices are given in a $|P_x\rangle$, $|P_y\rangle$-basis.

II.2 AN ELLIPTICAL POLARIZER

The matrices below correspond to an elliptical filter which produces and/or transmits unchanged the pure state:

$$|E\rangle = \cos\theta\, e^{-\frac{i\phi}{2}}\, |P_x\rangle + \sin\theta\, e^{+\frac{i\phi}{2}}\, |P_y\rangle,$$

$$\begin{bmatrix} \cos^2\theta & e^{-i\phi}\cos\theta\sin\theta \\ e^{i\phi}\cos\theta\sin\theta & \sin^2\theta \end{bmatrix}_{XY},$$

$$1/2\begin{bmatrix} 1+\sin 2\theta\sin\phi & \sin 2\theta\cos\phi + i\cos 2\theta \\ \sin 2\theta\cos\phi - i\cos 2\theta & 1-\sin 2\theta\sin\phi \end{bmatrix}_{RL},$$

$$1/2\begin{bmatrix} 1 & \cos 2\theta & \sin 2\theta\cos\phi & \sin 2\theta\sin\phi \\ \cos 2\theta & \cos^2 2\theta & \frac{1}{2}\sin 4\theta\cos\phi & \frac{1}{2}\sin 4\theta\sin\phi \\ \sin 2\theta\cos\phi & \frac{1}{2}\sin 4\theta\cos\phi & \sin^2 2\theta\cos^2\phi & \frac{1}{2}\sin^2 2\theta\sin 2\phi \\ \sin 2\theta\sin\phi & \frac{1}{2}\sin 4\theta\sin\phi & \frac{1}{2}\sin^2 2\theta\sin 2\phi & \sin^2 2\theta\sin^2\phi \end{bmatrix}.$$

Special cases of these matrices follow.

II.3 CIRCULAR POLARIZERS

The matrices for right and left circular polarizing filters are obtained from the elliptical filter matrices by setting $\phi = \pm\pi/2$ and $\theta = \pi/4$. In the following the plus sign is associated with right circular polarization states and the minus sign with left circular polarization states:

$$1/2\begin{bmatrix} 1 & \mp i \\ \pm i & 1 \end{bmatrix}_{XY},$$

$$1/2\begin{bmatrix} 1\pm 1 & 0 \\ 0 & 1\mp 1 \end{bmatrix}_{RL},$$

$$1/2\begin{bmatrix} 1 & 0 & 0 & \pm 1 \\ 0 & 0 & 0 & 0 \\ 0 & 0 & 0 & 0 \\ \pm 1 & 0 & 0 & 1 \end{bmatrix}.$$

II.4 THE PLANE POLARIZER

The matrices for a plane polarizing filter are obtained from those for the general elliptical filter by setting $\phi = 0$. The angle θ then corresponds to the angle between the line of the P-state produced by the filter and the x-axis of the coordinate system.

$$\begin{bmatrix} \cos^2\theta & \cos\theta\sin\theta \\ \cos\theta\sin\theta & \sin^2\theta \end{bmatrix} XY,$$

$$1/2\begin{bmatrix} 1 & ie^{-i2\theta} \\ -ie^{i2\theta} & 1 \end{bmatrix} RL,$$

$$1/2\begin{bmatrix} 1 & \cos 2\theta & \sin 2\theta & 0 \\ \cos 2\theta & \cos^2 2\theta & \frac{1}{2}\sin 4\theta & 0 \\ \sin 2\theta & \frac{1}{2}\sin 4\theta & \sin^2 2\theta & 0 \\ 0 & 0 & 0 & 0 \end{bmatrix}.$$

II.5 LINEAR RETARDERS

In the following the retardance, δ, is given by

$$\delta = kt(n_s - n_f).$$

The quantities n_s and n_f are the indices of refraction for the slow and fast P-states which are transmitted unchanged by the plate. The thickness of the plate is t. The angle θ is the angle between the x-axis of the coordinate system and the line of the *fast* P-state. The symbol k is related to the wavelength λ through $k = 2\pi/\lambda$.

$$\begin{bmatrix} \cos^2\theta e^{-i\delta/2} + \sin^2\theta e^{i\delta/2} & -2i\sin(\delta/2)\cos\theta\sin\theta \\ -2i\sin(\delta/2)\cos\theta\sin\theta & \cos^2\theta e^{i\delta/2} + \sin^2\theta e^{-i\delta/2} \end{bmatrix} XY,$$

$$\begin{bmatrix} \cos\delta/2 & \sin\delta/2 \\ -\sin\delta/2 & \cos\delta/2 \end{bmatrix} RL,$$

$$\begin{bmatrix} 1 & 0 & 0 & 0 \\ 0 & \cos^2 2\theta + \sin^2 2\theta\cos\delta & \begin{matrix} +\cos 2\theta\sin 2\theta \\ -\cos 2\theta\sin 2\theta\cos\delta \end{matrix} & \begin{matrix} +\sin 2\theta\sin\delta \\ \times \end{matrix} \\ 0 & \begin{matrix} +\cos 2\theta\sin 2\theta \\ -\cos 2\theta\sin 2\theta\cos\delta \end{matrix} & \sin^2 2\theta + \cos^2 2\theta\cos\delta & -\cos 2\theta\sin\delta \\ 0 & -\sin 2\theta\sin\delta & \cos 2\theta\sin\delta & \cos\delta \end{bmatrix}$$

The special cases of quarter-wave plates and half-wave plates are obtained from the above by setting $\delta = \pi/2$ and $\delta = \pi$.

II.6 CIRCULAR RETARDERS

In the case of circular retarders or optically active materials right and left circular polarization states travel with different velocities and possess different indices of refraction. The retardance β is given by

$$\beta = kt(n_l - n_r).$$

A circular retarder or retardance β rotates a P-state through angle $\beta/2$ away from the x-axis in the counterclockwise direction:

$$\begin{bmatrix} \cos\beta/2 & -\sin\beta/2 \\ \sin\beta/2 & \cos\beta/2 \end{bmatrix} XY,$$

$$\begin{bmatrix} e^{-i\beta/2} & 0 \\ 0 & e^{+i\beta/2} \end{bmatrix} RL,$$

$$\begin{bmatrix} 1 & 0 & 0 & 0 \\ 0 & \cos\beta & -\sin\beta & 0 \\ 0 & \sin\beta & \cos\beta & 0 \\ 0 & 0 & 0 & 1 \end{bmatrix}.$$

APPENDIX III

CONVENTIONS FOR POLARIZATION STATES

Right and left elliptical polarization states are physically distinguished by their helicity. At a given instant of time the tip of the electric vector traces out either a right- or left-handed helix advancing in the direction of propagation. From a quantum mechanical point of view the expectation value of the component of spin angular momentum in the direction of propagation is either positive or negative.

In practice one must decide which helicity shall be called right and which left. When these polarization states are then represented by certain mathematical forms, other choices must be made. It should be apparent that if a set of conventions is adopted for the particular case of right circularly polarized light, all other cases are also decided. The four conventions which must be decided for a right circular state are: the sign in the time dependence of a plane wave, $e^{+i\omega t}$ or $e^{-i\omega t}$; the helicity; the form of the Jones vector; and the sign of the Stokes parameter, P_3. Only two of these decisions may be made independently. Thus, there are four convention systems which may be used. The system used throughout this text is given in the first column of the table opposite, and is equivalent to the 1942 conventions of the Institute of Radio Engineers.

Conventions for right circular polarization

Exponential form of a plane wave	$e^{i(\boldsymbol{k}\cdot\boldsymbol{r}-\omega t)}$		$e^{-i(\boldsymbol{k}\cdot\boldsymbol{r}-\omega t)}$	
Helix of the plane wave				
Expectation valve of spin angular momentum, S_z	$+\hbar$	$-\hbar$	$+\hbar$	$-\hbar$
Form of the Jones vector, $\lvert R\rangle$	$\frac{1}{\sqrt{2}}\begin{pmatrix} e^{-i\pi/4} \\ e^{+i\pi/4} \end{pmatrix}$	$\frac{1}{\sqrt{2}}\begin{pmatrix} e^{+i\pi/4} \\ e^{-i\pi/4} \end{pmatrix}$	$\frac{1}{\sqrt{2}}\begin{pmatrix} e^{+i\pi/4} \\ e^{-i\pi/4} \end{pmatrix}$	$\frac{1}{\sqrt{2}}\begin{pmatrix} e^{-i\pi/4} \\ e^{+i\pi/4} \end{pmatrix}$
Sign of the Stokes parameter, P_3	$+1$	-1	-1	$+1$

APPENDIX IV

THE DIRAC DELTA FUNCTION

We examine in this appendix how one may interpret and utilize a symbol $\delta(x - x_0)$, which is defined by

$$\int_{-\infty}^{+\infty} f(x)\delta(x - x_0)dx = f(x_0), \qquad \text{(AIV.1)}$$

where $f(x)$ is a continuous function of x. The symbol $\delta(x - x_0)$ is known as the Dirac delta function.

Consider two functions $f(x)$, as above, and $\Delta(x)$, illustrated in Fig. AIV.1.

The function $\Delta(x)$ is considered to be nearly zero everywhere except in some small region around x_0. Let us examine the integral

$$\int_{-\infty}^{+\infty} f(x)\Delta(x)dx.$$

By stating that $\Delta(x)$ is nearly zero outside of the interval $x_0 - a < x < x_0 + a$ we mean that we may approximate as follows:

$$\int_{-\infty}^{\infty} f(x)\Delta(x)dx \simeq \int_{x_0-a}^{x_0+a} f(x)\Delta(x)dx.$$

(a)

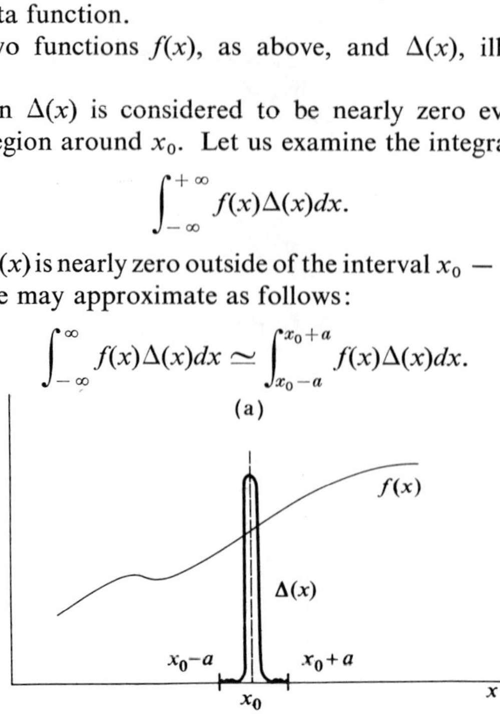

(b)

Figure AIV.1

Imagine now that the magnitude of the interval is greatly exaggerated in Fig. AIV.1 and that a is so small that $f(x)$ is essentially constant over the interval. Then $f(x)$ may be carried through the integral sign:

$$\int_{-\infty}^{+\infty} f(x)\Delta(x)dx \simeq f(x_0) \int_{-\infty}^{+\infty} \Delta(x)dx. \tag{AIV.2}$$

If, in addition,

$$\int_{-\infty}^{\infty} \Delta(x)dx = 1, \tag{AIV.3}$$

Eq. (AIV.2) becomes

$$\int_{-\infty}^{\infty} f(x)\Delta(x)dx \simeq f(x_0). \tag{AIV.4}$$

The difference between Eqs. (AIV.1) and (AIV.4) is, of course, that Eq. (AIV.1) implies a strict equality rather than an approximation.

By a judicious definition of $\Delta(x)$ we can, however, make the two equations equivalent. Consider the integral

$$\int_{-\infty}^{\infty} dx \int_{-\infty}^{\infty} dk f(x) e^{ik(x-x_0)}.$$

The integral over k is an indeterminate form and must be interpreted as

$$\begin{aligned}\int_{-\infty}^{\infty} dk e^{ik(x-x_0)} &\equiv \lim_{L\to\infty} \int_{-L}^{+L} dk e^{ik(x-x_0)} \\ &= 2\pi \lim_{L\to\infty} \frac{\sin L(x-x_0)}{\pi(x-x_0)}.\end{aligned}$$

Two graphs of this function for $L = L_1, L_2$ with $L_1 \lll L_2$ are shown in Fig. AIV.2. If one determines the area under this curve, it is independent of L and is given by

$$\int_{-\infty}^{\infty} \frac{\sin L(x-x_0)}{\pi(x-x_0)} dx = 1. \tag{AIV.5}$$

By comparing Fig. AIV.2 and AIV.1 and Eqs. (AIV.3) and (AIV.5), it is seen that a possible choice for $\Delta(x)$ is

$$\frac{\sin L(x-x_0)}{\pi(x-x_0)},$$

and that $\delta(x - x_0)$ may be realized by

$$\delta(x-x_0) = \lim_{L\to\infty} \frac{\sin L(x-x_0)}{\pi(x-x_0)}. \tag{AIV.6}$$

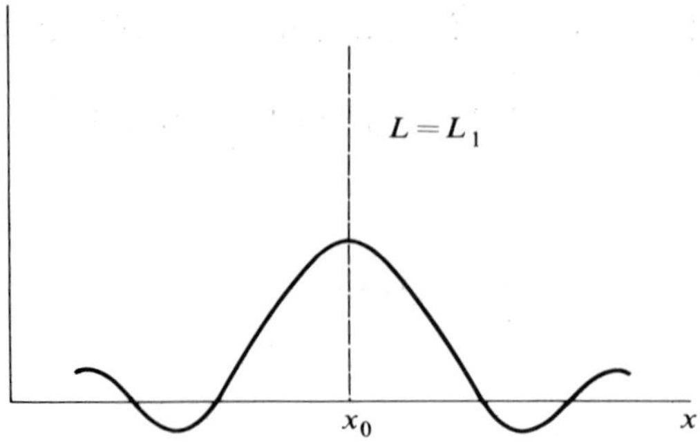

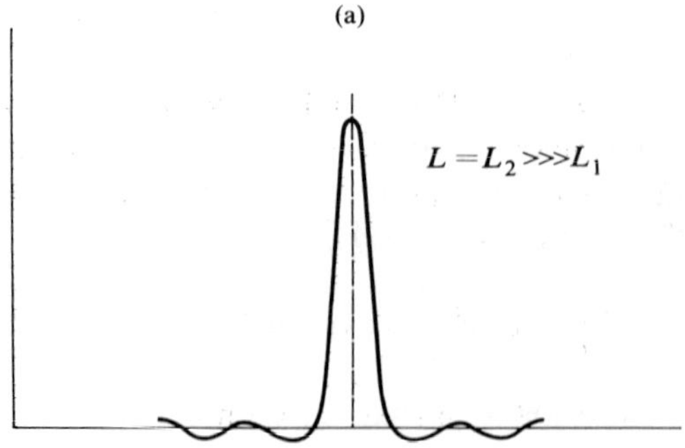

Figure AIV.2

If Eq. (AIV.6) is interpreted as a limit in the usual sense, no limit exists since the sine oscillates. However, when placed within an integral, it yields Eq. (AIV.1). Therefore, we may write

$$\int_{-\infty}^{\infty} e^{ik(x-x_0)} dk = 2\pi\delta(x - x_0). \tag{AIV.7}$$

A number of other useful representations of $\delta(x - x_0)$ may be obtained, examples of which are given without proof.

$$\delta(x - x_0) = \lim_{c\to\infty} \frac{\sin^2 c(x - x_0)}{c\pi(x - x_0)^2},$$

$$\delta(x - x_0) = \lim_{c\to\infty} \sqrt{\frac{c}{\pi}}\, e^{-c(x-x_0)^2}.$$

A commonly occurring integral is

$$\int d\mathbf{k} e^{i\mathbf{k}\cdot(\mathbf{r}-\mathbf{r}_0)}. \tag{AIV.8}$$

The integration be taken over all space where

$$\boldsymbol{k} = k_x\boldsymbol{e}_x + k_y\boldsymbol{e}_y + k_z\boldsymbol{e}_z$$

and

$$\boldsymbol{r} - \boldsymbol{r}_0 = (x - x_0)\boldsymbol{e}_x + (y - y_0)\boldsymbol{e}_y + (z - z_0)\boldsymbol{e}_z.$$

Writing out Eq. (AIV.8) in full, we have

$$\int_{-\infty}^{\infty} dk_x e^{ik_x(x-x_0)} \int_{-\infty}^{\infty} dk_y e^{ik_y(y-y_0)} \int_{-\infty}^{\infty} dk_z e^{ik_z(z-z_0)},$$

which from Eq. (AIV.7) may be interpreted as

$$\begin{aligned}\int_{-\infty}^{\infty} d\boldsymbol{k} e^{i\boldsymbol{k}\cdot(\boldsymbol{r}-\boldsymbol{r}_0)} &= (2\pi)^3\delta(x - x_0)\delta(y - y_0)\delta(z - z_0)\\ &= (2\pi)^3\delta(\boldsymbol{r} - \boldsymbol{r}_0).\end{aligned}$$

Equation (AIV.7) may be used to give a proof for the inversion formula for a Fourier transform. Consider the integral

$$\frac{1}{\sqrt{2\pi}}\int_{-\infty}^{\infty} g(k)e^{ikx}dk, \tag{AIV.9}$$

where $g(k)$ is given by

$$g(k) = \frac{1}{\sqrt{2\pi}}\int_{-\infty}^{\infty} f(x')e^{-ikx'}dx'. \tag{AIV.10}$$

Placing the form of Eq. (AIV.10) in Eq. (AIV.9) yields

$$\frac{1}{2\pi}\int_{-\infty}^{\infty} dk \int_{-\infty}^{\infty} dx' f(x')e^{ik(x-x')}.$$

Interchanging the order of integration and recognizing

$$\int_{-\infty}^{\infty} dk e^{ik(x-x')} = 2\pi\delta(x - x'),$$

we obtain

$$\frac{1}{2\pi}\int_{-\infty}^{\infty} f(x')2\pi\delta(x - x')dx' = f(x).$$

Thus, if $g(k)$ is defined by

$$g(k) = \frac{1}{\sqrt{2\pi}}\int_{-\infty}^{\infty} f(x)e^{-ikx}dx,$$

$f(x)$ is given by

$$f(x) = \frac{1}{\sqrt{2\pi}} \int_{-\infty}^{\infty} g(k) e^{ikx} dk.$$

The following useful properties of the delta function are given without proof

$$\int_{-\infty}^{\infty} F(t) \frac{d}{dt} [\delta(t - a)] dt = - \frac{d}{dt} F(t) \Big|_{t=a}, \tag{AIV.11}$$

$$\int_{-\infty}^{\infty} F(t) \delta[b(t - a)] dt = \frac{1}{|b|} F(a). \tag{AIV.12}$$

APPENDIX V

USEFUL VECTOR IDENTITIES

In the following, ϕ is taken as a scalar function of position. Vector functions of position are represented by $\boldsymbol{A}$, $\boldsymbol{B}$, $\boldsymbol{C}$, and $\boldsymbol{D}$. The operator ∇ expressed in Cartesian coordinates is given by

$$\boldsymbol{e}_x \frac{\partial}{\partial x} + \boldsymbol{e}_y \frac{\partial}{\partial y} + \boldsymbol{e}_z \frac{\partial}{\partial z} \cdot$$

Symbols such as $(\boldsymbol{B} \cdot \nabla)\boldsymbol{A}$ are to be interpreted as vectors whose i-th component is

$$B_x \frac{\partial}{\partial x} A_i + B_y \frac{\partial}{\partial y} A_i + B_z \frac{\partial}{\partial z} A_i .$$

1. $\boldsymbol{A} \cdot (\boldsymbol{B} \times \boldsymbol{C}) = \boldsymbol{C} \cdot (\boldsymbol{A} \times \boldsymbol{B}) = \boldsymbol{B} \cdot (\boldsymbol{C} \times \boldsymbol{A})$
2. $\boldsymbol{A} \times (\boldsymbol{B} \times \boldsymbol{C}) = \boldsymbol{B}(\boldsymbol{A} \cdot \boldsymbol{C}) - \boldsymbol{C}(\boldsymbol{A} \cdot \boldsymbol{B})$
3. $(\boldsymbol{A} \times \boldsymbol{B}) \cdot (\boldsymbol{C} \times \boldsymbol{D}) = (\boldsymbol{A} \cdot \boldsymbol{C})(\boldsymbol{B} \cdot \boldsymbol{D}) - (\boldsymbol{A} \cdot \boldsymbol{D})(\boldsymbol{B} \cdot \boldsymbol{C})$
4. $\nabla \times (\nabla\phi) = 0$
5. $\nabla \cdot (\nabla \times \boldsymbol{A}) = 0$
6. $\nabla \cdot (\phi \boldsymbol{A}) = \nabla\phi \cdot \boldsymbol{A} + \phi \nabla \cdot \boldsymbol{A}$
7. $\nabla \times (\phi \boldsymbol{A}) = \nabla\phi \times \boldsymbol{A} + \phi \nabla \times \boldsymbol{A}$
8. $\nabla(\boldsymbol{A} \cdot \boldsymbol{B}) = (\boldsymbol{A} \cdot \nabla)\boldsymbol{B} + (\boldsymbol{B} \cdot \nabla)\boldsymbol{A} + \boldsymbol{A} \times (\nabla \times \boldsymbol{B}) + \boldsymbol{B} \times (\nabla \times \boldsymbol{A})$
9. $\nabla \cdot (\boldsymbol{A} \times \boldsymbol{B}) = \boldsymbol{B} \cdot (\nabla \times \boldsymbol{A}) - \boldsymbol{A} \cdot (\nabla \times \boldsymbol{B})$
10. $\nabla \times (\boldsymbol{A} \times \boldsymbol{B}) = \boldsymbol{A}(\nabla \cdot \boldsymbol{B}) - \boldsymbol{B}(\nabla \cdot \boldsymbol{A}) + (\boldsymbol{B} \cdot \nabla)\boldsymbol{A} - (\boldsymbol{A} \cdot \nabla)\boldsymbol{B}$
11. $\nabla \times (\nabla \times \boldsymbol{A}) = \nabla(\nabla \cdot \boldsymbol{A}) - (\nabla \cdot \nabla)\boldsymbol{A}$

In the following, integrals including the differential symbol dV are to be taken over a finite volume of space. Such volumes are bounded by a closed surface whose differential area is represented by dA. The differential of an open area is written dS and the directed segment of the boundary of such an area is written dl. The symbol $\boldsymbol{n}$ represents a unit vector normal to a surface, directed outward from a closed surface, and $dl \times \boldsymbol{n}$ directed away from an open surface.

12. $$\int dS(\nabla \times \boldsymbol{B}) \cdot \boldsymbol{n} = \oint \boldsymbol{B} \cdot dl$$

13. $$\int dV(\nabla \times \boldsymbol{B}) = \int dA(\boldsymbol{n} \times \boldsymbol{B})$$

14. $$\int dV\, \nabla \cdot \boldsymbol{B} = \int dA\, \boldsymbol{B} \cdot \boldsymbol{n}$$

APPENDIX VI

MOMENTUM SPACE REPRESENTATIONS OF LINEAR MOMENTUM AND ANGULAR MOMENTUM

VI.1 LINEAR MOMENTUM

By Eq. (7.25)

$$g = \frac{S}{c^2} = \frac{1}{c^2} E(r) \times H(r),$$

$$p = \frac{1}{c^2} \int dr\, E(r) \times H(r),$$

which may be rewritten, by use of Eq. (8.7), as

$$\begin{aligned} p &= \frac{1}{(2\pi)^3 c^2} \int\int\int dk\, dk'\, dr\, \mathscr{E}(k) \times \mathscr{H}(k')\, e^{i(k+k')\cdot r} \\ &= \frac{1}{c^2} \int\int dk\, dk'\, \mathscr{E}(k) \times \mathscr{H}(k')\delta(k + k') \\ &= \frac{1}{c^2} \int dk\, \mathscr{E}(k) \times \mathscr{H}(-k). \end{aligned}$$

By Eq. (8.9),

$$\begin{aligned} ik \times \mathscr{H}(k) &= \dot{\mathscr{E}}, \\ ik \times [k \times \mathscr{H}(k)] &= \varepsilon k \times \dot{\mathscr{E}}(k), \\ i[k(k \cdot \mathscr{H}) - \mathscr{H}(k \cdot k)] &= \varepsilon k \times \dot{\mathscr{E}}. \end{aligned}$$

Since $k \cdot \mathscr{H} = 0$, by Eq. (8.11) we obtain

$$\mathscr{H}(k) = -\frac{\varepsilon}{i} \frac{k \times \dot{\mathscr{E}}}{k^2}, \tag{AVI.1}$$

so that

$$p = \frac{1}{c^2}\int dk\mathscr{E} \times \left[\frac{\varepsilon k \times \dot{\mathscr{E}}(-k)}{ik^2}\right].$$

Now,

$$\mathscr{E} \times [k \times \dot{\mathscr{E}}] = k(\mathscr{E} \cdot \dot{\mathscr{E}}) - \dot{\mathscr{E}}(k \cdot \mathscr{E}).$$

But $k \cdot \mathscr{E} = 0$, by Eq. (8.10), so

$$p = \frac{\varepsilon}{c^2}\int \frac{dk}{ik^2}\, k\mathscr{E}(k) \cdot \dot{\mathscr{E}}(-k).$$

By Eqs. (8.19), (8.20), and (8.27),

$$p = \tfrac{1}{2}\int dk\, \hbar k[-f^{+} \cdot f^{-} + f^{\star -} \cdot f^{\star +} - f^{-} \cdot f^{\star -} + f^{\star} \cdot f^{\star +}],$$

where the following definitions of symbols have been used:

$$f^{+} = f(+k), \qquad f^{-} = f(-k), \qquad \text{etc.}$$

Remembering that

$$\int dk = \int_{-\infty}^{+\infty} dk_x \int_{-\infty}^{+\infty} dk_y \int_{-\infty}^{+\infty} dk_z,$$

and that $\int_{-\infty}^{\infty} = \int_0^{\infty} - \int_0^{-\infty}$, it is possible by a simple change of variable, replacing k by $-k$, to show that the first two integrals vanish and that the last two are equal. We thus obtain

$$p = \int dk\, \hbar k\, f^{\star}(k) \cdot f(k)$$

or, for the α component of linear momentum,

$$p_\alpha = \int dk\, f^{\star}(k) \cdot [\hbar k_\alpha f(k)],$$

which is in agreement with Eq. (8.29).

VI.2 ANGULAR MOMENTUM

By Eq. (9.2)

$$J = \frac{1}{c^2}\int dr\, r \times [E(r) \times H(r)],$$

which may be rewritten, by use of Eq. (8.7), as

$$J = \frac{1}{(2\pi)^3 c^2} \int\int\int dk\, dk'\, dr\, [\boldsymbol{r} e^{i(\boldsymbol{k}+\boldsymbol{k}')\cdot \boldsymbol{r}}] \times [\mathscr{E}(\boldsymbol{k}) \times \mathscr{H}(\boldsymbol{k}')].$$

Now,

$$\int dr\, \boldsymbol{r} e^{i(\boldsymbol{k}+\boldsymbol{k}')\cdot \boldsymbol{r}} = -i\nabla_{K'} \int dr\, e^{i(\boldsymbol{k}+\boldsymbol{k}')\cdot \boldsymbol{r}}$$
$$= -i\nabla_{K'}\delta(\boldsymbol{k} + \boldsymbol{k}')(2\pi)^3,$$

so we may write

$$\boldsymbol{J} = -\frac{i}{c^2} \int\int dk\, dk'\, [\nabla_{K'}\delta(\boldsymbol{k} + \boldsymbol{k}')] \times [\mathscr{E}(\boldsymbol{k}) \times \mathscr{H}(\boldsymbol{k}')].$$

Treating $d\boldsymbol{k}'\nabla_{K'}\delta(\boldsymbol{k} + \boldsymbol{k}') = d\boldsymbol{u}$ as a perfect differential and integrating by parts over $\boldsymbol{k}'$, we obtain

$$\boldsymbol{J} = \frac{i}{c^2} \int\int dk\, dk'\, \delta(\boldsymbol{k} + \boldsymbol{k}')\nabla_{K'} \times [\mathscr{E}(\boldsymbol{k}) \times \mathscr{H}(\boldsymbol{k}')].$$

By Eq. (AVI.1)

$$\mathscr{H}(\boldsymbol{k}) = -\frac{\varepsilon}{i} \frac{\boldsymbol{k} \times \dot{\mathscr{E}}(\boldsymbol{k})}{k^2},$$

so that

$$\boldsymbol{J} = -\frac{\varepsilon}{c^2} \int\int dk\, dk'\, \delta(\boldsymbol{k} + \boldsymbol{k}')\nabla_{K'} \times \left\{\mathscr{E}(\boldsymbol{k}) \times \frac{\boldsymbol{k}' \times \dot{\mathscr{E}}(\boldsymbol{k}')}{k'^2}\right\}.$$

We expand the triple cross product, using

$$\boldsymbol{A} \times (\boldsymbol{B} \times \boldsymbol{C}) = \boldsymbol{B}(\boldsymbol{A} \cdot \boldsymbol{C}) - \boldsymbol{C}(\boldsymbol{A} \cdot \boldsymbol{B}),$$

and obtain

$$\nabla_{K'} \times \left\{\mathscr{E}(\boldsymbol{k}) \times \left[\boldsymbol{k}' \times \frac{\dot{\mathscr{E}}(\boldsymbol{k}')}{k'^2}\right]\right\} = \nabla_{K'} \times \left\{\boldsymbol{k}' \left[\mathscr{E}(\boldsymbol{k}) \cdot \frac{\dot{\mathscr{E}}(\boldsymbol{k}')}{k'^2}\right]\right\}$$
$$- \nabla_{K'} \times \left\{\frac{\dot{\mathscr{E}}(\boldsymbol{k}')}{k'^2} [\mathscr{E}(\boldsymbol{k}) \cdot \boldsymbol{k}']\right\}.$$

Each bracket in this expression is of the form

$$\boldsymbol{\nabla}(\phi \boldsymbol{A}) = \nabla\phi \times \boldsymbol{A} + \phi\boldsymbol{\nabla} \times \boldsymbol{A}.$$

The first term yields

$$\nabla_{K'} \left[\mathscr{E}(\boldsymbol{k}) \cdot \frac{\dot{\mathscr{E}}(\boldsymbol{k}')}{k'^2}\right] \times \boldsymbol{k}' + \left(\mathscr{E}(\boldsymbol{k}) \cdot \frac{\dot{\mathscr{E}}(\boldsymbol{k}')}{k'^2}\right) (\nabla_{K'} \times \boldsymbol{k}'),$$

but

$$\nabla_{K'} \times k' = 0.$$

The second term becomes

$$-\nabla_{K'} [\mathscr{E}(k) \cdot k'] \times \frac{\dot{\mathscr{E}}(k')}{k'^2} - (\mathscr{E}(k) \cdot k')\nabla_{K'} \times \frac{\dot{\mathscr{E}}(k')}{k'^2},$$

and, by using the identity

$$\nabla(A \cdot B) = (A \cdot \nabla)B + (B \cdot \nabla)A + A \times (\nabla \times B) + B \times (\nabla \times A),$$

we obtain

$$\nabla_{K'} \cdot [\mathscr{E}(k) \cdot k'] = \big(\mathscr{E}(k) \cdot \nabla_{K'}\big)k' + (k' \cdot \nabla_{K'})\mathscr{E}(k) \\ + \mathscr{E}(k) \times (\nabla_K \times k') + k' \times \big(\nabla_{K'} \times \mathscr{E}(k)\big).$$

The last three terms vanish identically and if the first term is written out explicitly, we obtain

$$\nabla_{K'}[\mathscr{E}(k) \cdot k'] = (\mathscr{E}(k) \cdot \nabla_{K'})k' = \mathscr{E}(k).$$

Collecting nonvanishing terms, we obtain

$$\nabla_{K'} \times \left\{\mathscr{E}(k) \times \left[k' \times \frac{\dot{\mathscr{E}}(k')}{k'^2}\right]\right\} = \nabla_{K'} \left[\mathscr{E}(k) \cdot \frac{\dot{\mathscr{E}}(k')}{k'^2}\right] \times k' - \mathscr{E}(k) \times \frac{\dot{\mathscr{E}}(k')}{k'^2} \\ - (\mathscr{E}(k') \cdot k')\nabla_{K'} \times \frac{\dot{\mathscr{E}}(k')}{k'^2}.$$

We then have

$$J = -\frac{\varepsilon}{c^2} \iint dk\, dk' \delta(k + k') \left\{\nabla_{K'} \left[\mathscr{E}(k) \cdot \frac{\dot{\mathscr{E}}(k')}{k'^2}\right] \times k' - \mathscr{E}(k) \times \frac{\dot{\mathscr{E}}(k')}{k'^2} \\ - (\mathscr{E}(k) \cdot k') \left(\nabla_{K'} \cdot \frac{\dot{\mathscr{E}}(k')}{k'^2}\right)\right\}.$$

If we now integrate over k', the last term vanishes, since $\mathscr{E}(k) \cdot k = 0$.

Some care must be taken with the first term, since $\nabla_{K'}$ does not operate on $\mathscr{E}(k)$ either before *or after* the integration over the δ-function. Thus, $\mathscr{E}(k)$ is treated as a constant in respect to $\nabla_{K'}$. We indicate this by writing $\mathscr{E}_c(k)$. After the integration, we obtain

$$J = -\frac{\varepsilon}{c^2} \int dk \left\{\nabla_K \left[\mathscr{E}_c(k) \cdot \frac{\dot{\mathscr{E}}(-k)}{k^2}\right] \times k - \mathscr{E}(k) \times \frac{\dot{\mathscr{E}}(-k)}{k^2}\right\}.$$

Now, using Eqs. (8.19), (8.20), and (8.27), we find

$$J = \frac{i\hbar}{2}\int dk\, \{\nabla_K[(f_c + f_c^{\star -})\cdot(f^- - f^{\star +})] \times k - (f^+ + f^{\star -})\cdot(f^- - f^{\star +})\}$$

$$= \frac{i\hbar}{2}\int dk\, \{\nabla_K[f_c^+ \cdot f^- - f_c^+ \cdot f^{\star +} + f_c^{\star -}\cdot f^- - f_c^{\star -}\cdot f^{\star +}] \times k$$
$$- [f^+ \times f^- - f^+ \times f^{\star +} + f^{\star -} \times f^- - f^{\star -} \times f^{\star +}]\}.$$

By a simple change of variable, replacing k by $-k$, one shows that

$$\int dk\, f^+ \times f^- = \int dk\, f^{\star +} \times f^{\star -} = 0$$

$$\int dk\, f^+ \times f^{\star -} = -\int dk\, f^{\star -} \times f^-,$$

so that the last four terms yield

$$-i\hbar \int dk\, f^\star(k) \times f(k).$$

Before proceeding, it is fruitful to examine integrals of the form

$$\int dk\, k \times \nabla_K(\mathscr{F}),$$

where $\mathscr{F}$ is any scalar function of k, $\mathscr{F}(k)$. We wish to show that these integrals vanish. Using the identity

$$\nabla \times (\phi A) = \nabla\phi \times A + \phi\nabla \times A$$

and $\nabla_K \times k = 0$, we obtain

$$\int dk\, k \times \nabla_K(\mathscr{F}) = \int dk\, \nabla_K \times (k\mathscr{F}),$$

and by using the identity

$$\int (\nabla \times B)\, dV = \int dA\, (n \times B),$$

we obtain

$$\int dk\, \nabla_K \times (k\mathscr{F}) = \int dA_K \left(\frac{k}{|k|} \times k\mathscr{F}\right),$$

where the surface of integration is taken over a sphere in K-space whose radius is allowed to become infinite so that the volume integral $\int dk$ may be taken over the whole of K-space. Since $k \times k = 0$, the integral vanishes. As an example, consider

$$\int dk\, k \times \nabla_K(f_c(k) \cdot f(-k)).$$

Our result does not apply directly because of the presence of $\boldsymbol{f}_c$ within the gradient. However, by applying

$$\nabla(\boldsymbol{A}\cdot\boldsymbol{B}) = (\boldsymbol{A}\cdot\nabla)\boldsymbol{B} + (\boldsymbol{B}\cdot\nabla)\boldsymbol{A} + \boldsymbol{A}\times(\nabla\times\boldsymbol{B}) + \boldsymbol{B}\times(\nabla\times\boldsymbol{A})$$

to

$$\nabla_K(\boldsymbol{f}_c^+\cdot\boldsymbol{f}^-),\quad \nabla(\boldsymbol{f}^+\cdot\boldsymbol{f}_c^-),\quad \text{and}\quad \nabla_K(\boldsymbol{f}^+\cdot\boldsymbol{f}^-),$$

one obtains

$$\nabla_K(\boldsymbol{f}^+\cdot\boldsymbol{f}^-) = \nabla_K(\boldsymbol{f}_c^+\cdot\boldsymbol{f}_c^-) + \nabla_K(\boldsymbol{f}^+\cdot\boldsymbol{f}_c^-),$$

and thus

$$\int dk\,\boldsymbol{k}\times\nabla_K(\boldsymbol{f}_c^+\cdot\boldsymbol{f}^-) = \int dk\,\boldsymbol{k}\times\nabla_K(\boldsymbol{f}^+\cdot\boldsymbol{f}^-) - \int dk\,\boldsymbol{k}\times\nabla_K(\boldsymbol{f}^+\cdot\boldsymbol{f}_c^-).$$

But we have shown that the first integral on the right vanishes. If we make the substitution of $-\boldsymbol{k}$ for $\boldsymbol{k}$ on the left, we obtain

$$\int dk\,\boldsymbol{k}\times\nabla_K(\boldsymbol{f}_c^+\cdot\boldsymbol{f}^-) = \int dk\,\boldsymbol{k}\times\nabla(\boldsymbol{f}^+\cdot\boldsymbol{f}_c^-).$$

By comparing these results, we must have

$$\int dk\,\boldsymbol{k}\times\nabla_K(\boldsymbol{f}_c^+\cdot\boldsymbol{f}^-) = 0.$$

By a similar argument we can show that

$$\int dk\,\boldsymbol{k}\times\nabla_K(\boldsymbol{f}_c^{\star-}\cdot\boldsymbol{f}^{\star+}) = 0.$$

However, the proof fails for the integrals

$$\int dk\,\boldsymbol{k}\times\nabla_K(\boldsymbol{f}_c^+\cdot\boldsymbol{f}^{\star+})\quad\text{and}\quad -\int dk\,\boldsymbol{k}\times\nabla_K(\boldsymbol{f}_c^{\star-}\cdot\boldsymbol{f}^-),$$

but does show that these integrals are both equal to

$$-\int dk\,\boldsymbol{k}\times\nabla_K(\boldsymbol{f}_c^{\star+}\cdot\boldsymbol{f}^+).$$

Finally, then, we obtain

$$\boldsymbol{J} = \int dk\,\{-i\hbar\boldsymbol{k}\times\nabla_K(\boldsymbol{f}_c^\star(\boldsymbol{k})\cdot\boldsymbol{f}(\boldsymbol{k}))\} + \int dk\{-i\hbar\boldsymbol{f}^\star(\boldsymbol{k})\times\boldsymbol{f}(\boldsymbol{k})\},$$

and taking the α component we obtain

$$J_\alpha = \int dk\sum_\beta f_\beta[-i\hbar\boldsymbol{k}\times\nabla_K]_\alpha f_\beta + \int dk[-i\hbar(\boldsymbol{f}^\star\times\boldsymbol{f})_\alpha],$$

which is that given by Eq. (9.51).

APPENDIX VII

INTEGRALS OF THE GAUSSIAN DISTRIBUTION

The following definite integrals are given without derivation:

$$\int_{-\infty}^{+\infty} e^{-ax^2}\,dx = \sqrt{\frac{\pi}{a}}, \tag{AVII.1}$$

$$\int_{\infty}^{+\infty} x^2 e^{-ax^2}\,dx = \frac{1}{2a}\sqrt{\frac{\pi}{a}}, \tag{AVII.2}$$

$$\int_{-\infty}^{+\infty} x^4 e^{-ax^2}\,dx = \frac{3}{4a^2}\sqrt{\frac{\pi}{a}}, \tag{AVII.3}$$

$$\int_{-\infty}^{+\infty} x^n e^{-ax^2}\,dx = 0 \quad n = \text{odd integer.} \tag{AVII.4}$$

VII.1 THE GAUSSIAN DISTRIBUTION

The Gaussian distribution $f(x)$ for the random variable x is given by

$$f(x) = \frac{1}{\sigma\sqrt{2\pi}}\, e^{-x^2/2\sigma^2}.$$

Using Eqs. (AVII.1) and (AVII.2), we easily show that

$$\int_{-\infty}^{+\infty} f(x)\,dx = 1,$$

$$\int_{-\infty}^{+\infty} x^2 f(x)\,dx = \sigma^2,$$

where σ is known as the standard deviation of the distribution and σ^2 as the variance.

VII.2 THE BIVARIANT GAUSSIAN DISTRIBUTION

The bivariant Gaussian distribution $f(x, y, \rho)$ is given by

$$f(x, y, \rho) = \frac{1}{2\pi\sqrt{1-\rho^2}\,\sigma_x\sigma_y} \exp\left\{-\frac{1}{2}\left[\left(\frac{x}{\sigma_x}\right)^2 + \left(\frac{y}{\sigma_y}\right)^2 - \frac{2xy\rho}{\sigma_x\sigma_y}\right]\right\}.$$

By completing the square of x in the exponent, we may cast this into the form

$$\begin{aligned} &f(x, y, \rho) \\ &= \frac{1}{2\pi\sqrt{1-\rho^2}\,\sigma_x\sigma_y} \exp\left\{-\frac{1}{2(1-\rho^2)}\left[\frac{x}{\sigma_x} - \rho\frac{y}{\sigma_y}\right]^2\right\} \exp\left[-\frac{1}{2}\left(\frac{y}{\sigma_y}\right)^2\right]. \end{aligned} \tag{AVII.5}$$

In this form it is easily seen from Eq. (AVII.1) that

$$\int_{-\infty}^{\infty} f(x, y, \rho)\, dx = f(y).$$

Likewise, one may show

$$\int_{-\infty}^{\infty} f(x, y, \rho)\, dy = f(x),$$

so that the bivariant distribution represents a Gaussian distribution for each of the variables x and y.

The correlation of the variables x and y is obtained by

$$\begin{aligned} \overline{xy} &= \int_{-\infty}^{\infty}\int_{-\infty}^{\infty} xyf(x, y, \rho)\, dx\, dy \\ &= \int_{-\infty}^{\infty}\int_{-\infty}^{\infty} \sigma_x\left(\frac{x}{\sigma_x} - \rho\frac{y}{\sigma_y}\right) yf(x, y, \rho)\, dx\, dy \\ &+ \int_{-\infty}^{\infty}\int_{-\infty}^{\infty} \sigma_x\rho\frac{y^2}{\sigma_y} f(x, y, \rho)\, dx\, dy. \end{aligned}$$

By use of the form of Eqs. (AVII.5) and (AVII.4), the integration over x in the first term is zero and the second term may be obtained as

$$\overline{xy} = \rho\sigma_x\sigma_y = \rho\sqrt{\overline{x^2}\,\overline{y^2}},$$

which identifies ρ as the correlation of the variables,

$$\rho = \overline{xy}/\sqrt{(\overline{x^2})(\overline{y^2})}.$$

The correlation between x^2 and y^2 is given through

$$\overline{x^2y^2} = \int_{-\infty}^{\infty}\int_{-\infty}^{\infty} x^2y^2f(x, y, \rho)\, dx\, dy.$$

Again completing the square and making the substitutions

$$z = \frac{x}{\sigma_x} - \rho\frac{y}{\sigma_y}, \qquad W = \frac{y}{\sigma_y},$$

we obtain

$$\overline{x^2y^2} = \frac{\sigma_x^2\sigma_y^2}{2\pi\sqrt{1-\rho^2}}\int_{-\infty}^{\infty}\int_{-\infty}^{\infty} dW\,dz\,(z^2 + 2zW\rho + \rho^2W^2)W^2$$

$$\exp\left(-\frac{1}{2}\frac{z^2}{1-\rho^2}\right)\exp\left(-\frac{W^2}{2}\right).$$

Integrating over z and using Eqs. (AVII.1) and (AVII.2), we obtain

$$\overline{x^2y^2} = \frac{\sigma_x^2\sigma_y^2}{\sqrt{2\pi}}\int_{-\infty}^{\infty} dW[1-\rho^2+\rho^2W^2]W^2\,e^{-\frac{w^2}{2}},$$

and the final integration yields

$$\overline{x^2y^2} = \sigma_x^2\sigma_y^2[1+2\rho^2],$$

or

$$\overline{x^2y^2}/(\overline{x^2})(\overline{y^2}) = 1 + 2\rho^2.$$

BIBLIOGRAPHY

GEOMETRIC OPTICS

General

JENKINS, F. S., and H. E. WHITE, *Fundamentals of Optics*, McGraw-Hill (1957).

SEARS, F. W., *Optics*, Addison-Wesley (1949).

Matrix Methods

BROUWER, W., *Matrix Methods in Optical Instrument Design*, Benjamin (1963).

HALBACH, K., "Matrix Representation of Gaussian Optics," *Am. J. Phys.* **32,** 90 (1964).

O'NEILL, E. L., *Introduction to Statistical Optics*, Addison-Wesley (1963).

POLARIZATION STATES

American Institute of Physics, *Polarized Light: Selected Reprints* (1963).

DIRAC, P. A. M., *The Principles of Quantum Mechanics*, Chap. 1, Oxford University Press (1958).

MCMASTER, W. H., "Matrix Representation of Polarization," *Rev. Mod. Phys.* **33,** 1 (1961).

O'NEILL, E. L., *Introduction to Statistical Optics*, Addison-Wesley (1963).

SCHURCLIFF, W. A., *Polarized Light*, Harvard (1962).

SCHURCLIFF, W. A., and S. S. BALLARD, *Polarized Light*, Van Nostrand (1964).

SCHURCLIFF, W. A., "Resource Letter PL-1 on Polarized Light," *Am. J. Phys.* **30,** 227 (1962).

QUANTUM MECHANICS

American Institute of Physics, *Quantum and Statistical Aspects of Light: Selected Reprints* (1963).

AKHIEZER, A. I., and BERESTETSKY, V. B., *Quantum Electrodynamics*, Part One, Chap. One, Interscience (1965). Available in paperback from Office of Technical Services, Dept. of Commerce, Wash., D.C., as AEC tr 2876.

CARRUTHERS, P., "Resource Letter QSL-1 on Quantum and Statistical Aspects of Light," *Am. J. Phys.* **31**, 5321 (1963).

FEYNMAN, R. P., *The Feynman Lectures on Physics*, Vol. III, Addison-Wesley (1965).

IKENBERRY, E., *Quantum Mechanics*, Oxford University Press (1962).

KAEMPFFER, F. A., *Concepts In Quantum Mechanics*, Academic Press (1965).

MCMASTER, W. H. "Matrix Representation of Polarization," *Rev. Mod. Phys.* **33**, 1 (1961).

MANDL, F., *Quantum Mechanics*, 2nd Edn, Academic Press (1957).

O'NEILL, E. L., *Introduction to Statistical Optics*, Addison-Wesley (1963).

ELECTROMAGNETIC FIELD AND COHERENCE

BORN, M., and E. WOLF, *Principles of Optics*, 2nd Edn. [the bible of non-quantum optics], Pergamon Press (1964).

FRANÇON, M., *Modern Applications of Physical Optics*, Interscience (1963).

JENKINS, F. S., and H. E. WHITE, *Fundamentals of Optics*, McGraw-Hill (1957).

MANDEL, L., "Fluctuations of Light Beams," *Progress in Optics*, Vol. III, E. Wolf (Ed.), Interscience (1963).

O'NEILL, E. L., *Introduction to Statistical Optics*, Addison-Wesley (1963).

REITZ, J. R., and F. J. MILFORD, *Foundations of Electromagnetic Theory*, 2nd Edn., Addison-Welsey (1967).

SCHURCLIFF, W. A., *Polarized Light*, Harvard (1962).

STONE, J. M., *Radiation and Optics*, McGraw-Hill (1963).

STRATTON, J. A., *Electromagnetic Theory*, McGraw-Hill (1941).

YOUNG, H., *Fundamentals of Optics and Modern Physics*, McGraw-Hill (1968).

INDEX

INDEX

ABCDE79876543210